Canada's Vegetation

Canada's Vegetation

A World Perspective

GEOFFREY A.J. SCOTT

McGill-Queen's University Press
Montreal & Kingston · London · Buffalo

© McGill-Queen's University Press 1995
ISBN 0-7735-1240-3 (cloth)
ISBN 0-7735-1241-1 (paper)

Legal deposit 1st quarter 1995
Bibliothèque nationale du Québec

Printed in Canada on acid-free paper

Canadian Cataloguing in Publication Data

Scott, Geoffrey A. J.
    Canada's vegetation: a world perspective
    Includes bibliographical references and index.
    ISBN 0-7735-1240-3 (bound) –
    ISBN 0-7735-1241-1 (pbk.)
    1. Phytogeography – Canada. 2. Botany – Canada –
    Ecology. 3. Phytogeography. I. Title.
    QK201.s37 1995   581.971   C94-900780-3

This book was typeset by Typo Litho Composition Inc.
in 10/12 Times Roman.

Contents

Figures

Tables

Preface

This book has evolved from my long exposure to biogeographic research around the world and to teaching biogeography and soils courses, where the emphasis is on soil-vegetation systems and the human disturbance of those systems. The absence of a suitable basic source on mid- and high-latitude environments that took a Canadian perspective has been the source of much frustration for me. The intention here is therefore to stress the description of Canada's ecoclimatic regions and their equivalents elsewhere, not simply to describe vegetation formations alone. While ecoclimatic regions include faunas along with their floras, the former will not receive the attention they deserve here.

The book has been written with two specific groups of people in mind. The first group includes anyone with a genuine love for Canada's incredible natural environment and our biodiversity heritage. The second group includes biogeographers, ecologists, and edaphologists, who by rights should also claim membership of the first group. The text is not designed to explain the workings of nature! While the emphasis centres on potential natural vegetation, I feel conscience-bound to mention examples of where the cover has been greatly modified, either intentionally or inadvertently, by humans. Though my support for preserving large portions of our natural environment is unequivocal, the point of the text is not to point fingers or list smoking guns, and I conform to the notion that there may yet be time for the true ideals of sustainable development to be realized. Despite this, ecosystem modifiers may say that I have sometimes gone too far, while conservers may say I have not gone far enough.

I readily admit to a personal point of view in the way that information is treated. This view is tempered, however, by the realization that ecosystems (biogeocenoses) are indeed holistic, integrated, and constantly changing entities in which *Homo sapiens* often plays an important, if not dominant, role. It also reflects my exposure to both North American and European terminologies and first-hand experiences gained while studying soil-vegetation systems throughout all of the Americas, in Southeast and East Asia, Melanesia, Polynesia, the Galapagos Islands, the Caribbean, Europe, and North Africa. In part, my point of view clearly reflects the outlook of those who have influenced my thinking about ecosystems. These people include Dieter Mueller-Dombois and the late John Street of the University of Hawaii; A.W. Küchler, professor emeritus of the University of Kansas; and the late Carl Sauer of Berkeley. In terms of a world vegetation overview, the greatest influence on me has been the writings of Heinrich Walter of the University of Hohenheim. In terms of the Canadian literature, my thoughts have been strongly influenced by a whole host of fine Canadian researchers too numerous to mention, whose published works form the cornerstone to this book.

I would like to acknowledge a large number of people who provided advice and encouragement and who helped with various aspects of text preparation. Most importantly, I am indebted to my wife, Nila, and my children, Troy, Gregory, and Sara, who have supported my interests and put up with my seclusion in the library and at the word processor, my absence on research trips, or with having to stop the car on a family outing merely to get a better impression of some roadside scene or examine a soil profile. A special thanks to Weldon Hiebert for his superb cartography, to Betty Harder, who kept draft after draft looking much more professional that I deserved, to Elizabeth Hulse, for her superb editing, and to Peter Tittenberger, whose skills help produce the selection of photographs. I am also indebted to Larissa McCutcheon, Darlene Taylor, Karen Scott, and Lori Sumner, who helped as research assistants. The help provided by the staff of the Churchill Northern Studies Centre is much appreciated. Ed Oswald, Gary Ironside, and Clay Rubec are thanked for their helpful comments on parts of the text, particularly the section on wetlands, as are two anonymous reviewers for their valuable input. Thanks are also extended to Mark Krawetz and my father, R. Ernest Scott, who provided some of the photographs. It was my father who gave me my interest in photography and my mother who instilled in me a lasting appreciation for my environment. Financial assistance from the University of Winnipeg is gratefully acknowledged.

Dwarf-shrub heath similar to that on Devon Island, but photographed here at Alexandra Fjord, Ellesmere Island. This mid-July photo shows the catkins of *Salix arctica* (Arctic willow) in full flower. White flowers seen on the lower left are *Dryas integrifolia* (white mountain avens). Photo by Mark Krawetz.

Typical polygon sedge meadow lowland on alluvial floodplain at Alexandra Fjord, Ellesmere Island. Attenuation causes fissuring along the ridge axis, and ice wedges fill the resultant crack. The ice wedge expands, forcing soil on either side into low parallel ridges. Numerous low polygon-margin ridges can be seen in the background. Photo by Mark Krawetz.

Coastal tundra on a raised beach slope at Bird's Cove, northern Manitoba. Herbaceous cover on the mid-slope in the left foreground consists primarily of the dwarf shrubs *Arctostaphylos alpina*, *Salix reticulata*, *Vaccinium uliginosum*, and *Rhododendron lapponicum*, together with light-coloured lichens. On the ridge crest to the right, cover is not continuous and is dominated by *Dryas integrifolia*, with their white seed tufts, and *Saxifraga oppositifolia*. The flagged *Picea glauca* in the background is about 2.5 m tall and exhibits layering.

Small fen palsas in a large palsa field near the Churchill Northern Studies Centre, Manitoba. These palsas have sedge fen between the 1–2 m tall mounds and consist of ice-filled fen peat, and mound surfaces exhibit a much droughtier tundra cover dominated by lichens and dwarf shrubs.

Fellfield alpine tundra at 2,450 m, The Whistlers, Jasper, Alberta. *Dryas octopetala* (mountain avens) is the white flowered dwarf shrub. Also common are saxifrages.

Flag forms in *Picea glauca* out on fen near the Twin Lakes, northern Manitoba. Note the basal skirts and lower branch die-back. Beyond the foreground spruce are less modified *Larix laricina*.

View of the bog–fen landscape in the forest subzone of the forest tundra, Hudson Bay Lowlands, northern Manitoba. Light-coloured open lichen woodland bog plateaux are expanding out over the darker fen. In the foreground, spruce form circles on the remnant ridges of palaeo-polygons.

Typical open lichen woodland stand at the boundary between forest tundra and the boreal, Twin Lakes, northern Manitoba. The soils on these better-drained sites have Eutric Brunisols with only two tree species, *Picea glauca* and *Larix laricina*. Open fen dominates the lowlands beyond.

iver, northern Manitoba. Trees are *Picea*
by *Sphagnum* moss, *Cladina stellaris*

Typical northern coniferous forest near Kenora, northwest Ontario. An earlier post-burning successional stage of *Pinus banksiana* and *Populus tremuloides* (canopy trees) is being taken over by successional *Picea glauca* and the occasional *Abies balsamea*.

Tamarack–spruce lakeshore forest near Dryden, northwest Ontario. This October photograph shows the deciduous *Larix laricina* beginning to shed its light-coloured needles, while *Picea mariana* retains its bold outlines. *Picea glauca* is also present on better-drained soils upslope to the right.

Pinus banksiana–dominated shield outcrop in northern coniferous forest, northwest Ontario. The rocky slopes are dominated by *Juniperus communis* (the low circular shrub) and *Betula papyrifera*. Spruce, birch, and aspen are common along the lower slopes.

Typical mixed-forest transition between northern coniferous forest and prairie, Riding Mountain National Park, early October. The dark conifers are *Picea glauca*, while the lighter fall-coloured broadleaf trees are *Populus* spp. and *Betula papyrifera*.

Mixed forest on deep mineral soils north of Parry Sound, Ontario, late October. The ridge-top community is dominated by *Pinus strobus*, while the lower slopes are dominated by a mixture of *Betula papyrifera*, *Populus tremuloides*, and *Picea glauca*.

Pinus strobus–dominated shield outcrop in the mixed-forest region near Parry Sound, Ontario.

Mixed forest on the southwestern escarpment on the Gatineau Hills, Gatineau Park, Quebec, overlooking the Ottawa River valley. *Pinus strobus* emergents are noticeable in an otherwise *Quercus-Acer*-dominated forest.

Old-growth mixed forest in Jobes' Woods, Presqu'ile Park, near Trenton, Ontario. *Acer* and *Fraxinus* deciduous species dominate, along with *Tsuga canadensis*. Many of the massive eastern hemlocks were windthrown during a storm in November 1992.

Xeric mixed-grass prairie in southern Saskatchewan. As a result of heavy grazing on this rangeland, the shrub *Artemisia frigida* (prairie sagewort) is common, as are numerous pads of *Opuntia polyacantha*. Because of the moist spring, many of the cacti are in flower. Bare earth mounds have been produced by burrowing mammals.

Bouteloua gracilis, Stipa comata, and *Artemisia* dominate xeric mixed-grass prairie along the Two Trees Trail, Grasslands National Park, Saskatchewan. Seventy Mile Butte dominates the horizon. Note the shrub cover along minor gullies and the arboreal growth along the Frenchman River on the valley floor.

Fescue prairie rangelands along the edge of the Rocky Mountain foothills west of Calgary, Alberta. A mixed forest of *Picea glauca, Pinus contorta,* and *Populus tremuloides* has encroached from the foothills during the last century.

Mixed-grass prairie in the aspen parkland, Brandon Hills, Manitoba. Terracettes on the slope in the middle distance have been produced by grazing cattle. This late September photo shows that the *Quercus macrocarpa* have already shed their leaves, while *Populus tremuloides* leaves are in the yellow stage.

Big bluestem (*Andropogon gerardii*) dominated tall-grass prairie in the Red River Valley southwest of Winnipeg. Because this never-tilled prairie has not been burned for many years, clonal aspen are now radiating out into it.

Riverine forest along the Assiniboine River, Spruce Woods Provincial Park, Manitoba. A willow-dominated shrub zone (*Salix interior* and *S. amygdaloides*) borders the river, while a zone of hardwoods (*Acer negundo, Fraxinus pennsylvanica, Ulmus americana*, and *Populus deltoides*) dominates on slightly higher ground. The *Picea glauca* in the foreground and on the distant hills are on the sandy, well-drained upland soils of the Carberry Desert and the Spruce Woods.

Canada's Vegetation

CHAPTER ONE

Introduction

1.1 Background

In 1993 the world's human population approached the 5.7 billion mark. While numerous species can boast that they surpass this figure by many fold, many can not, and still many more will soon become extinct. Recent calculations indicate a minimum of anywhere from 30 to 50 million plant and animal species on our planet, but of these, perhaps only 1.75 million have been scientifically classified and given binomials (Wilson, 1992). We have only recently discovered that this biological diversity is primarily due to the vastness of species variety, particularly among insects such as the microarthropods, in the more humid tropics. Tree and liana species richness is greatest in upper Amazonia, while non-tree species richness is greatest along the northern Andean foothills and southern Central America (Gentry, 1992). Unfortunately, this incredible biodiversity is in jeopardy due to the speed with which tropical forest clearing and associated species extinction is proceeding. One estimate is that fifty or more species become extinct daily, a rate that would see 10% of the world's biodiversity gone in the next twenty-five years (Senevirantne, 1992). Another is that 1 million plants and animal species will be exterminated during the next twenty years (Reid and Miller, 1989). Although species diversity outside the humid tropics is less in jeopardy, human influence on the areal extent and composition of vegetation there is just as, if not more, pronounced.

What can we as individuals do about the dilemmas of biodiversity, areal extent, and composition? The first step is to gain an appreciation for vegetation and the richness of its diversity, ecosystem variety, climatic and soils relationships, and dynamics. It is only through understanding some of the more important aspects of ecosystems that the seemingly esoteric discussions of, for example, biodiversity can become a reality that we can then attempt to handle. Even without a detailed understanding of the biodiversity dilemma, it would seem realistic that preserving such things as old-growth forests, undisturbed prairie, and even semi-natural biotic communities is the first step. To do so, however, will mean greater conservation efforts through wildlife reserve, wilderness area, and park establishment and protection; the greater adoption of agricultural practices such as "agroforestry," which maintains some tree cover while reducing the need to cut down more forest; promoting "extractivism" and "extractive reserves," where forest is otherwise earmarked for conversion to pasture; reforestation of logged-over sites, and abandoned pasture; improved environmental education; a greater adherence to the principles of sustainable development; and a greater control of both human population numbers and their degree of freedom to destroy remaining undisturbed plant communities – not a simple task! At least the problem is recognized as a real one and a start made to protect ecosystems and their diversity. At the international level, there has been a recent rush to "pledge to protect biodiversity, whatever it may be" (Gee, 1992, p. 639), and it is noteworthy that Canada was the first of 152 nations to sign the Biodiversity Convention in Rio de Janeiro on 11 June 1992. This convention came into effect on 29 December 1993.

Canadians "are the custodians of a substantial portion of the earth's northern latitude ecosystems" (Rubec, 1992, p. 16), but we must remember that the preservation of much of this area is not as yet guaranteed. We have the seasonal, but striking beauty of the tundra, the stillness of boreal forests with their pines, spruces, and birches, the grandeur of the old-growth Pacific mesothermal forests with their majestic firs, cedars, and hemlocks, and the brilliant flashes of fall colour from the beech, maple, aspen, and oak in our temperate deciduous forests. We have also had the destruction of natural vegetation on a vast scale due to such agencies as logging, forest fires, and clearing for both urban expansion and agriculture. The early history of the Red River valley in Manitoba was associated with the *Andropogon gerardii* (big bluestem) of the tall-grass prairie. Today, all we see are roads, houses, and fields, with some stands of aspen or oak; we have to search long and hard to find the remaining few tens of patches of true prairie, such as the small hectarage at the Living Prairie Museum in Winnipeg. We now have open conflicts between the users and the preservers. Occasionally, these fights are won by the preservers – case in point, the Haida, together with conservation groups, and their thirteen-year struggle against logging interests to establish a National Park on South Moresby in 1988. The Canadian Wilderness Charter and the 1990 Green Plan have set as a goal the protection of 12% of our national territory. By 1992, protected natural heritage areas of the federal and provincial/territorial governments constituted just over 7% of land and freshwater areas, of which 178,620 km^2 (or 1.8%) is in thirty-four national parks and national park reserves (Finkelstein, 1992). Much still remains to be protected, and a good deal of work still needs to be done to achieve this goal (Dearden and Rollins, 1993). It is hoped that the text which follows will give readers a better awareness of our vast terrestrial ecosystem heritage and make them much more disposed to its protection and even to contributing to that protection.

1.2 Scope of the Text

In the chapters that follow, the emphasis is on describing the natural vegetation cover of Canada's major ecoclimatic regions. To set a more global perspective, mention is made of similar biomes elsewhere. In essence, therefore, each major type of vegetation cover found in Canada is examined at the global scale but viewed from the Canadian ecoclimatic perspective. From the outset it must be recognized that at the local level, it may sometimes be considered unrealistic to define the natural vegetation of a site because of indigenous disturbances which prevent equilibrium from ever being established (Sprugel, 1991). Because the ecoclimatic approach is much more than simply describing vegetation cover types, a discussion of those environmental factors that impinge upon them is included to help account for observable plant-community differences within local areas. In essence, Canada's biomes are examined in terms of vegetation-soil-climate interactions.

The majority of vegetation maps used throughout this text represent the "potential natural vegetation" of the major biomes. This approach is taken because of the difficulty with generalizing on maps the mosaic approach that our own species takes in modifying cover and because of the difficulty of integrating environmental factors such as vegetation and climate. Land-use/land-cover maps may ultimately serve us better than maps of natural vegetation, but these are difficult to produce, few in number, and in need of constant update. In the case of Canada, an ecoclimatic map will also be used because it aids in relating climate to vegetation.

The study of Canada's vegetation is incomplete if there is not at least some attempt to integrate all aspects of ecosystems. To prepare readers for the discussion of major vegetation types, a brief review of some of the terminology used in the ecosystems-ecoclimatic approach to Canada's vegetation is given below. Emphasis is placed on terminology used in describing plant cover, relating climate to plant distributions, and demonstrating basic soil-vegetation relationships. To help accomplish this, existing Canadian classification systems for vegetation, soil, and ecoclimates are emphasized. Unfortunately, the scope of the text does not permit attention to similar detail on the faunal components of biomes. While it would also be of interest to place existing ecoclimates in the context of Canada's post-glacial (Holocene) vegetation history, such an approach is beyond the scope of the present study and has already been detailed by others (Pielou, 1991; Delcourt and Delcourt, 1991; Ritchie, 1987).

1.3 World Vegetation Types

Any discussion of continental or world-scale vegetation-cover types requires a classification system which is easy to use, has a practical number of categories, and has understandable terminology. Much of the literature uses either vegetation formations (biomes) (Figure 1.1) or the zonobiome system (Figure 1.2) when large areas of veg-

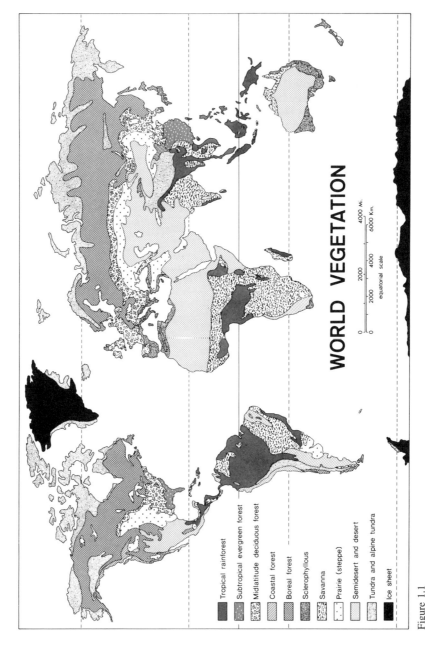

Figure 1.1
World vegetation formations (after Eyre, 1968)

Figure 1.2
Zonobiomes of North America (modified from Walter, 1979). See Table 1.1 for key to zonobiome numbers.

etation are discussed. While both these systems are used in the Canadian literature today, increasing attention is being given to the relatively new ecoclimatic regions approach in the study of Canadian vegetation. This ecoclimatic approach is to a large extent the local adaptation of the zonobiome approach since climate, as expressed through its control of soil and vegetation cover, is again the more important environmental determinant used in the classification. The Canada Committee on Ecological Land Classification has also recently developed a "first approximation" for the Canadian Vegetation Classification System. This new system should go a

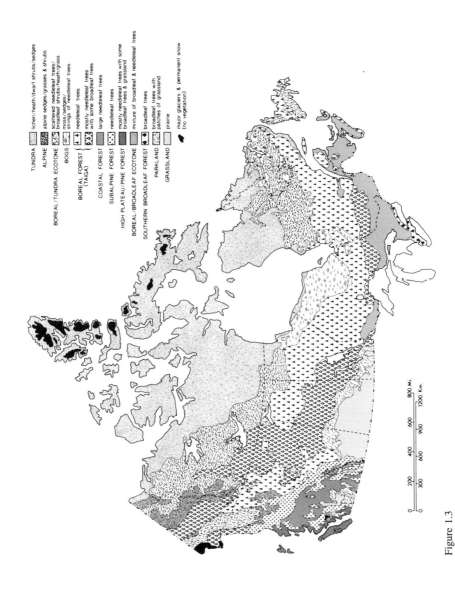

Figure 1.3

Vegatation Formations in Canada (after Canada, Department of Energy, Mines and Resources, 1973). See Energy, Mines and Resources Canada (1993) for cover types map based on satellite data.

long way towards helping standardize data reporting from new Canadian vegetation studies (National Vegetation Working Group, 1990).

1.3.1 Vegetation Formations and Zones

A vegetation formation is a world vegetation type dominated overall by plants of the same life form (Figure 1.1). The term "life form" refers to the growth form, or physiognomy, of the dominants. Examples would include prairie, a formation dominated by grasses, or boreal forest, where the needleleaf-tree life form dominates. There is generally no strict use of the term formation, and the number of classes used varies between authors, simply to reflect their interests and the degree of specialization that their writing requires. There is, however, an inherent relationship between the term formation and climate, because it seems that whatever the formation classes used by an author, each type occupies an area or region which possesses certain climatic characteristics to which a particular life form is most suitably adapted (Eyre, 1968). Acknowledging this climatic relationship, Dansereau (1957, p. 83) defines formation class as being "a geographic unit of vegetation which shows characteristic response to a particular climatic trend at a particular intensity." Figure 1.3 lists typical formation terminology as used in Canada. It is not uncommon to see subformation terminology such as tall-grass prairie and mixed-grass prairie being used on larger-scale vegetation maps. Likewise, transitional zones between two formations can be differentiated on larger-scale maps and are called "ecotones." The terms "zone" and "life zone" are often used interchangeably with formation, particularly if faunal elements are being considered (Furley et al., 1983).

1.3.2 Zonobiomes

A zonobiome system developed in Europe by Walter and his colleagues (Walter 1973) recognizes the strong relationship between climate and vegetation by categorizing together those biomes which are found under similar climatic conditions (Figure 1.2). A biome is simply a large and uniform environment where the vegetation is, for whatever reason, similar throughout. The term "biome" differs from formation or zone in that the concept includes the whole sociological unit of plants and animals both on and in the soil. A good example of a zonobiome (ZB) would be the Arid Temperate with a Cold Winter category (ZB VII), which, in terms of zonal or formation terminology, is practically synonymous with prairie. Figure 1.2 illustrates this "climatic determinant" approach to classifying vegetation. The simplicity of this classification lies in its limitation to nine categories (Table 1.1) and, as a world classification scheme, its avoidance of confusing regional terminologies. A zonobiome is therefore a grouping of biomes, but unlike the use of the term "formation," biomes within the parent zonobiome can be differentiated on the basis of life form (e.g., tall-grass prairie) or other environmental determinants, such as ele-

Table 1.1
Typical formation terminology with equivalent zonobiome numbers and verbal descriptions
(adapted from Walter, 1979)

Formation Examples	ZB	Zonobiome Description
Tundra	IX	Arctic, Antarctic
Boreal forest	VIII	Cold temperate (boreal)
Prairie (steppe)	VII	Arid temperate with cold winter (continental)
Temperate deciduous	VI	Typical temperate with short (nemoral) period of frost
Temperate evergreen	V	Warm temperate (maritime)
Mediterranean or sclerophyllous	IV	Winter rains and summer drought
Subtropical deserts	III	Subtropical arid (desert climate)
Savannas and tropical deciduous forest	II	Tropical with summer rains
Tropical rain forest (evergreen)	I	Equatorial (with diurnal climate)

vation (orobiomes), or peculiar soil conditions (pedobiomes), such as high salinity. In this system, transitional or ecotonal zones are called zonoecotones (ZE), an example of which would be ZE VIII–VI, the mixed-forest transition between boreal and temperate forest. For a more detailed discussion of zonobiomes, see Walter (1973, 1979).

1.3.3 Ecoclimates

Ecoclimatic regions can be defined as "broad areas on the earth's surface characterized by distinctive ecological responses to climate, as expressed by vegetation and reflected in soils, wildlife and water" (Ecoregions Working Group, 1989, p. 1). Strictly speaking, ecoclimatic regions are not based on specific climatic parameters, but as the regions do reflect macroclimate – as expressed using vegetation and soil development – as the defining criteria, the term "ecoclimatic" is therefore a practical one (Figure 1.4). To a large extent, this ecoclimatic approach is an extension of the zonobiome concept, and it is currently gaining acceptance in Canada as a valuable aid to the description of vegetation zones and in aiding land-use management decisions. The basic difference between the formation and zonobiome/ecoclimatic approaches can be summarized as the difference between description and quasi-explanation. As the thrust of this text is to describe, and in the ecosystem approach sense to explain, some aspects of Canada's vegetation mosaic, both approaches will be used.

   The ecoclimatic approach as applied to Canadian vegetation has one major advantage over the zonobiome approach in that it can legitimately take advantage of local terminology. It is also a hierarchial system with ten ecoclimatic provinces at the general level and seventy-two ecoclimatic regions (or ecoregions) at the specific level. Between these two are a variety of ecoclimatic region groupings found convenient

ARCTIC
HA High Arctic
MA Mid-Arctic
LA Low Arctic

SUBARCTIC
HS High Subarctic
MS Mid-Subarctic
LS Low Subarctic

BOREAL
HB High Boreal
MB Mid-Boreal
LB Low Boreal

COOL TEMPERATE
HCT High Cool Temperate
MCT Mid-Cool Temperate

MODERATE TEMPERATE
HMT High Moderate Temperate

SUBARCTIC CORDILLERAN
NSC Northern Subarctic Cordilleran

CORDILLERAN
NC Northern Cordilleran
MC Mid-Cordilleran
SC Southern Cordilleran

INTERIOR CORDILLERAN
ICb Boreal Interior Cordilleran
ICv Vertically Stratified I.C.

PACIFIC CORDILLERAN
NPC North Pacific Cordilleran
SPC South Pacific Cordilleran

GRASSLAND
Gt Transitional Grassland
Gs Subhumid Grassland
Ga Arid Grassland

Figure 1.4
Ecoclimatic regions of Canada (after Ecoregions Working Group, 1989)

Table 1.2
Ecoclimatic provinces in Canada and their equivalent formation and zonobiome terminology

Ecoclimatic Province	Formation Equivalent	ZB
1 Arctic	Tundra	IX
2 Subarctic	Tundra-boreal ecotone	ZE VIII–IX
3 Boreal	Boreal (taiga)	VIII
4 Cool Temperate	Boreal-broadleaf ecotone	ZE VIII–VI
5 Moderate Temperate	Southern broadleaf forest	VI
6 Grassland	Prairie grassland	VII
7 Subarctic Cordilleran	Subalpine-alpine	OB[1] VIII
8 Cordilleran	Boreal-subalpine-alpine	OB VIII
9 Interior Cordilleran	(vertically stratified)	OB VIII–VII
10 Pacific Cordilleran	Coastal forest	V

[1] OB is an orobiome (mountain biome) within a zonobiome.

in mapping. An example of the classification would be the Boreal Ecoclimatic Province (B), which is divided into three sub-provinces, the High, Mid-, and Low Boreal (HB, MB, and LB) respectively (Figure 1.4). In turn, sub-provinces such as the High Boreal are divided into ecoclimatic regions such as the Humid High Boreal (HBh), the Maritime High Boreal (HBm), and so on. Because of the scale used in Figure 1.4, every ecoclimatic region cannot be shown. For a map (1:7,500,000) showing all seventy-two ecoclimatic regions, see Ecoregions Working Group (1989). Ecoclimatic provinces, together with their formation (or zone) equivalents (from Figure 1.3) and zonobiome equivalents (from Table 1.1), are given in Table 1.2.

1.3.4 The Canadian Vegetation Classification System

The "first approximation" of this new system was published in 1990 and is designed to provide a nation-wide vegetation classification system for reporting relevé (plant species with cover) data. It provides a convenient mechanism for "reducing the complexity of natural vegetation to a small number of relatively homogenous, easily understood groups" (National Vegetation Working Group, 1990, p. 1). The system makes use of a combination of physiognomic, structural, floristic, and dominance criteria in a seven-level hierarchy. It does not use environmental criteria, it avoids connotative physiognomic "labels" such as grassland or shrubland (which bring with them inconsistent definitions), and it allows for easy mathematical analysis. Like many other vegetation classifications, the first level distinguishes between broad physiognomic types, but it restricts these to four categories: trees, shrubs, herbs (including ferns and their allies), and non-vascular species. At the second level, trees and shrubs are differentiated into evergreen, deciduous, and mixed, while herbs include forbs, graminoids (grasses, sedges, and rushes), and mixtures of both. The non-vascular groupings consist of bryophytes (mosses, liverworts, and hornworts), lichens, and mixes of the two. At the third level of classification, vegetation is differentiated on the basis of closed, open, and sparse, while the fourth level deals with height. The last three levels are differentiated on the basis of species dominants (National Vegetation Working Group, 1990).

This new vegetation classification system is designed for reporting vegetation-cover data in a standard way and is not intended to compete with traditional terms such as prairie, boreal, and tundra. At this data reporting level, it avoids connotative physiognomic labels such as savanna or grassland, which often lead to confusion among readers. Data users can then translate this standardized data into any connotative physiognomic description they prefer, provided they include clarifying descriptions. When the following discussion uses such physiognomic terms, they are normally defined to help reduce this confusion. Many of the terms used in this new classification system are both common and descriptively useful and will be used where appropriate in the text.

1.3.5 Ecozones

When it comes to the detailed description of vegetation cover in Canada, use will also be made of another system of classification. Because of Canada's size and valuable natural resources, it is important that land classifications be carried out not only at the macro level but also at the micro. To aid in this process, an Ecological Land Classification System has been established, with ecozones as the most general hierarchical unit used. Ecozones can be defined as "areas of the earth's surface representative of very generalized ecological units, based on the perception that the earth's surface is interactive and continuously adjusting to the mix of abiotic and biotic factors that may be present at any given time" (Wiken, 1986, p. 4).

A typical example of the fifteen Canadian terrestrial ecozones is the Boreal Shield, an area of boreal forest associated with acidic igneous bedrock and predominantly acidic morainic surficial deposits (Figure 1.5, number 9). Each ecozone can be subdivided into ecoprovinces, ecoregions, ecodistricts, ecosections, ecosites, and lastly, ecoelements. While the ecozone approach is clearly not a vegetation classification system, its descriptive/geographic value in vegetation studies cannot be overlooked.

1.3.6 Floristic Realms

In any discussion of vegetation distributions, it is inevitable that mention must also be made of the evolutionary affinities between species or genera located in quite different parts of the world. Plant distribution on a world scale can be looked at from this evolutionary point of view, and on the basis of historical factors, world vegetation can be classified into a number of floristic realms (Figure 1.6). These realms must not be confused with formations, zonobiomes, or ecoclimates, none of which

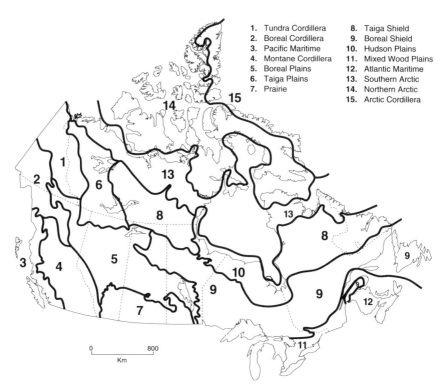

1. Tundra Cordillera
2. Boreal Cordillera
3. Pacific Maritime
4. Montane Cordillera
5. Boreal Plains
6. Taiga Plains
7. Prairie
8. Taiga Shield
9. Boreal Shield
10. Hudson Plains
11. Mixed Wood Plains
12. Atlantic Maritime
13. Southern Arctic
14. Northern Arctic
15. Arctic Cordillera

Figure 1.5
Terrestrial ecozones of Canada (after Wiken, 1986)

pay direct attention to evolutionary plant history but differentiate the species and life forms that are extant into cover types. Floristic realms say nothing about life forms; they simply distinguish areas of evolutionary integrity. As with the Holarctic, these floristic realms can include forested and non-forested regions.

The Holarctic is so large because of frequent and recent land connections between eastern Asia and Alaska through Beringia. This major land connection during glacial low sea level periods helps account for the amphi-Beringian similarities in structure and floristic composition of tundra sedge meadows (Ritchie, 1984). By the same token, the Austromalesian Realm is distinct because it has been separated for a long period of geologic time from the Holarctic and Palaeotropics by oceans, deserts, and mountains. The Holarctic is in turn divided into the Nearctic (essentially North America) and the Palaearctic (Europe and most of Asia). In turn, the Nearctic can be viewed as a number of floral regions which include the Boreal (the northern coniferous forest region) and the Carolinian (temperate deciduous forest region).

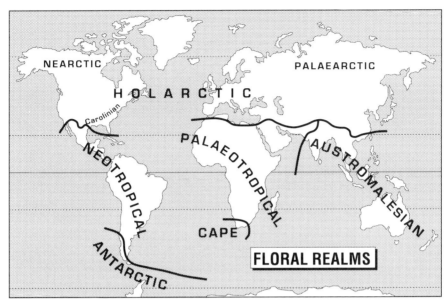

Figure 1.6
World floristic realms (modified from a number of sources, but primarily de Laubenfels, 1975)

1.3.7 Plant Species Nomenclature

As binomial plant nomenclature varies both over time and between authorities, Canadian vascular plant nomenclature follows *The Flora of Canada* (Scoggan, 1978), while most non-vascular nomenclature follows *Mosses, Lichens and Ferns of Northwest North America* (Vitt et al., 1988). Note, for example, the change from *Andropogon gerardi* (Scoggan, 1957) to *Andropogon gerardii* (Scoggan, 1978) for big bluestem. Also note that Canadian and American nomenclatures sometimes differ, with yellow birch usually being called *Betula lutea* in Canada, while in the USA it is *Betula alleghaniensis*. Likewise, in Canada the mid-grass little bluestem is called both *Andropogon scoparius* and *Schizachyrium scoparium*, while in the USA it is called only by the latter name. For regions outside of Canada, nomenclature follows general literature usage or the cited literature source.

Because of the large number of plant names used in the text, authorities for each binomial are not given. For these authorities, the reader should refer to Scoggan (1978) or the pertinent reference. Because the text uses binomials in preference to local names, all binomials, together with their most commonly used local name, are provided in the appendix.

1.4 Soil Classifications and Soil Systems

A soil classification is simply a convenient way of differentiating between soil types that vary from each other in certain profile characteristics. These characteristics could include such things as organic matter content, degree of oxidation or reduction, or effects of the leaching process as expressed by the development of eluvial or illuvial horizons. Soils do not vary suddenly or dramatically along a vegetation transect in which plant cover can suddenly change from one dominated by *Picea glauca* (white spruce) to one dominated by *Betula papyrifera* (white birch). Soil classifications are therefore often drawn up somewhat arbitrarily on the basis of their dominant profile characteristics and with convenience of use in mind.

A soil and its vegetation cover are normally closely geared to each other. This is not always the case, however, so it is important that we consider those soil conditions which clearly do influence cover. The relationship between a soil and its vegetation cover is not simply due to the fact that vegetation cover protects regolith sufficiently for soils to show horizon differences attributable to the cover. The moisture and nutrient status of soils is also important to the success of plant species in open competition with each other. It is therefore not surprising that certain soil conditions and soil types are associated with specific vegetation covers and life forms, and it is this relationship that will be considered in differentiating ecoclimatic regions. On a continental scale it is obvious that short-grass prairie is normally associated with Brown Chernozems, while tall-grass prairie is usually associated with Black Chernozems. Similarly, on a local scale in northwestern Ontario and southeastern Manitoba, it is expected that the *Larix laricina* (tamarack) and *Picea mariana* (black spruce) forest cover associated with Organic soils of fens and bogs differs from the *Picea glauca*, *Populus tremuloides* (trembling aspen), *P. balsamifera* (poplar), and *Pinus banksiana* (Jack pine) forests that frequently dominate the better-drained Luvisolic, Podzolic, and Brunisolic soils on the slopes of morainic ridges above.

In the chapters which follow, use will be made of three separate soil classification systems. These are (1) the traditional world soils terminology, (2) the New Comprehensive System (developed in the USA and used primarily there), and (3) the Canadian Soil Classification System. The traditional soils terminology and the Canadian systems are based primarily on important soil-forming factors such as climate, while the New Comprehensive System emphasizes actual profile characteristics. The reader should refer to Table 1.3 for basic terminology used by the Canadian system and its counterparts in the other two systems. Noteworthy is the fact that the Canadian classification was designed solely for use within Canada, so it is necessary to use other classifications to describe soil types found elsewhere.

Figure 1.7 is a map of world soils using the traditional terminology popular in many publications. Figure 1.8 is provided to show both how the New Comprehensive System terminology translates onto a world map and to provide a comparison with the map of Canadian soils (Figure 1.9), which uses only Canadian Soil

Table 1.3
Comparative soil classification terminologies used in this text (modified from a number of sources including Canada Soil Survey Committee, 1978)

Climate/ Vegetation Environment	Canadian System (Orders)	New Comprehensive System (US) (Orders)	Traditional System (Great Groups)
Tundra	Cryosolic	Pergelic subgroups	Tundra
High boreal	Brunisolic	Entisols	Arctic brown
Boreal	Podzolic	Spodosols	Podzols
Mixed forest	Luvisolic	Alfisols	Grey wooded/grey brown Podzolic
Moist/high water table	Gleysolic	Aqu-suborders	Gleys
Peatland	Organic	Histosols	Bog soils
Prairie	Chernozemic	Mollisols	Prairie, chestnut brown, Chernozem
Various	Solonetzic	Natric great groups of Mollisols and Alfisols	Solonetz
Various	Regosolic	Inceptisols	Regosols/Lithosols
Cracking clays	na	Vertisols	–
Hot deserts	na	Aridisols	Sierozems and red desert soils
Subtropical forest	na	Ultisols	Latosols and red-yellow Podzolics
Humid tropical forest	na	Oxisols	Latosols

Classification System terminology. When figures 1.8 and 1.9 are compared, it is clear that although terminologies differ sharply, distributions of soil types show noticeable similarities. Note, for example, how the Chernozems on the map of Canada are similar in distribution to the Mollisols on Figure 1.8. Even when the traditional terminology is used at the world scale (Figure 1.7), similarities can be noted between global soil zones and the maps using the other two systems. Similarities between these maps should be anticipated because, as with vegetation, the role of zonal climate (and the influence of vegetation cover itself) are of great importance in the development of soils profiles. Table 1.4 provides a breakdown of the area in Canada occupied by each of the nine soil orders of the Canadian Soil Classification System.

1.5 Climatic Parameters

We frequently read of the dramatic correlation between vegetation formations and climate. The reader should be cautious of such statements, for although they contain a basic truth, there are always exceptions to the rule. It can be clearly demonstrated,

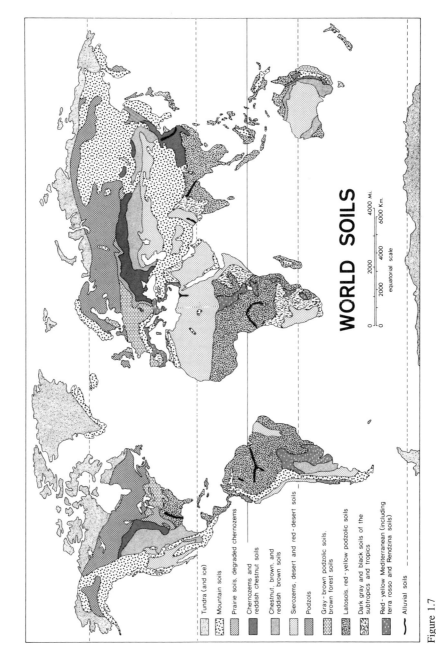

Figure 1.7
World soil zones using traditional terminology (after Basile, 1971)

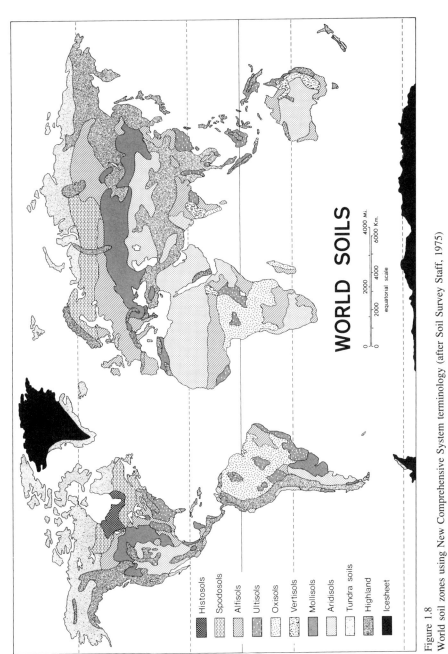

Figure 1.8
World soil zones using New Comprehensive System terminology (after Soil Survey Staff, 1975)

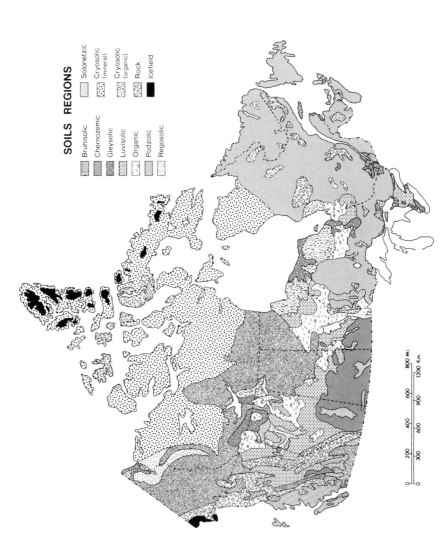

SOILS REGIONS

Brunisolic
Chernozemic
Gleysolic
Luvisolic
Organic
Podzolic
Regosolic
Solonetzic
Cryosolic (mineral)
Cryosolic (organic)
Rock
Icefield

0   200   400   600   800 Mi.
0   300   600   900   1200 Km.

Figure 1.9
Canadian soil zones using Canadian Soil Classification System terminology (modified from Soils Research Institute, 1972, to include Cryosols)

Table 1.4
Land area in Canada occupied by each soil order in the Canadian Soil Classification System
(from Editorial, 1985)

Soil Order	Area km²	% Land Area
Brunisolic	789,780	8.6
Chernozemic	468,190	5.1
Cryosolic	3,672,080	40.0
Gleysolic	117,143	1.3
Luvisolic	809,046	8.8
Organic	373,804	4.1
Podzolic	1,429,111	15.6
Regosolic	73,442	0.8
Solonetzic	72,575	0.7
(Rockland)	1,375,031	15.0

for example, that there are certain stands of tall-grass prairie found in areas with identical climates to temperate deciduous forests nearby. It can also be demonstrated that there are certain patches of conifers in the tundra which were established when climatic conditions were warmer but which survive even now, despite the fact that the climate has become harsher. Gallery forests found along rivers owe their presence to groundwater and not necessarily to the climate – savanna grassland patches can be found completely surrounded by forest. The list goes on, but these examples should suffice to point out that, as with soil-forming factors, more than one determinate is involved in explaining the distribution of vegetation formations. It would appear that in the case of vegetation, climate indeed plays a dominant role. Subordinate roles are, however, also played by such factors as soil conditions, influences of geomorphology, the fact that vegetation may only respond slowly to climatic change (the role of time), the influence of grazing animals and pests, and last but not least, the intentional and inadvertent influences of humans.

1.5.1 The Role of Climate

In the late nineteenth and early twentieth century, it became fashionable to stress the role of climate in explaining vegetation patterns. The work of Alexander von Humboldt (1805) and de Candolle (1855) on climate-vegetation relationships was taken one step farther by Köppen (1918) in his famous climatic classification. Köppen developed climatic indices so that climate zone boundaries matched those of vegetation. He even gave his climates vegetation names such as Tropical Savanna (Aw), Steppe (Bs), and Tropical Rain Forest (Af). Schimpfer (1903) and later Walter (1973) adapted the physiological approach to vegetation classification giving rise to the zonobiome (or ecophysiological) approach. Although this approach does not establish the exact mechanisms by which climate may control vegetation distribution, it provides a useful set of generalizations about distributions which is the basis of the ecoclimatic (regions) approach developed by the Ecoregions Working Group (1989). Holdridge (1967) and Box (1981) also base their recent vegetation classifications on an understanding of how climatic controls influence plant distribution (Woodward, 1987).

Our understanding of climate-vegetation relationships has developed to the point that Prentice et al. (1992) have been able to develop a climate-driven computer model that predicts which plant types can occur in a given environment. By choosing primary driving variables such as mean accumulated temperature over 5°C, mean coldest month temperature, and a drought index incorporating the available water capacity of the soil and the seasonality of precipitation, they have focused attention on those aspects of climate (and soils as moisture delivery systems) considered most germane to vegetation zonation. More important than simply noting that model predictions of global vegetation patterns are in agreement with actual ecosystem patterns is the appreciation that the model can be used in assessing impacts of possible future climatic changes (Prentice et al., 1992). Using predictor variables such as degree days, annual snowfall, and actual summer evapotranspiration, Lenihan (1993) has derived empirical ecological response surfaces for the eight dominant conifers and broadleaf deciduous trees in the Canadian boreal. He distinguishes five forest types that show a high degree of correspondence to extant zonation patterns, again confirming the strong climate-vegetation relationship.

Although the above authors stress the unique role of climate, they were also aware of vegetation patterns considered exceptions to the rule because they are influenced more by other factors. The impact of one such factor, a high water table, can be observed along the Nile floodplain near Luxor, Egypt, where mean annual precipitation is only 1 mm, yet the vegetation cover is evergreen broadleaf forest. A combination of factors such as excessive drainage promoted by low moisture retaining sands and the occasional natural fire help promote the physiological drought necessary to account for the presence of prairie–oak parkland in southern Ontario, a region otherwise dominated by temperate deciduous forest. In any discussion using the ecoclimatic regions approach, notice must also be taken of the exceptions.

1.5.2 Moisture Indices

While it is necessary to be familiar with the descriptive terminology associated with each climate-formation correlation, it is also useful to describe climates from the point of view of moisture availability. All ecosystems lose water to the atmosphere through evapotranspiration, so a measure of the success a climate has in satisfying the moisture requirements of a plant cover is useful in describing that cover. Thornthwaite (1948) developed the moisture index to allow for the comparison of the potential of individual climates to evaporate water (the potential evapotranspiration, or PE) with the actual precipitation (P). This index is written as $(P/PE - 1)\,100$, resulting in a negative moisture index (Im) if precipitation is less than PE and a positive value if precipitation exceeds PE. The mixed-grass prairie of the Regina region has a im of −40, while Winnipeg with its tall-grass prairie has a im of −12, and the corresponding value for Toronto, with its temperate deciduous forest, is +35.

It should be emphasized that this moisture index does not necessarily demonstrate when moisture stresses might occur and that even a region with positive values may experience severe moisture stress conditions during the hottest parts of the year. Nonetheless, researchers have found the use of moisture indices valuable in the discussion of vegetation distribution, and a recent analysis of vegetation distributions in China shows that desert, steppe, woodland, deciduous forest, and evergreen forest correspond to im values of <−40, −40 to −20, −20 to 0, 0 to +60, and >+60 respectively (Fang and Yoda, 1990).

1.5.3 Climate Diagrams

A practical way of summarizing the climate of any particular location is by using a climate diagram. Climate diagrams have been used since the time of Köppen, with the most comprehensive world collection having been produced by Walter and Lieth (1967). The example of the climate diagram for The Pas, Manitoba, given in Figure 1.10B is modified from the publication *Ecoclimatic Regions of Canada* (Ecoregions Working Group, 1989). All of the climate diagrams for Canadian stations used in this text are adapted from this source or from the original climatic normals data (Environment Canada, 1982), while all non-Canadian examples have been adapted from Walter et al. (1975) to fit this ecoclimatic format.

The advantage of using this type of graph to illustrate data for a Subhumid Mid-Boreal Ecoclimatic Region station (MBs) such as The Pas is not only that it gives a quick glimpse of the typical climate associated with the northern boreal forest vegetation but that it also allows for easy comparison with diagrams representing climates from other vegetation zones. In addition, it permits the presentation of important seasonal variations which can not be represented in the simple moisture index values detailed above, but which may be extremely important in influencing cover. Information at the top left of a typical climate diagram gives the vegetation-cover type, the ecoclimate, the station name, and the elevation of the station in metres above mean sea level (Figure 1.10). Mean annual temperature in degrees Celsius and mean annual precipitation in millimetres are given at the top right. The ordinate is divided into twelve months, beginning on the left with January for northern hemisphere stations and July for southern. Curve A represents mean monthly precipitation (mm), and curve B gives mean monthly temperatures (°C). Variations in seasonal temperatures are summarized in the horizontal bar attached below the ordinate. Here C represents months when the mean daily temperature is below 0°C, D gives months when the mean daily minimum is below 0°C, and E shows months with mean daily temperatures above 0°C. F represents the relatively humid season, and if a station was to have monthly precipitation values above 100 mm, the abscissa scale above 100 mm would be reduced to 1/10th (G). Likewise, if periods of relative drought occur, as in some grassland environments, the average monthly temperature

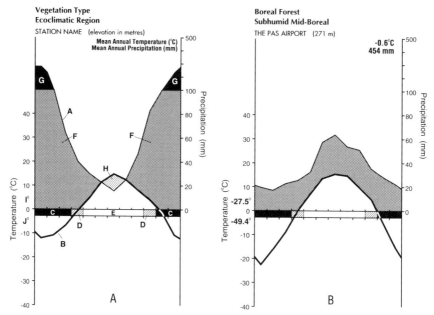

Figure 1.10
Examples of climate diagrams for (A) a theoretical station which illustrates diagram variety and (B) an actual Subhumid Mid-Boreal station (The Pas, Manitoba). For an explanation of how to read the diagram, see text. Modified from Ecoregions Working Group (1989).

curve rises above the curve for precipitation, so that the overlap area (H) represents the drought season. To the left of the abscissa are given two additional temperature values. For The Pas these are −27.5° and −49.4°C and represent mean daily minimum temperature of the coldest month (I) and the lowest recorded temperature (J) respectively.

1.6 Plant Strategies

Plants have evolved adaptations which permits them to occupy almost all terrestrial niches. Algae can live in the summer meltwater on Baffin Island's icefields, lichens can survive year round on the exposed granitic rock outcrops of the Canadian Shield or the dead branches of *Picea mariana* (black spruce), and *Opuntia fragilis* cacti seem to thrive on the sands of the Carberry Desert in Manitoba. Important strategies that help account for selected combinations of species making up formations include physiological adaptations to the environment and competitive strategies. The strategies of competition, moisture regulation, survival of stressful seasons, and leaf morphology and stress tolerance all deserve elaboration to see how each in turn helps account for the fact that at any particular location, only certain plant species survive and even fewer do well.

1.6.1 Competition

The floras of Saskatchewan and Manitoba each include in excess of 1,400 vascular plant species, the great majority of which they share in common (Fraser and Russel, 1944; Scoggan, 1957). When one actually visits a particular location in these provinces to examine the species composition present, the count is always much lower than this. The reasons are simple. Plant species are usually only adapted to a specific range of climatic/edaphic conditions, so many could not live at that spot anyway. Their distribution is therefore regulated by physiological requirements which, if the environment does not provide them, automatically lead to their exclusion. Perhaps more important is the fact that competition restricts or excludes a species from a site even though, from a physiological standpoint, it could survive there. In effect, a species must be physiologically suited to a site and from the ecological point of view, be competitive enough to survive the desire of other species also to live there.

We see in our gardens many perennial species surviving quite happily because they are within their physiological range (which to some extent may be maintained artificially by watering and fertilizer applications). The one important ecological difference between these garden species and those growing under natural conditions is that in the garden, weeding and spacing reduce or eliminate competition. The same can be seen in forestry plantations, where pure stands of *Pinus resinosa* (red pine) and *P. strobus* (white pine) often thrive in the controlled absence of competition from other woody species. Under natural conditions, many of our garden perennials would die out quickly due to this competition from "weeds" or "volunteers," and if poorly maintained, the plantation pines would perhaps succumb to natural competition from other volunteer (naturally seeded) conifers and/or hardwoods.

Figure 1.11 illustrates that the physiological and ecological response of individual species varies, but the various optima in combination help explain why so few species are found at any particular site. Figure 1.11B illustrates the relationship between the physiological optimum range through which a species such as *Pinus banksiana* can survive, as well as the ecological optimum that results when competition from species such as *Picea glauca* comes into play. In effect, *Pinus banksiana* is restricted to sites less than ideal for optimal growth, but to locations which give it the competitive edge. These sites include rocky outcrops on the Shield at one extreme and deep, sandy well-drained soils on the other. Jack pine can still be found occasionally on much better soils following forest fires, but as these soils are also more ideal for other species, the pine will succumb to competition and eventually be excluded. Figure 1.11C illustrates how the relationship between physiological optimum and ecological optimum would work for a species such as *Larix laricina,* which can tolerate high soil moisture contents for long periods. A comparatively strongly competitive species such as *Picea glauca* may in fact have identical optima for physiological and ecological distributions (Figure 1.11D). The coexistence of two or more plant species on the same site, however, demonstrates that they are able to avoid competitive exclusion (Aarssen, 1989).

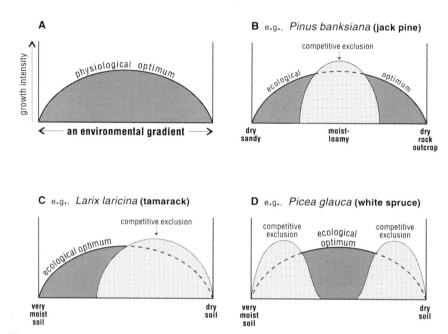

Figure 1 11
Physiological and ecological optima. See text for explanation. Modelled after Walter (1979, Fig. 17).

1.6.2 Hydrature and Moisture Regulation

All plant species both use water and need some form of water supply. The cells of higher (vascular) plants have central vacuoles which store water and supply it to the cell's cytoplasm when needed. Any water given up to the cytoplasm is replaced by inputs from the soil via vascular tissue. Vascular plants are therefore described as being "homeohydric" because they maintain a constant (or homeo) hydrature of the cytoplasm using vacuolar water. In non-vascular species such as bacteria, algae, fungi, and lichens, there is no guarantee of easy water replacement. In addition, these non-vascular groups lack cell vacuoles. These species are therefore described as being "poikilohydric" (or poikilohydrous) because their cytoplasm experiences fluctuations in hydrature as water is respired but not replaced due to the absence of vacuoles and vascular conducting tissue. The ecological significance of these differences is considerable. During drought periods, when no water can be supplied by the relatively sophisticated water-obtaining root system of vascular species, their cells may experience vacuolar shrinking, collapse of the cells, wilting, and death. The poikilohydric cells in lichens, on the other hand, having no vacuoles, can only lose water from their cytoplasms, shrink little, do not wilt or experience cell collapse, and essentially become dormant; on rewetting, cell activity is restored.

These strategies allow vascular species easy use of water in environments where considerable amounts are available in the soil or air during at least part of the year. Non-vascular species, such as lichens, have the advantage of being able to colonize the dead branches on living trees or even rock outcrops because their physiology allows them to be active during the short periods when moisture is available, or atmospheric humidity is high, and then become dormant when moisture is limiting. While these poikilohydric species have the advantage of being able to colonize environments too inhospitable for most species, they are not strong competitors when conditions more suitable for the homeohydric are provided. Along with species such as *Acer saccharum* (sugar maple), which are relatively well watered year-round and require little in the way of water conserving adaptations, are many vascular species that exhibit morphological adaptations to help them reduce water losses during water stress periods. *Opuntia polyacantha*, the prickly pear cactus typical of southwestern Saskatchewan prairie, is a good example of a vascular species which stores water in its succulent stems as a strategy that permits it basically to ignore the drought period. Other species, such as true xerophytes (which need some water at all times but do not have special water storage organs), limit water loss by developing hairy leaves, thick leaf cuticles, and so on. Some vascular species have evolved to make the most of droughty climates by producing extensive rooting systems to aid in water acquisition. *Populus tremuloides* (trembling aspen) have been recorded with roots exceeding 30 m in length in North American prairie, while some Australian eucalypts possess roots exceeding 60 m in length (Stone and Kalisz, 1991).

Between true vascular species and the poikilohydric groups are a number of species groups such as ferns and bryophytes (mosses and hepatics), many of which possess intermediate characteristics. Ferns have poorly developed vascular tissue and are more at the mercy of changes in atmospheric humidity than true vascular species, so as a consequence they are more frequently encountered in low respiration–high relative humidity ecosystems such as forest floors. On the other hand, those bryophytes such as *Sphagnum* spp. which have even less developed vascular systems than those of ferns are usually confined to perennially moist ecosystems such as the poorly drained wetlands of the Canadian and Eurasian boreal forest. Bryophytes found inhabiting dry sites, especially those in polar regions, are poikilohydric (Longton, 1988), however; so as a group, mosses can tolerate a broad range of conditions.

1.6.3 Life Forms

All higher plant species can be classified using a simple life-form classification based on the way in which individual species position their regenerating parts so as to survive the unfavourable season. This system was developed by Raunkiaer (1934) and essentially differentiates the ways in which the perennating buds of plants respond to the impact of climatic extremes (Scoggan, 1978; Pears, 1985). Generally, the closer buds are to the ground, the better protected they are. The Raunkiaer system

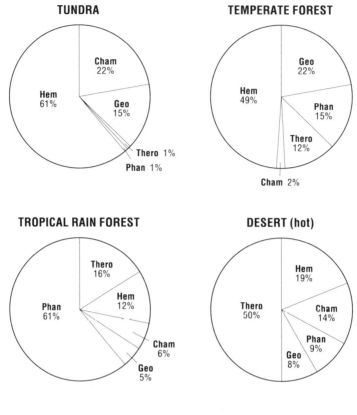

Phan - Phanerophytes (woody plants/trees with buds >25 cm)

Cham - Chamaephytes (woody herbaceous with buds <25 cm)

Hem - Hemicryptophytes (herbs with buds at soil level)

Geo - Geophytes (herbs with buds below soil level)

Thero - Therophytes (survive unfavourable season as seed)

Figure 1.12
Breakdown of four major climatic environments according to percentage vascular species using each life form strategy (data from Polunin, 1969)

is useful in that it allows us to divide up the species of a given area into groups and then compare the occurrence of these groups in different climatic regions. Such groupings can be illustrated as "spectra," as in Figure 1.12. Raunkiaer recognized ten groups of life forms, but later researchers such as Polunin (1969) have generalized these into five easily differentiated classes of land species and one category of aquatic species.

Phanerophytes are perennial woody plants (such as trees and tall shrubs) which have buds at least 25 cm above the soil surface. Because of their size, they exert a profound influence on the local ground cover even though their species numbers may be low. While phanerophytes comprise only about 12% of the species of the boreal and mixed forests of Canada, the discussion of these vegetation types will emphasize the contribution made by trees. Phanerophyte species are much more abundant in the tropics, so a normal worldwide spectrum would have 46% phanerophytes (Raunkiaer, 1934). Chamaephytes are herbaceous and woody species with buds above the soils surface but below a height of 25 cm. Hemicryptophytes are herbs which have perennating buds at or just below soil surface during the dormant season, while cryptophytes (or geophytes) are herbs where the buds (e.g., bulbs) survive well below the soil surface. Hydrophytes refer to all aquatic plants other than plankton (but this definition cuts across some of the previous terms). Therophytes survive unfavourable seasons as seed, and they differ from the previous categories (which are also usually seed producers) in that they are annuals which die off when the unfavourable season arrives. An example would be *Avena fatua* (wild oats), which is missing from the Canadian prairie agricultural landscape in fall but germinates from seed again the following spring. This therophytic strategy is particularly advantageous in deserts, where normally about half the species survive as seed until the next rains come. Surprisingly, this adaptation is of little advantage in cooler climates such as the tundra, where one growing season is usually too short to permit the complete cycle from germination to seeding. Raunkiaer (1934) reports therophyte values of 42% for desert in Death Valley, California, while Scoggan (1978) gives a range of 0–2% for tundra in the Northwest Territories. Raunkiaer (1934) reports an average world spectrum of 46% phanerophytes, 9% chamaephytes, 26% hemicryptophytes, 6% cryptophytes, and 13% therophytes.

Two additional Raunkiaer terms are mentioned because of their descriptive value, even though they are normally omitted from "spectra" diagrams such as Figure 1.12. Stem succulents are characteristic of many cacti, such as the *Opuntia polyacantha* mentioned above, but unfortunately this term also cuts across several of the definitions for other life-form terms. The term "epiphyte" will also be used to describe plants living on other plants. An example of an epiphyte would be *Arceuthobium pusillum* (dwarf mistletoe), which frequently parasitizes *Picea mariana* (black spruce), and as such, its perennating organs are not therefore restricted to any particular height above ground. Epiphytes are most frequently associated with the larger and longer-lived phanerophytes.

1.6.4 Leaf Morphology and Adaptation

A number of commonly used terms such as deciduous, evergreen, broadleaf, and needleleaf not only are useful for describing leaf morphology, but also provide insights into tree survival strategies. Angiosperm trees such as the *Acer saccharum* (sugar maple) have evolved the broadleaf habit, which provides them with a large photosynthetic surface somewhat at the expense of water regulation. In the warmer parts of Canada, on moist sites, these species do particularly well, but since they lose

their leaves in the fall to avoid winter freezing and moisture stress, they are described as deciduous. If the growing season for broadleaf trees is year-round, as for
example in humid temperate and tropical rain forests, severe moisture stress and
freezing are rare, leaves can remain year-round, and they are described as being
broadleaf evergreen. *Arbutus menziesii* (Pacific madrone) from southwestern British
Columbia is Canada's only broadleaf evergreen tree species.

In Canada we have many evergreen tree species, but these are, with the exception
of *Arbutus*, needleleaf conifers (Gymnosperms), not broadleaf species. Conifers
have thick, needle-like leaves with small photosynthetic area but good moisture
regulating properties. This ability to control moisture losses permits most conifers
living in cold climates, or on sandy soils prone to producing physiological drought,
to retain their needles year-round. The deciduous habit therefore favours broadleaf
species where moisture indices (Im) are above zero and winters are not very severe.
Where severe winters do not favour the broadleaf adaptation, evergreen conifers
compete successfully. In even more severe conditions, but conditions which still permit tree growth, the evergreen habit may again be unsuitable and deciduous
needleleaf species such as *Larix laricina* (tamarack) are about the only tree species
which will survive. In extremely harsh winter climates, such an adaptation is not
competitively advantageous, nor does it have to be, but it is a physiological necessity
(Gower and Richards, 1990). The fact that Canada's boreal forest is dominated by
conifers, with only a few common deciduous broadleaf species such as *Populus
tremuloides* and *Betula papyrifera,* and that forests in warmer, moist regions in
lower latitudes are dominated by broadleaf deciduous species is again a reflection
of the relationships between physiological requirements and competition.

1.7 Biomass and Net Primary Productivity

To provide additional ecosystem information useful in differentiating between vegetation types, data on plant biomass (phytomass), litter and soil organic-matter biomass, and net primary productivity are frequently included. Phytomass values are
usually expressed in tons per hectare (t ha^{-1}) or grams per square metre (g m^{-2}).
Useful comparative phytomass measures include (1) green plant phytomass, (2)
woody plant phytomass, (3) above-ground phytomass, and (4) below-ground phytomass. Values of this kind are useful when it comes to the comparison of community types, where ratios between above-ground and below-ground biomasses give
insight into community survival strategies. Living phytomass and litter and soil
organic-matter contents can also be viewed as nutrient reservoirs; therefore, when
appropriate, additional biomass data will be included for litter and soil organic matter. When the discussion is extended beyond the immediate ecosystem biomass to
such topics as atmospheric CO_2 and carbon sinks, it is again important to provide
total ecosystem biomass storage. In wetlands, peat buildup can also trap vast
amounts of carbon in the form of dead organic matter, and it must be appreciated
that one hectare of typical boreal forest wetland contains a much larger carbon store

Table 1.5
Mature ecosystem biomass,[1] soil organic matter,[2] and net primary productivity (NPP)[1] for selected vegetation types

Vegetation Type	Mature Biomass t ha^{-1}	Soil Organic Matter t ha^{-1} (inc. Litter)	NPP Approx. Mean g m^{-2}yr^{-1}
Tundra	1–30	408	140
Boreal forest	200–250	412	500
Temperate forest	420–460	268	1,000
Temperate grassland	?–30	378	500
Desert scrub	1–40	116	70
Tropical grassland	?–50	82	700
Tropical lowland rain forest	450	96	2,000

[1] From Lieth (1975).

[2] Modified from Schlesinger (1984) on the basis of soil/litter organic carbon representing 50% of the biomass.

than an equivalent area in well-developed Amazonian rain forest on freely drained soil.

While climate is considered to influence community structure and composition, it is also of great importance in net primary productivity (NPP), which is the remainder of annual photosynthetic production (gross primary productivity) after taking into account losses through respiration. It is therefore an important measure of an ecosystem's success in harnessing insolation and soil nutrients to produce energy storage and increase tissue production. Many ecosystems may not appear to increase in phytomass from one year to the next, even where net primary productivities are high, because of losses due to grazing animals and litter production. In plant communities undergoing secondary succession, net primary productivity exceeds these other losses and ecosystem phytomass increases. The same can be said for most bog ecosystems, where peat biomass increases because net primary productivity rates exceed the slow rates of decomposition. Net primary productivity values are generally expressed in grams per metre square per year (g m^{-2}yr^{-1}).

One measure of the harshness or suitability of a climate in terms of growth conditions on freely drained soils is the ratio between the soil organic-matter content and annual net primary productivity of an ecosystem (both expressed in kg m^{-2}). Known as the "K factor," this ratio normally decreases as soil organic-matter decomposition is slowed (Moore, 1978). As examples, lichen-heath tundra would have a decomposition K factor of 0.03, while boreal forest values would range from 0.1 to 0.2. By comparison, the large net primary productivities and the more rapid soil organic-matter decomposition rates of the humid tropics provide K factors as high as 4.0 (Moore, 1978).

CHAPTER TWO

Tundra

2.1 Tundra Distribution

Tundra vegetation consists of the low moss-herbaceous and sometimes shrubby ecosystems characteristic of areas between ice deserts and forest. The term "tundra" is derived from the Finnish word *tunturi* meaning "completely treeless heights" (Chernov, 1985). Primarily a circum-Arctic formation, tundra (ZB IX) extends across the northern edge of North America and Eurasia, with major areas in Alaska, the Northwest Territories, and northern Russia. An area close to nearly three million square kilometres is found in Siberia alone. Smaller areas are located in Quebec, Greenland, and Iceland and in some of the sub-Antarctic islands such as Kerguelen, Macquarie, South Georgia, and the South Sandwich Islands. Tundra communities are also found on the northern edges of the Antarctic continent itself, but only two vascular species are indigenous there. Tundra-like communities can also be found in alpine locations. Due primarily to the effects of latitude, these alpine tundra areas possess many differences from true high-latitude tundra, so they will be discussed separately at the end of this chapter.

2.2 Climate

Tundra vegetation is found in climates which favour a herbaceous cover not usually dominated by grasses and where the true tree life form is absent. It is the product of a harsh growing season more than the extremes of winter cold. In Arctic regions, winters can be bitterly cold, and the growing season has usually less than four months above 0°C and no months above 10°C (Figure 2.2). In the short, harsh growing season, productivity is more limited by available nutrients than by temperature. Plant physiological processes are, however, less temperature sensitive than would be the case among their warmer climate counterparts (Chapin, 1983).

Although mean annual precipitation is generally less than 200 mm, summer cli-

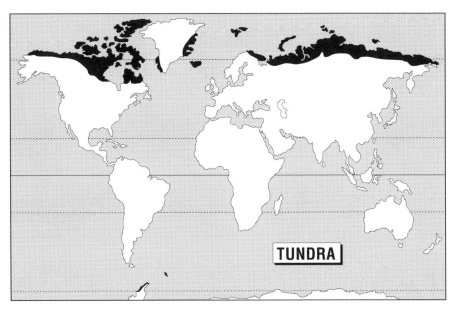

Figure 2.1
World distribution of tundra

mate is normally relatively moist since evapotranspiration is also low and drainage is often impeded by permafrost. Locally, well-drained regolith produces effectively arid ecosystems, while increasing aridity is the norm in the High Arctic, where large areas could be better described as Arctic or polar "desert" (Billings, 1974; Bliss, 1988) or by the more colloquial term "barrens." Wind is an important ecosystem modifier due to its role as a desiccating agent and its ability to distribute seed and spores. Water stress occurs more usually in winter, when plant parts stick out through the snow cover into desiccating and ice-crystal blasting winds. During the winter, soil moisture is frozen so plants must reduce or eliminate transpiration. Woody species are therefore discouraged from taking on true tree life form because the microclimate above snow cover, being effectively arid, promotes more rapid dessication of emergent parts. The short Arctic growing season is somewhat compensated for by longer days and a thinner atmosphere through which the sun's rays must penetrate, but even when the daylight period exceeds twenty hours, air temperatures do not rise significantly. This is because of the large amounts of energy taken up in the form of latent heat as soil ice melts and water evaporates. In contrast to the Arctic, sub-Antarctic tundra vegetation may be found in almost isothermal conditions, with all months above freezing. This results from the extreme expression of maritime influences.

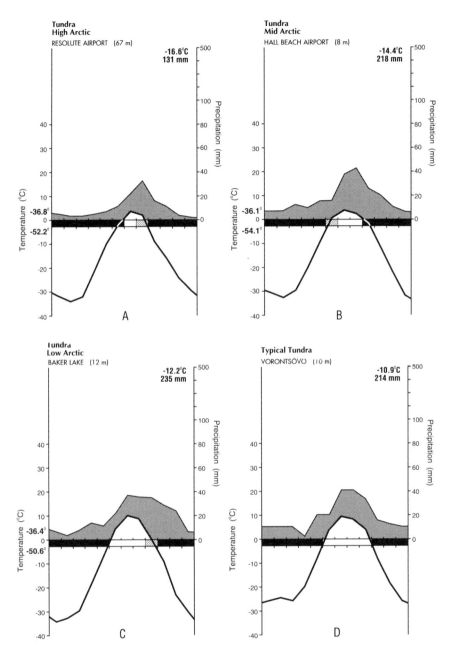

Figure 2.2
Climate diagrams from the Canadian and Eurasian tundra. A = Resolute Airport, NWT, a High
Arctic station; B = Hall Beach, NWT, a Mid-Arctic station; C = Baker Lake, NWT, a Low Arctic station;
D = Vorontsovo on the Arctic Ocean coast, Russia.

Figure 2.3
Distribution of permafrost in Canada (modified from Geological Survey of Canada, 1969)

2.3 Soils

Except in the western Arctic, which was not glaciated, North American soils usually develop on regolith derived from glacial activity. To this must be added the occurrence of frost-shattered debris, alluvium, raised beachs, and marine sediments, which are particularly common on the islands experiencing isostatic uplift. A common characteristic of surficial regolith in the Canadian Arctic is the presence of permafrost (Figure 2.3). Permafrost is soil and subsoil, as well as rock and sea floor, where temperatures remain below 0°C for two or more years (Bliss, 1988). In soils, this occurs where moisture has been allowed to collect in fine-textured or poorly drained regolith or in peat. Elsewhere, permafrost may be missing on deep sandy or gravelly substrates or on rock outcrops (non-soil). An often related phenomenon occurring in tundra soils is cryoturbation, a process whereby freeze-thaw action gives rise to vertical disturbances within soil profiles.

One consequence of having permafrost, with its potential for associated cryoturbation, is the frequent occurrence of relief features that result from selective freezing and thawing, ice wedging, and associated soil heaving or subsidence. Micro-relief land-forms include dominant landscape features such as polygons and hummocky terrain, as well as the less frequently encountered sorted and unsorted stone circles, stone stripes, and solifluction lobes. The self-organizing group of polygons, circles, and stone stripes are collectively known as patterned ground and are associated with water-laden soils subject to diurnal or seasonal refreezing (Krantz, 1990). Polygons are common in the Low Arctic, occurring most frequently in organic deposits and unconsolidated marine sediments, and they are associated with frost wedging, which develops when water collects in cracks formed by the winter contraction of frozen tundra (Hobbie, 1980). It may be that all of these patterned ground features owe their origin to the one mechanism, the effects of free convection induced by unstable density stratification in the soil (Krantz, 1990). The ice wedges that form in polygons can vary in breadth from a few centimetres to 8 m across, allowing the resultant wedge network to produce polygons 20–50 m across. Hummocky terrain, such as that found north of Inuvik in the Northwest Territories, results primarily from the effects of cryoturbation on mineral soil profiles (Tarnocai, 1972).

Two macro-relief features associated with the warmer tundra are pingos and thaw lakes. Pingos are mounds up to 25 m tall formed by massive ice growth in thick deltaic sediments, such as in the Mackenzie Delta, Northwest Territories. The localized melting of ground ice results in the development of a number of land-form types (thermokarst), one of which consists of depressions known as thaw lakes (Walker et al., 1987). On the central plains near Barrow, Alaska, up to 80% of the top 3–4 m of permafrost ground is actually ice, so disturbances or changing heat transfer leads to thaw lake development (Hobbie, 1980). The shorelines of these lakes vary due to permafrost melt and erosion but are generally elliptical in shape and provide important migratory waterfowl habitat. A dramatic change in the concentration of soluble salts occurs in these aquatic environments each year due to snowmelt dilution in spring, followed by summer runoff enrichment (Douglas and Bilgin, 1975). One micro-relief feature of interest is water tracks. These tracks have surfaces 10–20 cm lower than adjacent ground and function as subsurface downslope drainage channels (Chapin et al., 1988). What makes them of special interest is that they support distinct bands of vegetation, often 5–15 m wide.

All tundra soils freeze at the surface in winter and melt out again in summer. This happens whether or not permafrost is found below this seasonal freeze-thaw layer. Tundra soils are called Cryosols if permafrost is within two metres of the surface and the effects of cryoturbation are present or if permafrost is within one metre of the surface with no cryoturbation. Classification based on the presence or absence of ice creates an all-encompassing soil order that includes both mineral and organic soil profiles. Mineral Cryosols possess little real profile development other than where cryoturbation causes horizon disruptions (Figure 2.4A). Recognizable mineral soil profile horizons include thin, peaty layers (L-H), organic stained Ah hori-

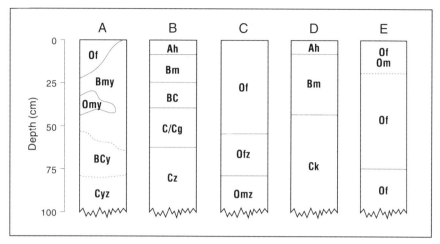

Figure 2.4
Typical soil profiles of the Canadian tundra (after Canadian Soil Survey Committee, 1978).
A = Turbic Cryosol (horizons disrupted by cryoturbation); B = Static Cryosol; C = Organic Cryosol;
D = Eutric Brunisol; E = Fibrisol. The lower-case letter y represents a horizon disturbed by cryotur-
bation, while z refers to permanently frozen.

zons, and gleyed Bg and Cg horizons overlying permafrost. In summer, the upper
15–25 cm of many poorly drained tundra mineral topsoils resemble the peaty or
Humic Gleysols more characteristic of warmer climates where permafrost is absent.
Another consequence of poor drainage in areas of carbonate-bearing sediments is
that average pH values range from only mildly acid to alkaline (Douglas and Bilgin,
1975).

Three great groups of the Cryosolic order are readily differentiated. The Turbic
and Static Cryosols are mineral soils with or without cryoturbation effects respectively, while the Organic Cryosols are of peaty origin (Figure 2.4). At the subgroup
level, these mineral Cryosols show typical Gleysolic, Brunisolic, or Regosolic characteristics (e.g., Gleysolic Turbic Cryosols, Brunisolic Static Cryosols, and Regosolic Static Cryosols). Applying New Comprehensive System terminology, which
does not use the presence of permafrost as a central theme at the order level, these
tundra soils would be poorly developed mineral soils such as Cryaquepts or in the
case of the Organic soils, Pergelic Histosols.

Brunisols can also be found in the tundra, particularly towards the forest-tundra
transition zone. Brunisols develop on well-drained sites where permafrost is absent
or at such depth that the effects of cryoturbation do not influence the soil profile.
These produce some of the more productive tundra soils in terms of plant growth,
particularly if they are Eutric Brunisols (pH >5.6, with parent material usually glacial or outwash deposits derived from sedimentary rock), as opposed to the more

acid Dystric Brunisols (pH <5.6, and frequently derived from glacial deposits of igneous or metamorphic origin). As one moves toward the transition zone with the boreal, particularly around James Bay, soils of the Organic order can be found. At the great group level, these soils are classified as Fibrisols if little decomposition of the peat has occurred or Mesisols if decomposition has been modest. However, because of the influence of permafrost, the great majority of tundra organic soils are Cryosols.

2.4 Tundra in North America

2.4.1 Ecoclimatic Sub-Provinces and Regions

While the great percentage of North American tundra is in Canada, sizeable areas are also found in Alaska. In terms of land area, Arctic tundra is Canada's major vegetation formation type, and the Arctic Ecoclimatic Province covers 26.09% of Canada (see Figures 1.3 and 1.4 and zones 13, 14, and 15 in Figure 1.5). Subzoning of this region is necessary not only because tundra includes a number of diverse areas ranging from the polar desert, cryptogam (lichen and mosses), and "impoverished" cushion plant-lichen community types on the colder side of the biome to continuous herbaceous low-shrub communities on the warmer, but because of the very important influences these differences have on wildlife, particularly migratory ungulates and waterfowl (Figure 2.5). The Arctic Ecoclimatic Province is subdivided into three sub-provinces and five regions, as given in Table 2.1.

In the High Arctic Sub-Province, such as at Resolute in the Northwest Territories (Figure 2.2A), mean annual precipitation is low (approximately 130 mm yr^{-1}) while mean daily temperatures exceed 0°C only in July and August and daily winter temperatures average below −30°C, except where strong maritime conditions are experienced. In the Mid-Arctic, conditions are frequently humid (300 mm yr^{-1}), with a marginally longer growing season. In the Low Arctic Sub-Province, such as at Baker Lake, Northwest Territories (Figure 2.2B), summers can be described as being cool and moist, with as many as four months averaging above 0°C, but no months above 10°C.

It is important to note that while the High Arctic is generally found at higher latitudes than the Low Arctic, it is growing season conditions, not latitude, which dictate their location. In the northwest and north-central parts of the Queen Elizabeth Islands, vascular species diversity is least, while the more diverse and dense vascular plant assemblages are found on the protected eastern side of Axel Hieberg and Ellesmere islands (Edlund and Alt, 1989). On Banks Island, all three ecoclimatic sub-provinces are located at the same latitude in close juxtaposition. Of significance is the possibility that Holocene recolonization of the tundra may have been aided by distribution from refugia. *In situ* sedge-peat deposits from Arctic Bay in northwestern Baffin Island have been radiocarbon dated to 16,849 ±860 BP (Short and Andrews, 1988). It is also important to note that post-glacial climate has fluctuated

Table 2.1
Zonation within the Arctic Ecoclimatic Province

Ecoclimatic Sub-Province	Ecoclimatic Region
High Arctic	1 High Arctic (HA)
	2 Oceanic High Arctic (HAO)
Mid-Arctic	3 Mid-Arctic (MA)
Low Arctic	4 Low Arctic (LA)
	5 Moist Low Arctic (LAm)

considerably across this vast zone. While studies in the Queen Elizabeth Islands to the north indicate present temperatures might actually be warmer than for much of the last few thousand years (Bradley, 1990), others to the south show *Picea mariana* open spruce woodland reverting to shrub tundra during the last 5,000 years (Cwynar and Spear, 1991; Nichols, 1967).

In the Canadian tundra, typical ecosystems would include (1) exposed bedrock and/or coarse-textured glacial till or outwash deposits on which few species other than lichens exist; (2) gentle to steep slopes where drainage is poor to fair and where a discontinuous to continuous herbaceous community, perhaps with some woody shrubs, is found; and (3) boggy/marshy sites where bryophytes (mosses), sedges, and reeds dominate. In addition to these vegetated ecosystems, there are areas without any cover which can best be described as polar deserts. On occasion, contrasting exposed polar desert rock outcrops and continuous herbaceous cover are seen almost side by side, because in this harsh environment substrate (soil), characteristics and microclimate can vary dramatically within only a few metres. One of the more important and almost unique characteristics of the tundra biome is that here minor microclimatic differences are sufficient to differentiate between sites where vascular species are physiologically adapted, and can thrive, and sites where higher plant growth is simply not possible. It would seem that many tundra species can withstand long periods with frozen tissues, but they must have a reasonable growing season or they can not survive. In most other biomes, variations in microclimate normally only produce variations in cover dominants. The primary controller of plant establishment in the Arctic is therefore the physical environment, with biological constraints such as competition only becoming important if succession is progressing (Billings, 1987).

A number of other important characteristics of the Arctic flora should be noted. One is the fact that the primary soil nutrient deficiency in the biome is often nitrogen and that while other mineral nutrients may be in adequate supply, many tundra soils are essentially oligotrophic in terms of available nitrate-nitrogen (Flint and Gersper, 1974). It would appear that plants not only have more difficulty assimilating nitrogen due to the low soil temperatures, but that available nitrate-nitrogen is limited in any event because recycling via litter decomposition is greatly retarded as a result of high moisture contents and low temperatures. Another contributing factor to low nitrogen availability is the relatively high demand for this nutrient by microbial decomposers. Research along Alaska's North Slope reveals that both nitrogen and phosphorus mineralization rates are low due to high microbial demands for both, while respiration release of carbon remains high (Nadelhoffer et al., 1991). Exceptions to this nutrient limitation can be found around uplifted boulders used as perches by seagulls and other birds, where the soil is enriched with guano (E. Oswald, personal communication) and where leguminous forbs such as *Oxytropis maydelliana, O. arctobia,* and *Astragalus alpinus*, or nitrogen-fixing lichens such as *Stereocaulon arenarium* and *S. rivolorum* are common (Karagatzides et al., 1985). Likewise, fertilization studies in Alaska show that fellfield alpine tundra vegetation is nutrient limited (Fox, 1992).

A second point is that a higher percentage of all vascular species survive as perennials and biennials compared to other biomes, where larger percentages of the local flora survive to the next growing season only as seed (i.e., as therophytes). It would appear that germination, plant maturation, flowering, and then seed production are simply too many stages to be completed by most tundra vascular species in one growing season, given the harsh climatic conditions, low levels of soil nitrogen, and so on. As only one to two per cent of the Arctic flora are therophytes (Scoggan, 1978), it would appear that a better life form adaptation is one which allows seeding in the second or a later growing season. It is important to also note that the actual flowers themselves are often very large in proportion to plant size, an adaptation which may well aid in attracting or accomplishing fertilization (E. Oswald, 1990, personal communication).

Arctic species are primarily geophytes (15%), chamaephytes (22%), and hemicryptophytes (61%) (Figure 1.12). Woody chamaephytes, especially prostrate and dwarf shrubs, often exhibit branch layering and have the larger proportion of their phytomass below ground (Billings, 1974). Chamaephytes are in particular harmony with Arctic and alpine environments where both cold and desiccating winds eliminate phanerophytes (Scoggan, 1978). Different shrub forms can also be considered as general bioclimatic indicators. As mean July temperatures increase, one passes from polar desert to a non-woody herbaceous zone, then to a prostrate shrub–herbaceous zone, a prostrate shrub zone, a dwarfed and prostrate shrub zone, a low, erect shrub zone, and finally into forest-tundra (Edlund, 1990).

Phanerophytes are normally too tall to survive the severe climate, except in the more sheltered hollows in the Low Arctic. In High Arctic tundra, phanerophytes are absent, while they only comprise one per cent of the species closer to the forest-tundra transition. Arctic plants therefore have to be hardy, are usually perennial, often have well-developed underground storage organs, and normally survive the harsh winters because they seek protection in the soil, hug the ground, or at least stay beneath the surface of the mid-winter snow. Woody plants which penetrated through winter snow are often described by the term "krummholz" because of their contorted shape. This krummholz form (from the German for "crooked wood") is the result of branch die-back due to winter wind desiccation and ice-crystal abrasion. A frequent form of this krummholz is the "flag" shape that is more characteristic of many *Picea glauca* exposed to cold northwesterly winds along the southern shores of Hudson Bay in the forest-tundra transition (Scott et al., 1987a; see also chapter 3.5.3). Many of these flag forms also have "skirts" of basal branches, which are usually protected in winter below snow. Of the phanerophytes which do survive here, most have flattened or creeping "mat krummholz" forms that spread by vegetative propagation.

It is noted that here in this almost ice-desert situation, plant species are predominantly perennials, while in warm deserts exactly the opposite is the case, with the great majority surviving the adverse (hot, dry) season as therophytes. It is also noted that in comparison to most other biomes, nearly all tundra vascular species possess the C_3 photosynthetic pathway, and a greater percentage are also polyploids. In polyploids, each cell nucleus contains three or more sets of chromosomes instead of the usual two, an adaptation which appears to give enhanced tolerance levels. Packer (1969) reports that in the High Arctic, on islands such as Prince Patrick, over 70% of the Dicots and over 90% of the Monocots are polyploid. Others conclude there appears to be no direct and general causal connections between the ecology, habitat, and distribution of Angiosperms on the one hand and polyploidy on the other (Ehrendorfer, 1980). Ehrendorfer (1980) does conclude, however, that when diploids and polyploids within related species are compared, it would appear that the diploids are more common in the stable habitats, while polyploids are often associated with more successional or labile communities.

Despite the shortness of the growing season, with its low temperatures, some community types are quite xeric or polar desert–like. In sites such as rock outcrops, well-drained raised beach ridges, or fluvioglacial outwash deposits, the poikilohydric properties of lichens and some bryophytes allows most of even the harshest environments to have a plant component. Based on the structure and development of the fruiting body (ascocarp), lichens are classified into three basic groups: crustose, foliose, and fruticose (Vitt et al., 1988). Crustose types have low biomass, are thin, are attached over their entire thallus to the substrate, and are often primary colonizers. Foliose types are somewhat leafier but also tend to do well on rock outcrops, to which they become attached by means of short hairs (rhizines). Fruticose lichens are tufted and tend to have larger biomass, the primary thallus is foliose-like, and it produces fruiting stalks from secondary thalli. These fruticose types have generally greater water needs and tend to dominate mineral and organic soil material where moisture supplies other than direct precipitation are found in storage.

2.4.2 High and Mid-Arctic

The flora of Canada's High Arctic is characterized by discontinuous vegetation cover and low diversity and is dominated by cryptogams and herbaceous, rather than

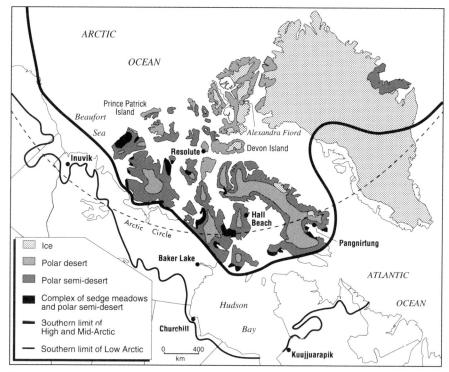

Figure 2.5
Zonation within the Arctic tundra of North America (adapted from Bliss et al., 1981). Note that many areas within the High and Mid-Arctic can be classified as polar desert as they are without any vegetation cover at all.

woody, species. Typical continental sites may have 15% cover but can increase up to 75% where oceanic influences are strong. The relative importance of cryptogams in this impoverished flora is seen on the calcareous erosional mountains in central Ellesmere Island, where a vegetation analysis revealed 156 plant species, of which 81 were lichen and 37 bryophytes, with only 38 being vascular species (Maycock and Fahselt, 1992). The majority of vascular species are the more resistant, hardier representatives of families and genera found in warmer habitats to the south. Strong positive correlations have been noted both between vascular species numbers and mean July temperature (Rannie, 1986) and between phytomass and growing season energy (Wielgolaski et al., 1981). Where drought promotes polar desert conditions, communities are often only patches of moss fed by summer meltwater from snow banks. The well-drained nature of coarse-textured regolith and high alkalinity combine to limit plant cover to only about 20% of Cornwallis Island and adjacent islands (Edlund, 1990). In effectively more humid sites, such as on Devon Island (Figure 2.5), percentage cover is much greater, and species diversity higher.

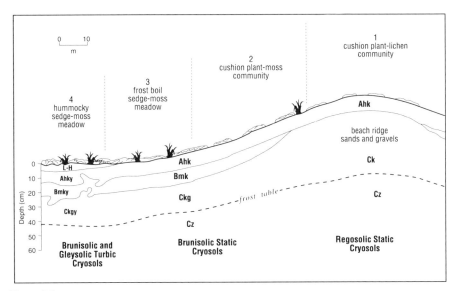

Figure 2.6
Typical beach ridge ecosystem diversity in the High Arctic, Truelove Lowlands, Devon Island,
NWT (modified from Bliss, 1977)

In the High Arctic, sedges such as *Eriophorum angustifolium* (cotton-grass), *Carex membranacea*, and *C. aquatilis* and mosses (but no *Sphagnum* spp.) dominate cover in very moist locations, while the dwarf shrubs *Dryas integrifolia* (white mountain avens) and the female plants of the dioecious *Salix arctica* (Arctic willow), together with sedges, tend to dominate on moist, but drained sites. Dryer sites have low cover and much bare ground, with *Dryas*, *Saxifraga oppositifolia* (purple sax-ifrage), male *Salix arctica* (Dawson and Bliss, 1989), and lichens such as *Alectoria pubescens* dominating. Cover height rarely exceeds 15 cm. In the Mid-Arctic, cover increases to 40–60% in typical continental sites, with *Saxifraga, Dryas*, and *Salix* spp. dominating, along with sedges, mosses, and lichens. More cover is present on drier sites than in the High Arctic, but it is still usually less than 15%. Wet sites nor-mally have a continuous cover of sedges (e.g., *Eriophorum*), saxifrages, and mosses. As with the High Arctic, mosses do not include *Sphagnum* spp. Soils throughout both the High and Mid-Arctic are typically Cryosols.

Figure 2.6 illustrates typical ecosystem differences in a High Arctic environment on the Truelove Lowlands, Devon Island, Northwest Territories, an International Biological Program (IBP) site. Floristically, these lowlands are species-rich com-pared with most High Arctic environments (Bliss, 1977, 1990) and are biologically close in diversity to the Mid-Arctic, except for fewer plant species, especially shrubs. The Truelove Lowlands flora includes 96 vascular plant species, 132 mosses, 30 hepatics, 182 lichens, and 92 species of fungi. For Devon Island in general, more than 50% of the Dicots and 80% of the Monocots are polyploids (Packer, 1969). While there are many more habitat types on Devon Island than shown in Figure 2.6, the four depicted here for the mesotopographic profile from hummocky sedge-moss meadow upslope to cushion plant-lichen – dominated raised beach ridges, along with a fifth type – the dwarf-shrub heath on bedrock outcrops – are the more typical. Because these lowlands contrast so noticeably from polar desert nearby, they are often called polar "oases," a name that takes on greater significance when it is appreciated these sedge meadows serve as valuable feeding habitat for muskox (Pearce, 1991). Descriptions of the five representative cover types noted above follow (primarily from Muc and Bliss, 1977, and Bliss, 1977). For a similar mesotopographic gradient on Bathurst Island, see Miller and Alpert (1984).

1 Cushion Plant–Lichen Community   This community dominates the beach ridge crest where soils are well-drained Regosolic Static Cryosols that exhibit little profile development or cryoturbation. At best they have an Ah horizon above calcareous parent material (ck) and permafrost (cz). This community type makes up some 6.4% of the Truelove Lowlands non-aquatic environment. Common vascular species include *Saxifraga oppositifolia, Carex nardina,* and the dwarf shrubs *Dryas integrifolia* and *Salix arctica.* Cryptogams include the lichen *Alectoria pubescens,* the community dominant, along with five other common lichen species and six common moss species, including *Encalypta rhaptocarpa* and *Distichium capillaceum.* Lichens cover 55–60% of these beach ridge crests, with approximately 30, 10, and 5% being crustose, foliose, and fruticose respectively (Richardson and Finegan, 1977).

2 Cushion Plant–Moss Community   This community type makes up almost 17% of the area. A somewhat better moisture regime is found here, and soils are usually Brunisolic Static Cryosols with modestly weathered Bm horizons. *Dryas integrifolia* is the dominant vascular species in this community, while *Saxifraga oppositifolia, Cassiope tetragona* (Arctic white heather), *Salix arctica,* and the sedges *Carex rupestris* (rock sedge) and *C. misandra* are common. Co-dominant with *Dryas* are the three moss species *Tortella arctica, Oncophorus wahlenbergii,* and *Racomitrium sudeticum.* At least five species of lichen are common and include *Cetraria cucullata* and *C. nivalis.*

3 Frost-Boil–Sedge-Moss Meadow   At these topographically lower sites, drainage is becoming more impeded and the possibility of some cryoturbation exists, but Brunisolic Static Cryosols are still the dominant soil type. This community type dominates 18.7% of the non-aquatic environment. Dominant vascular species are *Eriophorum triste* (a cotton-grass) and *Carex membranacea,* while *Carex stans, C. misandra, Salix arctica, Arctagrostis latifolia, Polygonum viviparum,* and *Juncus biglumis* are common. The dominant moss is *Drepanocladus revolvens,* while at least eight additional moss species are frequently encountered. High moisture levels limit the competitive strategies of lichens, so that only two species, *Xanthoria elegans* and *Solarina saccata,* are common, and they comprise less than one per cent of the cover.

4 Hummocky Sedge-Moss Meadow   Covering 26% of the Truelove Lowlands non-aquatic environments, this community type is typically found on fair to poorly drained soils, such as Brunisolic Turbic Cryosols and Gleysolic Turbic Cryosols. As the subgroup names imply, cryoturbation and the anaerobic conditions conducive to gleying and peaty surface-layer development are commonly encountered. *Carex stans* is the dominant vascular species, although *Eriophorum angustifolium, Salix arctica, Saxifraga hirculus* (yellow marsh-saxifrage), *Polygonum viviparum* (alpine bistort), and several other *Carex* spp. are common. *Cinclidium arcticum* and *Drepanocladus revolvens* are the dominant mosses, while lichens constitute less than one per cent of the cover. Polygonal peat plateaux can also develop in these lowland sites if substrates are formed from saturated sediments left as shallow lakes and ponds drain following isostatic rebound (King, 1991).

5 Dwarf-Shrub Heath   Covering 15.7% of the lowlands, these rock outcrop communities are the floristically richest lowland community type. Small pockets of soils associated with granite rubble and their related microclimatic diversity encourage this species variety. The dwarf shrub *Cassiope tetragona* (Arctic white heather) dominates, along with the moss *Racomitrium lanuginosum*. Other common vascular species include *Dryas integrifolia, Saxifraga oppositifolia, Salix arctica,* and *Carex misandra*. While a number of other moss and lichen species are common on soil patches, lichens such as *Rhizocarpon geographicum* and *Alectoria* spp. are common on exposed rock surfaces.

Cushion plants (especially the prostrate dwarf shrub *Dryas integrifolia*) and moss-lichen communities dominate vast areas on the southern Arctic islands in Canada's High and Mid-Arctic Ecoclimatic Regions. Where rock outcrops are found, the dwarf shrub *Cassiope tetragona* often dominates. It must be stressed that the harsh climate dictates that community species diversity is modest and plant height restricted. Vascular species distribution depends primarily on microclimate and soil conditions, particularly moisture content, with lichens common on drier sites and mosses on wetter. The instability of soils due to frost heaving also reminds us that most Arctic communities remain in a state of flux (Griggs, 1934).

2.4.3 Low Arctic

The Low Arctic is the single largest ecoclimatic sub-province in Canada (Figures 1.4, 2.5). Here plant species numbers have increased considerably over High Arctic equivalents, with the vascular flora for an area of 100–200 km^2 possibly including 100 to 150 species (Bliss, 1988). In addition, tundra vegetation becomes more noticeably woody, with shrubs such as *Salix* spp. and *Ledum palustre* (dwarf Labrador tea) dominating, together with sedges on wetter sites. On drier sites, shrubs such as *Betula glandulosa* (scrub birch), *Ledum*, and *Vaccinium* spp. join *Salix*. To these should be added krummholz and flag forms of both *Picea glauca* and occasionally *Larix laricina* in the southern tundra regions along the southwestern shores of Hudson Bay. South of the Brooks Range in north-central Alaska, *Alnus crispa* (alder) dominates many areas and creates a shrub tundra landscape in which resource-related competition on low-resource sites gives rise to a remarkably regular alder spacing (Chapin et al., 1989).

Although in favoured south-facing Low Arctic depressions, some woody species reach 2 m, cover normally does not exceed 30 cm. The fact that shrub tundra is so weakly developed or is virtually absent from much of the northeast Mackenzie District, across Keewatin, and into northernmost Quebec seems to be primarily due to the strong winds of the region, which cause snow abrasion (Savile, 1972). On some slopes, solifluction lobes result where summer thawing above permafrost encourages slow downslope creep of the saturated soil trapped between frozen subsoil and the matted surface-vegetation cover. This not only acts as a geomorphic agent in the denudation process but helps to create habitat differences, thereby encouraging ecosystem variety. On the well-drained south-facing slopes of pingos within the Central Plain of Alaska can even be found steppe tundra – like communities where cover is dominated by grasses and forbs (Walker et al., 1991). Another important ecosystem-modifying feature of level, coarse-textured mineral-soil, and organic-soil areas in the Low Arctic is the development of polygons. Here the freeze-thaw activities in mineral soils cause the finer material to collect at the centre of the polygon while the coarser material migrates to the edges. In organic landscapes, frost wedging is more important in the development of peat polygons. Polygons are particularly noticeable because any micro-relief created greatly influences soil moisture levels and profile drainage, and consequently vegetation cover. In the Old Crow Flats region of the Yukon, the lowlands are broken up by dozens of thermokarst lakes surrounded by marshes, willow thickets, and fields of low-centred peat polygons. These polygons are usually dominated by *Sphagnum balticum* and *Eriophorum* and *Carex* species (Ovenden, 1981).

Ecosystem variety typical of the Low Arctic ecoclimatic environment is illustrated in Figure 2.7, which is based on tundra studies performed near Baker Lake in the eastern Keewatin District of the Northwest Territories between 60° and 68° north (Zoltai and Johnson, 1979). When climate data for Baker Lake (Figure 2.2C) are compared to Hall Beach (Figure 2.2B) in the Mid-Arctic, it is noted that the mean annual temperature has risen over two degrees Celsius, and the growing season has increased to four months averaging above freezing. The effects of this climatic difference is sufficient to encourage greater percentage ground cover, increased plant productivity, more species diversity (particularly shrubs and graminoids), and taller plants. Occasionally, small *Picea mariana* (black spruce) stands are found in protected Keewatin lakeshore sites, but they take on krummholz forms, not true tree life

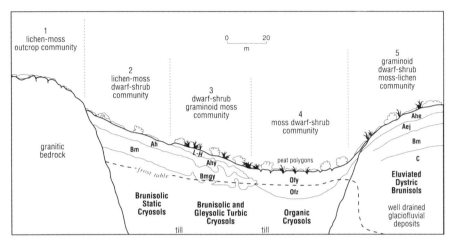

Figure 2.7
Idealized ecosystem variety in the Canadian Low Arctic (designed after descriptions given in Zoltai and Johnson, 1979)

forms, and are essentially outliers of the forest-tundra transition (boreal-tundra ecotone) of the High Subarctic Ecoclimatic Region. Likewise, in the Cape Churchill region on calcareous, till, raised beachs, and eskers, *Picea glauca* are frequently encountered (Scott et al., 1987a). As with cooler regions of the Arctic, community differences depend primarily on differences in microclimate, the presence or absence of soil, and soil moisture conditions (Figure 2.7). Permafrost and cryoturbation actively influence soil development. Cryosols dominate except where deep, well-drained glaciofluvial deposits are found. Variety typical of shrub tundra in the Baker Lake region is illustrated in Figure 2.7 and described below (after Zoltai and Johnson, 1979).

1 Lichen-Moss Outcrop Community   On exposed granite and quartzite bedrock non-soil ecosystems, crustose and foliose lichens dominate, with fruticose lichens and mosses becoming common in moist cracks, troughs, or basins. Important outcrop crustose lichen genera include *Lecidia, Rhizocarpon,* and *Lecanora*, while *Cetraria* and *Parmelia* are the commoner foliose genera. Due to lichen coloration, these outcrops can be strikingly beautiful. In the moister cracks, the mosses *Grimmia affinis* and *Racomitrium lanuginosum* dominate, along with fruticose lichens such as *Sphaerophorus fragilis*.

2 Lichen-Moss–Dwarf-Shrub Community   On the better-drained till sites, lichens and the moss species *Dicranum elongatum* dominate, with the more important vascular species being dwarf shrubs. Soils are Brunisolic Static Cryosols and Turbic Cryosols. To the south, where parent materials are mainly of granitic origin, soils are acidic (pH <5.0) with low base saturation, while pH rises to >6.5 on more calcareous parent materials to the north. Dominant lichens include the brightly colour-ed *Alectoria ochroleuca, Cetraria cucullata, C. nivalis* (all yellow), *Bryocaulon divergens* (black), and species of *Sphaerophorus, Cladina,* and *Stereocaulon.* The dominant moss is *Dicranum,* while *Aulacomnium turgidum, Tomenthypnum nitens,* and *Ptilidium ciliare* are common on moister sites. The common dwarf shrubs (in terms of decreasing order of cover) are *Empetrum nigrum, Ledum palustre, Vac-cinium vitis-idaea* (rock cranberry), *V. uliginosum* (alpine bilberry), and *Arcto-staphylos alpina* (alpine bearberry). Non-sorted circles resulting from frost boil activ-ity give variety to this community. Circle centres are virtually unvegetated due to contemporary frost action but may support occasional sedges such as *Carex capillaris.* The elevated peaty ring surrounding boils is usually dominated by the dwarf shrubs *Ledum palustre* (dwarf Labrador tea), *Betula glandulosa, Loiseleuria procumbens* (alpine azalea), and *Empetrum* and *Vaccinium* species. Only towards the High Arctic does the dwarf shrub *Dryas integrifolia* (white mountain avens) be-come important in this habitat type.

3 Dwarf-Shrub–Graminoid-Moss Community   On poorly drained tills, shrub, graminoids, and moss cover increase, while lichen cover decreases. Cryosols are Turbic, peaty surface layers (L-H) are the norm, and gleying is an important pedo-genic process. Here the dwarf shrub component is well developed and includes *Salix arbusculoides, S. glauca* (blue-green willow), *S. phylicifolia, Betula glandulosa, Ledum palustre* ssp. *decumbens,* and *Vaccinium uliginosum.* To these, add the dwarf shrubs *Dryas* and *Cassiope* towards the Mid-Arctic. Common graminoids include the grass *Arctagrostis latifolia* and several sedges (*Carex* spp. and *Eriophorum vaginatum*). Typical Low Arctic bryophyte dominants include *Dicranum* spp., *Polytrichum strictum, Hylocomium splendens,* and *Ptilidium ciliare,* while these are joined by more "boreal" species such as *Sphagnum* spp. and *Aulacomnium stricta.* While lichens comprise a smaller percentage of the cover, the bright yellow *Cetraria* and *Cladina* spp. are greatly in evidence.

4 Moss–Dwarf-Shrub Communities   Characteristic features of Keewatin organic terrain include high centre peat polygons and elevated peat mounds. In high centre polygons, peat attains thicknesses of 1.5–2 m. They are dissected by deep polygonal trenches often up to 70 cm deep and 1–2 m wide and underlain by frost wedges. This creates ecosystem diversity, with wet fen communities in depressions and better-drained plateaux or domed polygon centres. Wedging and cryoturbation ensure soils are Turbic Organic Cryosols. The elevated, better-drained peat areas are dominated by *Sphagnum* spp., with the mosses *Dicranum elongatum* and *Polytrichum strictum* being secondary, along with such forbs as *Rubus chamaemorus* (cloudberry). Dwarf shrubs include *Ledum, Andromeda polifolia* (bog rosemary), *Vaccinium* spp., and *Betula glandulosa,* while lichens have low cover. If fen surrounds the raised peat areas, they are completely dominated by the wetland sedges *Carex aquatilis, C. chordorrhiza, C. membranacea, Eriophorum angustifolium,* and *E. vaginatum* and the mosses *Drepanocladus aduncus, D. fluitans, D. revolvens*, and *Scorpidium scorpioides*. The dwarf shrub *Salix arctica* is found occasionally, while lichens tend to be absent.

5 Graminoid–Dwarf-Shrub–Moss-Lichen Communities   Found on the well-drained glaciofluvial deposits, this community has similar characteristics to the well-drained till community (no. 2), except that here graminoids comprise a significant portion of cover, herbs are rare, and the dominant dwarf-shrub species change. Permafrost is usually well below 2 m in depth and the well-drained, somewhat acid mineral soils are classified as Eluviated Dystric Brunisols. Cover dominants include *Hierochloe alpina* (holy grass), *Luzula confusa* (woodrush), *Poa* spp., and *Arctagrostis latifolia* (polar grass), while a thick moss turf dominated by *Dicranum elongatum* is widespread. Dwarfed versions of the dwarf shrubs *Ledum decumbens* and *Vaccinium uliginosum* are common, while *Betula glandulosa* is rare. In northern sites the dwarf shrubs *Vaccinium vitis-idaea, Empetrum nigrum, Arctostaphylos alpina,* and *Dryas integrifolia* become the dominant woody representatives. This community is noted for its striking colour, imparted by the yellowish impigmented lichens (*Cetraria* and *Cladonia* spp.).

In addition to the variety shown along the topographic profile given in Figure 2.7, there may also be variety associated with differences along the contour. An example of the latter is the impact of water tracks, which are subsurface drainage channels on slopes. These channels have similar species compositions to their surroundings (which are also moist), but the excess throughflow of subsurface water increases the nutrient flux, which may explain why productivity increases (Chapin et al., 1988). Communities of this type found along the northern foothills of the Brooks Range in Alaska are as much as 2.4-fold more productive than adjacent sites, primarily because of the sedge *Eriophorum vaginatum,* which experiences an almost tenfold increase in above ground phytomass (Chapin et al., 1988). Other micro-relief features that add variety to shrub tundra landscapes are stone stripes and solifluction lobes.

Shrub tundra communities are common east of Churchill along the shores of Hudson Bay. Here, winds off the pack ice cause such severe ice-crystal blasting that *Picea* and *Larix* are uncommon, and ridge-top communities are usually less than 20 cm tall. In sheltered areas behind gently curved, raised beach ridges, however, well-developed shrub communities can develop in which *Salix* spp. can reach heights of 3 m. Important members of these sheltered communities are *Salix brachycarpa* (short-capsuled willow), *S. candida* (silver willow), *S. planifolia* (flat-leaved willow), *S. lanata* (lime willow), *Betula glandulosa* (scrub birch), and *Shepherdia canadensis* (soapberry). On mid-slopes a low dwarf-shrub–lichen community ranging between heights of 20 and 35 cm is dominated by *Arctostaphylos alpina* (alpine bearberry), shorter *Betula glandulosa* shrubs, ground-hugging *Salix reticulata* (snow willow) and *Shepherdia canadensis*, tufts of *Dryas integrifolia* (white mountain avens), lichens such as the dark-coloured *Cetraria icelandica*, and several species of light-coloured *Cetraria* and *Cladina* lichen. Only occasionally are *Picea glauca* or *Larix laricina* encountered here, and when they are, they are usually less than 1.5 m tall and both species possess basal skirts and the flag form. *Larix*, being a deciduous conifer, generally seems less influenced by flag promoting die-back. On the windswept ridges, *Dryas integrifolia* and *Saxifraga tricuspidata* (three-toothed saxifrage) are common where vegetation cover is not continuous. Where cover is continuous, dwarf shrubs such as *Arctostaphylos alpina*, *Vaccinium uliginosum* (alpine bilberry), and krummholz forms of *Betula glandulosa* dominate, while some *Rhododendron lapponicum* (Lapland rose-bay) are present, together with light-coloured fruticose lichens such as *Cladina stellaris* and *C. rangiferina*.

It is important to stress that while plant communities in the Low Arctic frequently appear diminutive in terms of productivity and phytomass, they are indeed a relatively rich feeding ground for fauna, particularly migratory ungulates such as caribou (*Rangifer tarandus*). The variety of habitat type, the higher protein content of Arctic plants, and the variation in timing of community-type growth bursts all contribute to a reasonably continuous supply of fresh browse throughout the summer, while wetland habitats do likewise for migratory waterfowl. Localized tundra overgrazing, however, can occasionally lead to the removal of most of the macrolichen biomass, with lichen recovery taking several decades (Henry and Gunn, 1991). So important is tundra lichen production to the caribou, and caribou to the local inhabitants, that co-management projects, such as the one which oversees the Beverley and Kaminuriak caribou herds in the Northwest Territories, northern Saskatchewan, and northwestern Manitoba, have been set up to manage this important ecosystem resource. Wildlife protection is also aided by the establishment of parks such as the large Auyuittuq National Park north of Pangnirtung on Baffin Island.

It is also important to appreciate the vulnerability of this ecosystem to other natural and human disturbances and to take measures to mitigate damage where it is present. While many tundra species seem to be well adapted to such disturbances as fire, the effects of oil drilling or the loss of topsoil on the excavation of borrow pits along highways are more problematic. Because of the importance of fire, Rowe (1983) has categorized tundra species into five "fire-response" categories: invaders, evaders, avoiders, resisters, and endurers. In northwestern Alaska, Racine et al. (1987) found that seed and seedling fluxes of certain species increase dramatically after lightning-initiated fire and that the "resister" species *Eriophorum vaginatum* may well be fire dependent. In the Yukon and western Northwest Territories, Kershaw and Kershaw (1987) observed that of the 433 taxa found in 80 borrow pits abandoned for at least five years, 18 species had naturally colonized all of them. Important among these 18 were *Arctagrostis latifolia*, *Betula glandulosa*, *Empetrum nigrum*, *Epilobium* spp., *Poa* spp., and *Salix alaxensis*. In the high lime content borrow pits of the Churchill region of northern Manitoba, *Dryas integrifolia* is the dominant early colonizer.

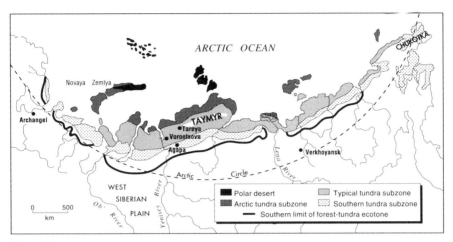

Figure 2.8
Tundra in Eurasia (modified from Chernov, 1985)

The Low Arctic shrub-dominated tundra gives way to shrub-dominated forest-tundra of the High Subarctic at the tree line. This tree line is where the true tree growth form for such species as *Picea glauca* and *P. mariana* exceeds a height of 3.5–5 m. Although exceeding this height, many of the trees at the edge of the zone, particularly where influenced by the cold winds off Hudson Bay pack ice, show deformed characteristics such as flag and skirted growth (Scott et al., 1987a). A fuller discussion of this transition is given in chapter 3.

2.5 Tundra in Other Northern Hemisphere Locations

Eurasia and some of the northern Atlantic and Arctic Ocean islands, such as Greenland, Iceland, and Spitsbergen, contain large expanses of tundra. Russia alone has four million square kilometres of treeless tundra (Chernov, 1985). The Eurasian tundra (exclusive of alpine tundra) is subdivided into the three subzones: Arctic Tundra, Typical Tundra, and Southern Tundra (Chernov, 1985; Figure 2.8). These essentially correspond to the Canadian High, Mid-, and Low Arctic Ecoclimatic Sub-Provinces, respectively. In both the Arctic and Typical Tundra regions, there is a particularly strong positive correlation between mean July temperatures and plant species diversity, more so than in either the polar deserts or in the Southern Tundra (Chernov, 1989).

2.5.1 Arctic Tundra

The Arctic Tundra extends along much of the Arctic Ocean shores of northern Asia and covers most of Novaya Zemlya. Bare ground dominates, with the mossy cover generally appearing secondary except when its herbaceous species are in flower. These herbaceous species produce minimal cover and are generally only 3–10 cm tall, with occasional shoots reaching 15 cm. At Maria Pronchitsheva Bay on the Arctic coast of the Taymyr Peninsula (75°30' N), annual average temperature is −15°C and annual precipitation is 220 mm. Matveyeva et al. (1975) distinguish four basic community types in the Maria Pronchitsheva Bay area: (1) mossy *Salix polaris* mesic frost-boil tundra, (2) mossy *S. polaris* dry frost-boil, (3) lichen-*Dryas* fellfield on skeleton soils, and (4) herb-grass meadows. Mosses dominate on moister sites, while lichens are common in the dry frost-boil and skeleton soil areas. Except for the occasional prostrate *Salix arctica* (Arctic willow) and ground-hugging sub-shrubs *Dryas octopetala* (dryas) and *Salix polaris*, shrubs are absent. The characteristic sedges of warmer tundra are absent in interfluve communities but can be found in stream valleys. Grasses such as *Alopecurus alpinus* (alpine foxtail) and *Poa alpigena* form meadow-like interfluve landscapes, especially on southerly slopes, where they are joined by *Luzula confusa* (woodrush), *L. nivalis,* and *Arctagrostis latifolia* (Metveyeva et al., 1975).

2.5.2 Typical Tundra

The Typical Tundra subregion contrasts with the Arctic subregion in that it is dominated by bryophytes. There are more than a dozen important moss species, which include carpet, cushion, and hummock formers. The waterlogging properties of moss cover inhibit grasses but encourage sedges such as *Carex ensifolia,* the most common of all flowering plants in the Eurasian tundra (Chernov, 1985). Shrubby thickets in snow-protected depressions are also common, with bushy *Salix* spp. and other dwarf shrubs dominating, along with herbs such as *Saxifraga* spp. Locally, in the moss cushions are found many species of lichen. The Typical Tundra can therefore be described generally as a sedge-moss tundra with variations resulting from any co-dominance with shrubs, *Dryas,* or lichens.

Tareya, at the confluence of the Pyasina and Tareya rivers in the Western Taymyr, can be considered a Typical Tundra site. Here mean annual temperature is −14.1°C and mean annual precipitation is 295 mm. This site is east-northeast of Vorontsovo, for which climatic data is illustrated in Figure 2.2. An important community type found here is the "mossy-*Dryas* sedge spotted tundra," where some 50% of the ground is in the form of hummocks, 20% troughs along the frosty cracks, and the remaining 30% in the form of bare ground. Mounds and troughs are completely vegetated, with *Carex ensifolia* and *Dryas octopetala* ssp. *punctata* being the dominants and *Cassiope tetragona, Salix arctica, S. polaris, Arctagrostis latifolia, Luzula nivalis, Polygonum viviparum,* and *Parrya nudicaulis* being important (Chernov et al., 1975). "Mossy-*Dryas* hummock tundra" is similar to the first community type, but bare ground is absent and *Salix reptans* and *Eriophorum angustifolium* become abundant. In "polygonal mire" sites the lower ground remains wet throughout the summer. Mosses cover the low ground completely, with *Carex stans,*

C. chordorrhiza, Eriophorum medium, Hierochloe pauciflora, and *Salix reptans* emerging through them. On polygon ridges, drier tundra species more typical of the first two community types dominate (Chernov et al., 1975).

2.5.3 Southern Tundra

The Southern Tundra subzone has many similarities to the Canadian Low Arctic, with 50 cm tall dwarf-shrub communities being variously dominated by species of *Salix, Betula,* and *Alnus* and with *Carex* spp. and *Eriophorum* joining herbaceous species to form ground cover (Chernov, 1985). This similarity is particularly apparent when one compares the *Eriophorum* and *Carex* southern tundra meadows of the Chukotka region in the Russian far east with equivalent communities in northwestern Canada (Ritchie, 1984). Typical southern tundra is found south of Tareya on the Pyasina River at Agapa (due east of Vorontsovo) where the mean annual temperatures are −12.3°C and annual precipitation is 344 mm. Here the flatter areas are dominated by *ernik* tundra (*ernik* is the dwarf birch, *Betula nana*), sphagnum bogs, and willow stands (Vassiljevskaja et al., 1975). As in the Canadian Low Arctic, shrubs do well in bogs and shrubby meadows, where they constitute some 30% of the above-ground plant biomass. Common shrubs include *Betula nana, Salix glauca, S. pulchra* (or *S. phylicifolia*), *S. lanata, S. reptans,* and *S. hastata.* The biomass of sedges is also highest on the bogs, with *Carex stans, C. rariflora, Eriophorum medium,* and *E. scheuchzeri* being important. On higher ground are found communities more characteristic of Typical Tundra.

Towards the south of the Southern Tundra are encountered areas of *redina* (very open *Larix dahurica* woodland) and *redkolesja* (open larch woodland), which represent a transition to forest-tundra (Norin and Ignatenko, 1975). In the equivalent Canadian biome, trees encountered are more likely to be *Picea mariana* (black spruce). To the west, in northern Finland, Southern Tundra *Calluna-Cladina* heath communities are typically associated with *Betula tortuosa* and the occasional stand of *Pinus sylvestris* (Kallio, 1975). Overgrazing of the northwestern Finnish Lapland pine-lichen heaths by semi-domesticated reindeer causes the climax ground-cover dominant, *Cladina stellaris*, to give way to other *Cladina* spp. Only horn lichen species (*Cladina* spp.) and small-bodied mosses can survive under the heaviest grazing pressure (Helle and Aspi, 1983).

2.5.4 Tundra on Arctic Islands

Just as Greenland has a similar tundra flora to that of North America, so Novaya Zemlya shares the Eurasian flora (Aleksandrova, 1988). Iceland, on the other hand, is more isolated in the North Atlantic and before settlement was at least 65% vegetated, with tundra-like communities dominated by dwarf shrubs such as *Salix* spp. found above 300–400 m (Arnalds, 1987). Today, following 1,100 years of livestock grazing, only 25% is considered vegetated. At lower elevations, the *Betula pub-*

escens woodland, which once covered some 25% of Iceland, has been reduced to only 1%. Spitsbergen, which is located much farther north than Iceland, can boast only an impoverished marine High Arctic flora in scattered *Saxifraga oppositifolia–Cetraria delisei* communities (Klokk and Ronning, 1987). The mosses *Hylocomium splendens* and *Sanionia uncinata,* which dominate cover in little auk (*Alle alle*) colonies on Spitsbergen, have been found to contain 1.5–2 times higher nutrient concentrations due to the fertilizing effect of guano (Godzik, 1991).

2.6 Tundra in the Southern Hemisphere

It is difficult to equate the treeless ecosystems on and around Antarctica with those of the northern hemisphere because of different climatic conditions. Antarctica includes ice-free areas much closer to the pole than in the Arctic, and most of this continent lies at latitudes poleward of continental Arctic ecosystems. In the southern hemisphere, at equivalent latitudes to North American and Eurasian continental tundra, there are no continental landmasses, only small, scattered islands. This allows for the southern hemisphere tundra to be readily classified into two bioclimatic subregions: the Antarctic and Sub-Antarctic. The Antarctic Subregion includes a Maritime Antarctic Subzone, which consists of a few islands such as the South Orkneys. The intensely cold Antarctic continent itself has basically a cryptogam cover in vegetated areas. The Sub-Antarctic Subregion consists of islands distant from the continent, where strong oceanic influences promote tussock grass communities.

2.6.1 The Antarctic Subregion

Despite its size of 14 million km^2, Antarctica has only 4% of its territory ice free. Reddish, yellow, and green patches of cryoplankton (ice algae) from the *Chlamydomonas, Scotiella, Chodatella,* and *Chlorosphaera* genera can be found on ice in summer (Mackenzie Lamb, 1970), while land areas which are vegetated are subdivided into (1) Antarctic non-vascular cryptogam tundra and (2) Antarctic herb tundra (Gimingham and Lewis Smith, 1970). Cover in the non-vascular cryptogam tundra is frequently discontinuous. In terms of decreasing cover abundance are found lichens, mosses, algae, fungi, and bacteria. Lichens can survive in the moist condition at temperatures of –75°C for prolonged periods and are even more resistant in the dry condition (Mackenzie Lamb, 1970). About 400 species of lichen have been recorded from Antarctica, with many having bipolar distributions (Mackenzie Lamb, 1970), while 72 species of moss are also present, with the bipolar species *Polytrichum alpinum* being the most common. Moss individuals have been found as close to the South Pole as 87°17' south latitude. Because of their drought tolerance, lichen species are not only more numerous than mosses, but are more widespread and in some places constitute the only land vegetation (Steere, 1967).

The Antarctic herb tundra includes two indigenous species, the grass *Deschampsia antarctica* and the herb *Colobanthus crassifolius*, together with two introduced species of grass, *Poa pratensis* (Kentucky bluegrass) and *P. annua* (annual bluegrass). Scottsberg (1954) considers these two indigenous species to be post-glacial, having arrived via wind or birds (anemochory or biochory). These vascular species combine with bryophytes and lichens to provide patchy cover in ice-free areas and coastal strips. Many of the non-vascular species could have remained in Antarctica during the Pleistocene, with bipolar species having arrived in the late Tertiary via "stepping stones" (Mackenzie Lamb, 1970). While continental isolation and low summer temperatures have traditionally been considered the reasons for this impoverished flora, moisture underavailability could well be the primarily limiting factor. This suggests that in Antarctica, dehydration resistance may be more important to survival than cold resistance (Kennedy, 1993).

The Maritime Antarctic Subzone includes the South Orkneys, which are essentially an extension of the Antarctic Peninsula. Here Signy Island (60°S, 45°W) is typical of the subzone and, as with the continental portions of the Antarctic bioclimatic subregion, vegetation is primarily cryptogamic (Collins et al., 1975). Lichens such as *Caloplaca* and *Verrucaria* do well in the salt spray zone along the coast, while *Biatorella* and *Buellia* colonize rock where bird colonies provide nitrogen-rich excrement, and *Lecidia* and *Lecanora* do well where nitrogen is limited. On more stable lithosols (Regosols), fruticose lichens and the mosses *Andreanea, Grimmia,* and *Dicranowesia antarctica* dominate, along with the moss hummock species *Brachythecium austrostamineum*. Grass swards of *Deschampsia antarctica* and *Colobanthus quitensis* occur on favoured north-facing slopes in the coastal zone (Collins et al., 1975). Recent increases in the fur seal population on Signy have unfortunately given rise to major destruction of coastal cryptogam-dominated vegetation (Lewis Smith, 1988).

2.6.2 The Sub-Antarctic Subregion

The floras of Antarctic islands such as South Georgia and Macquarie are much better developed than on the continent, although they are quite impoverished when compared to their northern hemisphere tundra counterparts. On South Georgia (54–55°S, 36–38°W) there are 56 species of vascular plant, 24 of which are indigenous (Mackenzie Lamb, 1970), and the environmental pattern in the vegetation is related mainly to the moisture status of the substratum. The most widespread cover type consists of dense stands of the tussock grass *Poa flabellata*, which can exceed 2 m in height. Tussock growth can be locally improved where seals and birds leave excrement (Lewis Smith and Walton, 1975), although grazing by introduced Norwegian reindeer reduces biomass (Leader-Williams and Ricketts, 1981), and disturbance by expanding fur seal populations has led to the eradication of *Poa flabellata* over large areas of coastal breeding territory (Bonner, 1985). *Deschampsia antarctica* is common on wet level ground, while oligotrophic mires are dominated by the rush *Rostkovia magellanica*, along with mosses and liverworts. Eutrophic seepage slopes are dominated by *Juncus scheuchzerioides*, the moss *Tortula robusta*, and the dwarf shrub *Acaena magellanica*. This shrub forms dense, pure stands on drier, stonier slopes, while on moister slopes the tall turf-forming mosses *Chorisodontium aciphyllum* and *Polytrichum alpestre* develop deep banks. *Festuca contracta* forms extensive closed stands in sheltered situations (Lewis Smith and Walton, 1975).

Macquarie Island (54°30'S, 158°57'E) is located south-southeast of Tasmania. Below 230 m are found pedestalled *Poa foliosa* tussock grasslands up to 1.5 metres tall, while throughout the island are herb fields dominated by *Pleurophyllum hookeri*, with *Stilbocarpa polaris* occasionally co-dominating (Jenkin, 1975). Fens are dominated by *Juncus scheuchzerioides*. Only 35 species of vascular plant are found on Macquarie (Jenkin and Ashton, 1970). Some tundra-like areas are also found in the Falklands, but technically this heath-dominated island group lies outside the tundra region.

2.7 Alpine Tundra

Tundra-like communities can be found in mountainous areas well outside the Arctic and Antarctic regions. The term "alpine" used along with the word "tundra" to identify these non-Arctic herbaceous communities is derived from the Latin name for the white-topped Alps, *Alpes* (Love, 1970). Found at increasingly higher elevations as one approaches the tropics, these communities owe their existence to the effects of elevation on climate. Essentially they are herbaceous communities where frost and not drought is usually responsible for the absence of trees. They are therefore found above the climatic limit of upright trees (Bliss, 1985), a limit which is often referred to as the "tree line" or "timberline." Many temperate-latitude alpine tundra areas resemble Arctic tundra because many alpine species are also found there. Alpine species which are also circumpolar constitute a group known as "arctic-alpine" (Scoggan, 1978). Alpine tundra found at lower latitudes is usually quite different; so different in fact that the name "tundra" is often excluded in its description. These differences should be expected because, although low-latitude species face frequent frost, they do not have to tolerate the extreme temperatures so characteristic of the long sub-polar winters. They also have a quite different photoperiod regime, have to tolerate more intense UV radiation, and may also be able to grow for most of the year because their "winter" often occurs only at night. In recognition of these environmental differences, information on alpine communities is divided into two sections: (1) temperate-latitude alpine tundra and (2) low-latitude (equatorial) alpine tundra.

2.7.1 Temperate-Latitude Alpine Tundra

Alpine tundra is more commonly encountered on mountain systems in temperate latitudes than in the tropics simply because tundra climatic conditions are encountered in the former at much lower elevations. Alpine tundra is particularly widespread throughout the western cordillera of North America (Figures 6.1 and 6.5). It is also common in Scandinavia, the Urals, and other European mountain systems such as the Alps, as well as in central and east Asia. From a photoperiod standpoint, these regions have 15–18 hours of light during the growing season, and average summer temperatures can be considerably warmer that in true high-latitude tundra. As a result, vegetation cover just above the tree line can be quite meadow-like. Because of exposure to strong winds, many alpine surfaces have either a thin winter snow cover or it is missing altogether. Here, impoverished tundra is encountered along with bare ground, locally called "golzy" in Europe (from the Russian *golly,* meaning bare; Walter, 1979). "Fellfield," a term derived from the Norwegian *fell,* meaning open mountainside, is often used to describe alpine tundra where vegetation covers less than half the area and is the alpine equivalent of the Arctic terms "polar desert" and "barrens."

2.7.1.1 North America

In eastern Canada, small alpine communities are found in the Gaspé Peninsula in Quebec, the Torngat Mountains of Labrador, and the Long Range in Newfoundland. Most of these mountain tops possess low, dry fellfield-type alpine tundra which show remarkable Arctic affinities in their species composition. Billings (1988) notes that 62 per cent of the vascular species in the alpine portion of Gerin Mountain in Labrador are also found in Finnish tundra, and shared species include *Poa alpina, Eriophorum vaginatum*, and *Carex aquatilis.*

In western Canada and Alaska, alpine tundra is found at elevations as low as 900 m in the British and Richardson Mountains north of the Arctic Circle and 950 m on Eagle Summit northeast of Fairbanks, Alaska. Here in the Alpine Northern Subarctic Cordilleran Ecoclimatic Region (NSC), vegetation cover frequently consists of crustose and fruticose lichens, mosses, *Alnus* spp., *Betula glandulosa* (scrub birch), and ericaceous dwarf shrubs. In the Rocky Mountains to the south, moist and wet alpine meadows are characterized by mesophytic species such as *Eriophorum* spp., *Erythronium grandiflorum* (yellow avalanche lily), *Phleum alpinum* (alpine timothy), *Caltha leptosepala* (marsh marigold), and *Carex* spp. (Scoggan, 1978). In less mesic sites, characteristic species include both woody and herbaceous types, with *Vaccinium, Dryas, Cassiope, Poa, Carex, Epilobium, Saxifraga, Eriogonum* (umbrella plant), *Salix*, and *Luzula* (woodrushes) spp., *Sedum roseum* (roseroot), *Empetrum nigrum*, and *Ledum glandulosum* (trapper's tea) being among the more common. A typical alpine heath cover in the Jasper National Park region would be *Cassiope tetragona, Dryas octopetala*, and *Salix reticulata* var. *nivalis* (snow willow) (Holland and Coen, 1982). Although the true tree life form is not frequently encountered above the tree line (or timberline) in the cordillera of western North America, *Abies lasiocarpa, Picea engelmannii*, and *Salix* spp. are particularly adapt at taking on a mat krummholz form. At timberline, needle death from year-round dehydration due to wind exposure is the major factor limiting woody forms to krummholz mats and the occasional flagged tree (Hadley and Smith, 1986). In alpine tundra, these tree and shrub mat krummholz forms are generally found in the lee of rocks or rock outcrops.

In the USA many tree species also take on mat krummholz forms. In the Sawatch Range, Colorado, mat krummholz patches are overwhelmingly dominated by *Picea engelmannii*, while *Pinus aristata* (Rocky Mountain bristlecone pine) and *P. contorta* var. *latifolia* (lodgepole pine) are also found (Hartman and Rottman, 1988). Because of decreasing latitude, communities here are somewhat less "Arctic" than their Canadian alpine tundra equivalents; however, community diversity is still evident. Typical community representatives for the northern Front Range in Colorado would include (1) *Kobresia*-turf, (2) *Carex-Trifolium*-turf, (3) *Paronychia*-fellfield, (4) *Dryas*–mat shrub, 5) *Sibbaldia-Selaginella*-snowbed, and (6) *Deschampsia*-meadow (Eddleman and Ward, 1984). Throughout the Rocky Mountains, woody heath species such as *Cassiope tetragona* and *Vaccinium uliginosum* are Arctic tundra–like, while cushion and rosette-forming species are particularly abundant, especially on the more exposed sites. Typical cushion plants include *Silene acaulis* (moss-campion), *Trifolium nanum, Paronychia pulvinata, Minuartia obtusiloba* (sandwort), *Saxifraga oppositifolia*, and *Diapensia lapponica* (Bliss, 1985). Likewise, the most common growth form of alpine tundra plant in the Northern San Juan Mountains in Colorado is the rosette (Rottman and Hartman, 1985).

Alpine meadows are typical of sheltered areas just above tree line. Often in drier meadows, forbs and grasses dominate, while in moister sites, sedges are common. In the Sawatch Range, drier meadows are commonly dominated by *Dryas octopetala* var. *hookeriana* (mountain avens), *Kobresia bellardii* (cobresia), *Carex elynoides,* and *Salix reticulata* var. *nivalis*, while in wetter meadows, *Deschampsia caespitosa, Carex nigricans*, and prostrate sub-shrubs such as *Salix reticulata* dominate (Hartman and Rottman, 1988). To the west, in the Olympic Peninsula, drier sites are dominated by *Festuca idahoensis* and *Lupinus lepidus* (lupin), while on late snowmelt sites and in moister swales, *Carex spectabilis* is the most common species (Loneragan and del Moral, 1984).

A number of important generalizations can be made about alpine floras. Billings (1988) notes that Arctic affinity decreases with latitude, with only 19% of the alpine tundra vascular species above Piute Pass in the Sierra Nevada also being found in the Arctic. Likewise, therophytes, which still comprise a very small percentage of alpine tundra species, are both more numerous than in Arctic tundra and also tend to increase with decreasing latitude. Jackson and Bliss (1984) found as many as fifteen annual species in subalpine open meadow in the central Sierra Nevada, while Spira (1987) found thirteen in the White Mountains in eastern California. As might be expected, an analysis of the floras of isolated alpine tundras throughout the western cordillera supports the conclusion that the size of tundra "islands" best explains the number of species found there and that latitude best explains floristic similarity (Hadley, 1987). Mount St Helens, the volcanic peak which exploded violently in 1980, might be considered an exception to this last statement. Here the balance between episodic local extinction caused by eruptions and presumably only gradual colonization has produced a monotonous alpine flora limited in richness (del Moral and Wood, 1988b). In addition, while the short-term effects of modest tephra deposition does not greatly reduce primary production in previously well developed subalpine meadows, the long-term consequence is for community simplification (Pfitsch and Bliss, 1988).

2.7.1.2 Eurasia

In Scandinavia, alpine tundra has many similarities to north European lowland tundra, with dwarf shrub *Empetrum*-heath species common on mesic and dry sites, while *Betula nana*–dominated communities do better on moister ground (Jonasson and Skold, 1983). *Betula pubescens* ssp. *tortuosa* (mountain birch) is an important tree-line species which has been used to study recent retrogression of the forest-alpine ecotone in the Scandes Mountains, Sweden (Kullman, 1989). In the Alps of southern Europe, vegetative propagation by *Picea abies* is particularly common along the tree line. While *Larix decidua* also layers here, it does so decidedly less frequently than the spruce (Tranquillini, 1979). These layering differences between *Picea* and *Larix* were previously noted from the Canadian shrub tundra south of Hudson Bay.

Just above the tree line, krummholz forms of *Pinus cembra* (Aroila pine) are common in drier tundra dominated by dwarf shrubs and cushion and rosette species. In the Vanoise Massif in the northern French Alps, *Carex curvula* and *Festuca halleri* dominate alpine grasslands on acid substrates on south-facing slopes, while on strongly calcareous south-facing slopes, *Sesleria coerulea* dominates (Gensac, 1990). In the central Alps in Switzerland, *Nardus stricta* tends to dominate on the acid sites, while *Sesleria* again dominates the calcicole (alkaline) substrate grasslands (Gignon, 1987).

Islands of alpine tundra are also common in the Urals and Himalayas. In the Urals, woody species include *Betula nana* (dwarf birch), *Salix glauca,* and *Juniperus sibirica* and occasional patches of elfin woodland comprised of krummholz forms of *Betula tortuosa,* along with *Pinus sibirica, Abies sibirica,* and *Picea obovata* (Siberian pine, fir, and spruce respectively) (Famelis and Nikanova, 1989). Typical lichen, lichen–dwarf shrub, moss-lichen-sedge, shrub-moss, and grass-moss tundras are also present. To the south, in the central Himalayas, areas of alpine tundra, or *bughiyals,* as they are known locally, are found above the tree line at about 3,000 m to just below the snowline at 5,400–5,600 m. Here six families – Brassicaceae, Asteraceae, Ranunculacea, Poaceae, Fumariaceae, and Caryophyllaceae – dominate in terms of numbers of species present (Rawat and Pangtey, 1987). As already noted for North American alpine tundra, the effects of lower latitude permit a somewhat larger percentage of therophytes (5%), while the hemicryptophyte life form (61%) dominates. In terms of phenological characteristics, such as length of growing season and timings of growth initiation, flowering, and fruiting, these *bughiyals* have

more in common with the low-latitude (equatorial) alpine regions than with the Arctic tundra (Ram et al., 1988). Even so, typical Tibetan alpine grassland genera above 4,860 m include the sedges *Kobresia* and *Carex* and the grasses *Stipa*, *Poa*, *Calamagrostis*, and *Koeleria*, and most communities experience some grazing pressure from domesticated yak, sheep, and goats (Cincotta et al., 1991).

2.7.1.3 Southern Hemisphere Alpine Tundra

There are few major areas of temperate-latitude alpine tundra in the southern hemisphere other than those in the southern Andes of South America and in New Zealand. The flora in alpine areas of New Zealand is particulary interesting because of the fact 93% of the species are endemic to the New Zealand Biological Region, which includes the sub-Antarctic islands (Mark and Adams, 1973). In New Zealand, low alpine tundra is separated from subalpine forest by stunted (but non-krummholz) forms of *Nothofagus solandri* (southern hemisphere beech) that result from the extremes of climate and subsequent vegetative proliferation (Norton and Schonenberger, 1984). Just above tree line are found the so-called snow tussock grasses and shrub communities dominated by *Chinochlora flavescens* (broad-leaved snow tussock) and the shrubs *Podocarpus nivalis* (snow totara) and *Phyllocladus alpinus* (mountain toatoa). These communities grade into tussock–herb fields and then herb fields dominated by species from such genera as *Aciphylla, Astelia, Anisotome,* and *Gingidium*. Bogs are also common and are characterized by *Sphagnum* mosses, sedges (*Carex* spp.), rushes (*Juncus, Carpha,* and *Schoenus* spp.), cushion plants, creeping heath plants, and carnivorous herbs such as *Drosera* spp. (sundews). The high alpine belt is typically made up of fellfield, scree, and cushion plant communities (Mark and Adams, 1973) •

2.7.2 Low-Latitude (Equatorial) Alpine Tundra

Although equatorial alpine tundra regions are found from New Guinea to Africa and South America, and vast distances separate communities and promote extreme endemism, they still retain many striking similarities, with their tussock grasses and perennial rosette-form herbs (Deshmukh, 1986). In the Neotropics, alpine tundra communities are found in parts of the Sierra Madre and also on volcanoes on the Mexican Plateau, in Guatemala, and in Costa Rica, as well as along much of the spine of the Andes, especially in the moist *paramo* of Venezuela, Colombia, Ecuador, Peru, and Bolivia. While alpine tundra in Mexico and Guatemala show clear North American alpine affinities, those in Costa Rica and south are more typically South American *paramos*-like (Billings, 1988). Typically, *paramo* regions have a year-round growing season characterized by low mean temperatures (0–5°C) and pronounced diurnal temperature fluctuations (Pfitsch, 1988). Occasionally, this diurnal climate is interrupted by an infrequent *friagem*, or cold spell, which can last several days (Winterhalder and Thomas, 1978). Temperatures vary somewhat with latitude, as do the amount of precipitation, the length of dry season, and the frequency of dry spells. In general, these conditions mean that unlike Arctic tundra, sites which get warmest during the day will have the greatest advantage in terms of growth potential, provided that they also have a supply of moisture.

The perennial rosette species is one of the most distinguishing features of low-latitude alpine tundra. Frequently, dead leaves form a thick cylinder around the stem, and such forms have evolved independently in alpine tundra from the silverswords of Hawaii (*Argyroxiphium sandwicense*) to the lobelias of East Africa (Smith, 1981). Large numbers of species of the genus *Espeletia*, which is endemic to the Andes of Venezuela, Colombia, and northern Ecuador, are also of the rosette type. Experiments with *E. schultzii* shows that the buds are better protected from frost because the rosette closes at night and that the sheath of dead leaves acts as an insulator (Deshmukh, 1986). In Venezuela the rosette species *Draba chinophila* occurs on the cooler, frost-heaved soil, where plant cover and species richness are low. The subshrub *Draba arbuscula* and other shrub species occur preferentially on the rock outcrops, where thinner soils allow for the soil surface temperatures to be 2–4°C warmer. These rock outcrop sites have somewhat higher cover and species richness (Pfitsch, 1988).

In the Altiplano region of Peru and Bolivia, greater moisture stress is experienced than in typical *paramo* grasslands to the north and east, and the alpine communities are known as the *puna*. In the *puna* of southern Peru, species are perennial and dwarf and come primarily from the Compositae, Gramineae, Leguminosae, Solanaceae, and Verbenaceae families. Bunch grasses dominate, with *Stipa ichu* (ichu), *Festuca dolybergetum,* and *Calamagrostis densiflora* being the most prominent at lower elevations and *Calamagrostis antonianus, Festuca heterophylla,* and *Stipa obtusa* dominating on higher slopes and rocky areas (Winterhalder and Thomas, 1978). Grazing, agriculture, and burning over several thousand years has greatly reduced the woody component of the cover and has lowered the tree line. Small pockets of the slow-growing tree *Polylepis incana* (quenua) can still be found in ichu grasslands isolated from humans and their domesticated animals, and these attest to a former, once more forest-like cover. At higher elevations, where somewhat more natural alpine tundra is encountered, woody phanerophytes are more common than in Arctic tundra. This is primarily because they can grow year-round and do not have the severe krummholz-producing winters typical of higher latitudes. Nonetheless, woody species tend to be evergreen and less than one metre tall. The great majority of species take the cushion form as an adaptation to both temperatures and drying winds. Because of these harsh conditions, therophytes constitute only 2% of the species.

In the African Palaeotropics, alpine communities and valley glaciers can be found close to the equator only on several of the isolated peaks in the Ruwenzori Range and on isolated volcanos such as Mount Kenya and Mount Kilimanjaro. Above the tree line is a *paramos*-like vegetation of shrub, tussock grassland–moorland, mire, and desert communities. Composite shrubs (*Helichrysum* spp., or everlastings) are frequently found just above the tree line with ericaceous woodland. Where humidity is high, branches on these shrubs are commonly covered with lichens and bryophytes (Lind and Morrison, 1974). Tussock grasslands may be of natural origin on the drier peaks or derived from burned ericaceous woodland in wetter areas. Thirty-four grass species form part of the Afro-alpine flora, with *Festuca pilgeri* being the dominant bunch grass on the high moorlands of Mounts Kenya and Elgon, while *F. pilgeri* var. *supina* dominates on Mount Kilimanjaro, along with the large tussocks of *Pentaschistris minor* on drier sites. Many colourful forbs, such as *Kniphofia thommonii* (red-hot poker), *Carduus nyssanus* (thistle), and the clover *Trifolium cryptopodium*, also form part of this community. In places, tussock-forming *Carex* spp. form acidic mires in poorly drained valley floors and depressions, especially in the Ruwenzories. Close to the upper limit of plant growth, such as close to glaciers or above about 5,000 m, the grass *Poa ruwenzoriense* is the commonest vascular plant (Lind and Morrison, 1974).

In Southeast Asia, alpine tussock grasslands are also found near the tops of some of the higher volcanic peaks, such as Mount Wilhelm and Mount Giluwe in Papua New Guinea. In Papua New Guinea, mid-height tussock grasses such as *Deschampsia klossii* and *Poa keyessii* ssp. *sarawagetica* dominate on the relatively well drained sites at and just above tree line. Associated with these tussocks are smaller grasses such as *Danthonia archiboldii, D. vestita, D. schneideri,* and *D. semiannularis* (Robbins, 1961). On poorer sites and at higher elevations, low stunted tussock grasses and dwarf grasses such as *Poa, Festuca, Monostachya,* and *Danthonia* dominate. Alpine forbs such as *Gentiana, Potentilla, Centrolepis,* and *Hydrocotyle* species, many of which form cushions, are found among the tussocks in a mat of lichens and moss (Robbins, 1961). Occasionally, the fern *Gleichenia vulcanica* (false staghorn fern) is found, and quite noticeable in more sheltered sites are tree ferns (*Cyathea* spp.), which often have fire scars recording recent burning episodes. While the presence of an alpine flora at the higher elevations is distinctive, the role of human disturbance through the use of fire for hunting is an important ecological factor nearer the tree line and has contributed to the formation of secondary grasslands. Palynological studies carried out on Mount Wilhelm suggest that the shrubs currently invading grasslands there at 3,900 m are doing so because of reduced burning and that they are the precursors of forest regeneration (Corlett, 1984). Near the peak of Mount Giluwe at 4,200 m, species are typically alpine, with bryophytes, dwarf shrubs, and herbaceous species such as gentians and sedges, many of which are endemic.

2.8 Primary Production and Phytomass in Tundra

Net annual plant production in the High Arctic is understandably low (Webber, 1974). Figures as low as $1-2$ g m^{-2} are typical of High Arctic polar desert sites, while rates increase to $20-40$ g m^{-2} in cushion plant–lichen communities such as found on Devon Island (Wielgolaski et al., 1981) (Table 2.2). Values as high as

Table 2.2
Living phytomass (g m^{-2}) and net annual plant production (g m^{-2} yr^{-1}) in tundra (from various sources as summarized by Wielgolaski et al., 1981)

Community Type	Location	Vascular Plants		Cryptograms	Net Annual Production
		Above Ground	Below Ground		
Cushion plant–moss	Disko, Greenland	66	32	33	46
Moss-herb	Truelove, Devon Is	15	10	232	8
Cushion plant–moss	Truelove, Devon Is	126	50	623	54
Wet sedge–moss	Truelove, Devon Is	78	691	1,097	281
Dwarf shrub–heath	Kevo, Finland	329	195	84	–
Wet sedge-moss	Point Barrow, Alaska	13	1,305	30	144
Wet sedge-Moss	Cambridge Bay, NWT	372	462	671	123
Dwarf shrub–heath	Harp, Russia	1,285	2,037	95	–

200–300 gm^{-2} have been determined for High Arctic wet sedge-moss communities. Net annual production in high-latitude tundra is normally below 300 g m^{-2}, while in lower-latitude alpine tundra, values can exceed 500 g m^{-2}. Lieth (1975) gives an approximate mean net primary productivity value of 140 gm^{-2}yr^{-1} for tundra in general, but rates are very variable. One study comparing four contrasting tundra sites near Toolik Lake, Alaska, showed that vascular plant primary productivity (not including root production) varied from 32 to 305 g m^{-2}yr^{-1} (Shaver and Chapin, 1991). Among the highest annual net production values for tundra is the figure of 6,025 g m^{-2} determined for a wet grassland community on sub-Antarctic South Georgia (Lewis-Smith and Walton, 1975), while values as high as 800 g m^{-2} are reported for waterlogged meadow at Agapa in the southern tundra, Taymyr, Russia (Vil'chek, 1987).

For total living phytomass, values range from near zero in the polar deserts of the High Arctic to 3,000 g m^{-2} (30 t ha^{-1}) in the Low Arctic, with Lieth (1975) reporting a range of 1–30 t ha^{-1}. Above-ground phytomass (including cryptogams) increases by an order of magnitude from polar deserts to semi-deserts and from there to shrub tundra in the Low Arctic. Increase is greater for vascular plants than cryptogams. Small above-ground to below-ground phytomass ratios are characteristic of tundra ecosystems, with the top-to-root ratio generally decreasing with increasing mean annual temperatures (Table 2.3). In a hummocky sedge-bog polar oasis at Alexandra

Table 2.3
Average ratios of tundra phytomass in selected community types (after Wielgolaski et al., 1981)

Ecoclimate	Community Type	Shoot-to-Root (Live)	Green–to–Non-Green Live Vascular Plants	Live-to-Dead Above-Ground Vasc. Plants
High Arctic	Desert and semi-desert	1:0.9	1:2.3	1:1.9
	Wet sedge meadow	1:21	1:23	1:1.6
Mid-Arctic	Mesic-dry meadow	1:5.0	1:7.7	1:0.8
	Dwarf shrub tundra	1:3.1	1:12	1:0.6
Low Arctic	Low shrub tundra	1:2.0	1:19	1:0.2

Fjord, Ellesmere Island, Henry (1987) reports a total vascular species standing crop of 1,400–3,200 g m^{-2} (both living and dead), of which 85% is below ground. In this community live-to-dead ratios were 1:4 to 1:7 for above-ground phytomass and 1:0.8 to 1:6 for below-ground, while net production is only 100–200 g m^{-2}yr^{-1}.

Small annual net primary productivities and relatively high soil organic-matter contents produce K factor values often between 0.03 and 0.06 (Moore, 1978). Green-to-non-green ratios generally increase as increasing temperatures improve woody shrub growth. In the Low Arctic, stem secondary growth in woody shrubs becomes an important component of NPP. Shaver (1986) determined that 13–49% of total above-ground NPP for individual shrubs in an Alaskan shrub tundra went into secondary growth. Harsh growing conditions in the High Arctic result in greater amounts of dead, than living, phytomass for above-ground vascular species parts (Wielgolaski et al., 1981). These same harsh climatic conditions have also promoted a modest accretion of carbon into humus and peaty layers. Oechel et al. (1993) report that recent warming on the North Slope in Alaska has lowered water tables and promoted decomposition to the point the soil has switched from a carbon sink to a carbon source.

CHAPTER THREE

Forest-Tundra or Boreal-Tundra Ecotone

3.1 Definitions

The term "boreal-tundra ecotone" (ZE VIII–IX) is defined in a number of different ways in the literature (Timoney et al., 1992; Larsen, 1989; Payette, 1983). As used here, it describes that environment found between areas of tundra (usually shrub tundra, where the true tree growth form is absent) and, on its warmer side, continuous boreal forest (Figure 1.3). Many authors consider this whole region as an ecotone or transitional zone between tundra and true boreal forest, because throughout the region is found a mosaic or matrix of tundra-like communities coexisting with stands of coniferous trees (Sirois, 1992; Larsen, 1989; Arno and Hammerly, 1984). Others might describe it as a "landscape boundary," but current usage shows no compelling reasons for differentiating between the terms "landscape boundary," "transitional zone," and "ecotone" (Gosz, 1991). Nevertheless, the term "ecotone" is considered a misnomer by some because the extent of this community matrix is zonal in nature. The expression "forest-tundra" is frequently substituted for "boreal-tundra ecotone" to signify this zonal recognition. Although rarely seen in the literature, the term "ecocline" might also be an appropriate one to use here as it signifies an environmentally more stable condition than that of the original ecotone concept (van der Maarel, 1990). Regardless of which expression is used to identify this vegetation community, authors agree that in terms of geographical area, it is necessary to give this cover type effective zonal status.

The forest-tundra includes two important subzones; the shrub subzone and the forest subzone (Payette, 1983). In the shrub subzone, the lichen–heath–dwarf birch communities so typical of the southern Canadian tundra dominate on uplands, exposed sites, and slopes; only in the more protected basins are found small stands of up to a few hundred conifers which clearly possess boreal affinities and are dominated by *Picea* spp. In the forest subzone, the dominant community type becomes reversed, with coniferous forest (open or closed) dominating the landscape except on the more exposed slopes and uplands, where lichen–heath–dwarf birch persists.

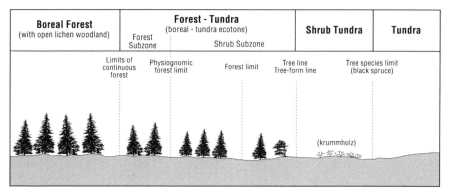

Figure 3.1
A typical profile across the Canadian forest-tundra (modified from a number of sources including Ball and Scott, 1987, Payette, 1983, and Atkinson, 1981; from *Earthscapes*, by W.M. Marsh, with permission of the publisher, John Wiley & Sons, Inc. © 1987)

Figure 3.1 illustrates a typical profile, from boreal forest across this forest-tundra to true shrub tundra, for the Canadian Subarctic. Terms used in Figure 3.1 are defined in Table 3.1. It is recognized that many authors consider other definitions or explanations for some of these terms. Atkinson (1981) considers the tree line to be identical with the "limits of continuous forest" line shown in Figure 3.1, while Hare and Ritchie (1972) equate forest limit with the break-point between closed crown boreal forest and open woodland. Timoney et al. (1992) use more quantifiable criteria, with the northern and southern boundaries being located where the 1:1,000 and 1,000:1 tree:upland tundra cover isolines occur respectively. As Timoney et al. (1992) consider this boundary with shrub tundra to be where trees are at least 3–4 m tall, their 1:1,000 tree:upland tundra boundary is therefore almost identical to the tree form line definition (Payette, 1983) of 5 m tall trees. Reflecting on Larsen's (1989, p. 2) comment that "there will probably never be universal agreement as to where the Arctic begins, and there will be a continual need for redefinition to suit individual requirements," the discussion below will therefore restrict itself primarily to the terms given in Table 3.1. For fuller reviews of terminology associated with the forest-tundra, see Timoney et al. (1992), Payette (1983), Elliott-Fisk (1983), and Atkinson (1981).

3.2 Distribution

Extensive areas of forest-tundra are found in Alaska, northern Canada (Figures 1.3 and 3.2), and northern Eurasia (Figure 2.8). Although there are no boreal forests in the southern hemisphere and therefore no sub-Antarctic equivalent of the Canadian forest-tundra, there is some forest-tundra where broadleaf forest and tundra are in

Table 3.1
Terminology associated with the forest-tundra (boreal-tundra ecotone or transition) in Canada (primarily after Payette, 1983). These terms are used in Figure 3.1.

	Definition
Tree species limit	A line separating areas where tree species are totally absent from areas where tree species may be found only in the shrubby and/or krummholz form and never in the true growth form (life form) or as individuals over 5 m tall. This limit is normally located between non-shrub and shrub tundra and refers here to the most likely encountered tree species, *Picea mariana*.
Tree-form line	The boundary between tundra (where tree species do not take true growth form) and areas where trees actually take true tree growth form (with mature individuals exceeding 5 m in height and restricted to favourable topographic locations only). The tree form line is therefore coincident with the cooler side of the forest-tundra (boreal-tundra ecotone) and separates the tundra from the shrub subzone of the forest-tundra.
Forest limit	A line representing the northern limit of forest stands in the shrub subzone of the forest-tundra. On the colder side of this line can be found only the occasional individual tree exhibiting true growth form, while on the warmer side are found small forest stands of perhaps a few hundred individuals occupying catchments with favourable microclimates.
Tree line	Used here synonymously with tree-form line.
Physiognomic forest limit	On the warmer side of this line, forest is extensive in lowlands but sparse on rock outcrops and exposed sites, and regeneration is episodical. On the colder side of this line, small forest stands may be found in more favourable sites but lichen–heath–dwarf birch communities dominate. This physiognomic forest limit is therefore a useful line to take as the boundary between the shrub subzone and the forest subzone of the forest-tundra.
Limit of continuous forest	Water-free landscapes covered by boreal forests in valleys and on hills. This includes both open and closed forest growing on xeric and mesic sites respectively. The northern limit of continuous boreal forest coincides with the southern limit of the forest-tundra.

juxtaposition. For those who prefer the expression "boreal-tundra ecotone," the problem of a southern hemisphere forest-tundra transition zone does not apply. This chapter will therefore restrict itself to discussing the forest-tundra regions of the northern hemisphere.

Because of difficulties inherent in defining the terms "continuous forest" and "tundra" and due to our lack of knowledge about the area and the temporal effects of recent climatic change, it is difficult to find two maps which agree on the boundaries for the forest-tundra zone. The distribution of forest-tundra shown in Figure 3.2 is based on a number of sources, particularly Timoney et al. (1992) for the region west

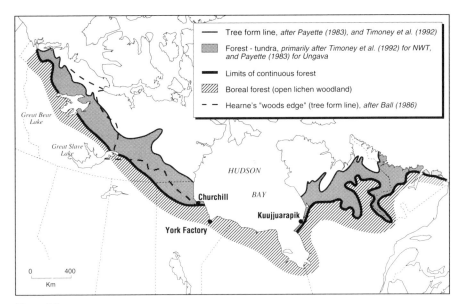

Figure 3.2
Distribution of forest-tundra (boreal-tundra ecotone) in Canada

of Hudson Bay and Payette (1983) for northern Quebec and Labrador. Figure 3.2 also shows possible responses to historical climatic changes in western Canada (Ball, 1986). The accuracy of Hearne's 1772 "woods edge" is difficult to confirm even although he used the tree line as his "woods edge" boundary. Our lack of knowledge about these changes underscores both the problems associated with the effects of climatic change in this harsh environment and the dearth of studies on vegetation performed by early explorers and scientists. Fortunately, palynological studies provide evidence for Holocene fluctuations in the tree line west of Hudson Bay (Moser and MacDonald, 1990; Nichols, 1967) and in northern Quebec (Gajewski et al., 1993; Ritchie, 1987).

3.3 Climate

The Canadian forest-tundra region is synonymous with the High and Mid-Subarctic Ecoclimatic Sub-Provinces (Figure 1.4). A typical climatic station for the tundra margin of the foresttundra region is Churchill, a port located on the southwestern shore of Hudson Bay (Figure 3.3A). Here mean annual temperatures are −7.2°C, which is some 4–6 degrees warmer than for typical Low Arctic tundra regions such as in the Keewatin District, located just to the north. July is the warmest month (11.8°C), with almost five months averaging above 0°C. In warmer parts of the forest-tundra, such as at Kuujjuarapik (Poste-de-la-Baleine; Figure 3.3B) on the

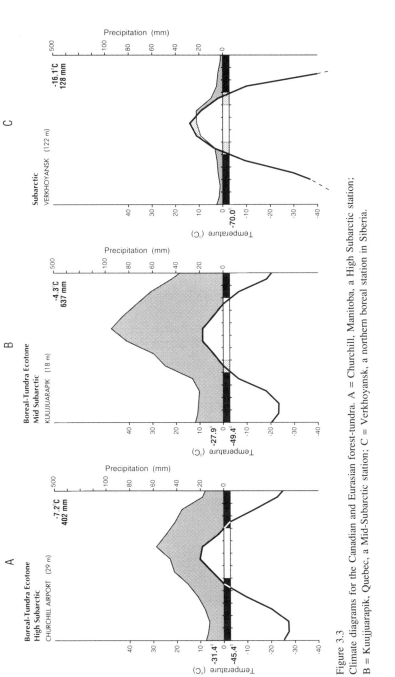

Figure 3.3

Climate diagrams for the Canadian and Eurasian forest-tundra. A = Churchill, Manitoba, a High Subarctic station; B = Kuujjuarapik, Quebec, a Mid-Subarctic station; C = Verkhoyansk, a northern boreal station in Siberia.

eastern shore of Hudson Bay, mean annual temperatures are 2–4 degrees higher and as many as six months average above 0°C, while the climate remains humid.

To a certain extent, the forest-tundra region in Canada coincides with the average summer position of the polar front. The strong southward thrust of this summer polar front over central Canada, a feature which is primarily due to the cold air masses spawned over Hudson Bay (Rouse, 1991), helps account for the arced shape of this zone so apparent in Figure 3.2. The particularly sharp southward plunge of the ecotone between Great Bear Lake and Great Slave Lake may also be related to both soil nutrient differences and a shift of tree dominancy from *Picea glauca* to *P. mariana*, which occurs as summer germination temperatures improve to the southeast (Timoney at al., 1992, 1993a). Hare and Ritchie (1972) stress that the pronounced north-south temperature gradient produced at the average position of the polar front is sufficient to account for the transition from a harsh, essentially non-arboreal ecosystem on the colder side to boreal forest stands on the warmer. Between the Cordillera and Hudson Bay, this zone is narrower than in northern Quebec, averaging 131 km (using the 1,000:1 and 1:1,000 tree:upland tundra cover isolines of Timoney et al., 1992). As frontal activity is also associated with precipitation events, it is not surprising that precipitation values are higher than that of the tundra, with maxima occurring in summer (Figure 3.3). Positive moisture indices throughout the zone, combined with poor natural drainage, help explain why large areas of wetlands (primarily bog and fen) are associated with this environment, as well as within the boreal forest along its warmer margins. While moisture may be abundant, Hustich (1983) is of the opinion that annual temperature variation is more significant than the variation in annual precipitation. In addition, others stress the significance of climatic forcing by the trees themselves, which may well modify a strictly polar front determination of the northern boundary (Lafleur et al., 1993).

In Eurasia, particularly in the forest-tundra and northern boreal (taiga) of Siberia, conditions are more severe in winter but warmer in summer than their tundra equivalents. Climatic data for Verkhoyansk given in Figure 3.3C indicate a mean annual temperature of −16.1°C with extreme conditions in winter, yet a number of summer months above 10°C. It is not surprising that bogs and deciduous conifers dominate under such conditions. The genus of deciduous conifers, *Larix*, is capable of survival under extreme growing conditions beyond the normal tolerance of evergreen conifers but is restricted to only ten species worldwide (Gower and Richards, 1990). It would appear that in Canada the moist, mild summers and cold winters encourage bryophyte–*Picea* spp. dominated landscapes where deciduous conifers such as *Larix laricina* are normally only secondary. In western Europe, the mild, moist summers and cold winters encourage *Betula pubescens* var. *tortuosa,* while in eastern Europe, with its more severe winter, the dominant conifers become *Picea obovata* and *Larix russica* (Siberian larch; formerly *L. sibirica*). In more continental Siberia, particularly in northeastern Siberia, the only conifer that readily tolerates the extremely severe winters is the deciduous *Larix gmelinii* (Asian larch; includes *L. dahurica*).

3.4 Soils

Increased species diversity in the forest-tundra, as compared to within tundra, is not simply a reflection of an improved climate and latitudinal variation. This increased diversity also reflects a broader range of soil conditions, which include such factors as the presence or absence of permafrost and soil nutrient status (Timoney et al., 1993a). The forest-tundra region is associated with both the continuous and discontinuous permafrost zones (Figure 2.3). A comparison of Figures 2.3 and 3.2 shows that there is a significant extension of continuous permafrost into this zone in the Hudson Bay lowlands. As a consequence of the presence of permafrost and the possible effects of cryoturbation, it is normal to find typical Organic, Turbic, and Static Cryosols. Associated with these are found non-Cryosolic soils where permafrost is either absent or at sufficient depth that it does not influence the soil profile.

Well-drained mineral soils include Brunisols, which are quite common in Western Canada in areas where parent materials are alkaline, being derived primarily from carbonate-rich sands and tills (Figure 1.9). In the northwest of the Northwest Territories, parent materials are generally more calcareous than in the eastern parts of the Territories, where parents materials are derived from acid Canadian Shield Precambrian rock. This again changes abruptly in the southeastern Hudson Bay lowlands, where Palaeozoic limestones and dolomites are the sediment sources for the ground moraine, esker, kame, and raised beach land-forms. Due to isostatic rebound, many of the coastal features in the Hudson Bay lowlands east of Churchill have only had a few thousand years for profile development, so this nutrient-rich parent material strongly influences the young wetland soils of the region, with most wetlands within 20 km of the coast being alkaline fens and not acid bogs. To the east, in the Quebec-Labrador forest-tundra region, Podzols dominate on well-drained glacial and outwash deposits derived from acid Shield bedrock. Soils of the Organic and Turbic Cryosol and Gleysolic great groups are also common in the Hudson Bay Lowlands where natural drainage is impeded. For typical Gleysolic and Podzolic soil profiles, see Figure 3.4, and for Cryosols and Brunisols, see Figure 2.4.

While Fibrisols and Mesisols are encountered in the High and Mid-Subarctic, the frequent presence of permafrost, or the cryoturbation effects of permafrost, means that Organic Cryosols are the dominant organic soil type throughout the region. Slow decomposition rates ensure many organic profiles are fibric, while moderately decomposed organic profiles are common only where the moisture regime allows moderate decomposition of *Sphagnum* peat in acid bog environments or in fen peat, where less acid conditions prevail. Because of the cryoturbic effects of permafrost, most gleyed profiles are actually Gleysolic Turbic Cryosols, often capped by a peaty layer. Where permafrost is at depth, Gleysolic soils are typically Peaty Gleysols. With both gleyed types, reduction conditions promoted by a high water table inhibit oxidation throughout all but the upper few centimetres of the soil. This high moisture content, combined with the cool climate, reduces litter decomposition so that a peaty

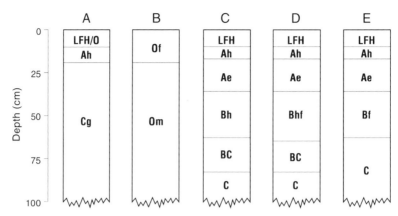

Figure 3.4
Typical soil profiles of the forest-tundra zone in Canada (modified from Canada Soil Survey Committee, 1978). A = Gleysol; B = Typic Mesisol; C = Humic Podzol; D = Ferro-Humic Podzol; E = Humo-Ferric Podzol.

layer can be expected. These conditions also discourage faunal mixing of mineral and humic materials to form thick Ah horizons. If litter decomposition was to be even less effective, this peaty layer would simply thicken until the profile required reclassification. A Peaty Gleysol would be reclassified as a Mesisol if the moderately decomposed peat layer exceeds a thickness of 40 cm or as a Fibrisol if fibrous peat depth exceeded 60 cm.

   On acid, well-drained mineral deposits, Podzols (or Spodosols using the New Comprehensive Soil Survey System terminology) form distinctive soil profiles, with sharp contrasts between the wood ash colour of their Ae horizons and the bright brown-to-red colours of the B horizon. The terms Podzol and Spodosol are both derived from the respective Russian and Greek words for wood ash. A high leaching potential, resulting from a combination of (1) a positive moisture index, (2) acids from the acidophilous vegetation cover, (3) the soils' natural acidity, and (4) good drainage, produces this distinctive ash-coloured Ae eluvial horizon. This same acidity almost totally limits mesofaunal populations so that humus is not necessarily mixed with the eluvial zone to form an Ah. While the Podzolic profiles shown in Figure 3.4 indicate the presence of a thin Ah layer, this horizon may be totally absent. In moist climates a strong leaching potential does, however, promote the eluviation of iron, aluminum, and humus compounds from the A horizon. Some of these compounds are illuviated in the B horizon, so at the great group level, Podzols are differentiated on the basis of the degree to which humus or iron dominates in this illuvial zone. Humic Podzols (Bh) form when the illuviated material is primarily humus with little iron, Ferro-Humic Podzols (Bhf) develop when iron joins with the humus, and Humo-Ferric Podzols (Bfh) result when iron dominates over humus in the B horizon. In parts of the drier Subarctic in northern Quebec, sandy substrates sometimes exhibit patchy Podzol development in otherwise little eluviated Regosols. Patchy Podzol development results from greater rates of eluviation-illuviation caused by the melting of snow trapped within the krummholz layering around scattered spruce (Payette and Filion, 1993). As shown in Figure 1.9, Podzolic soils are particularly important in both the forest-tundra and boreal forest in Quebec.

As with other vegetation covers, soil forming factors dictate frequent recurrent soil type patterns (associations). On better-drained slopes where permafrost is absent, Podzolic or Brunisolic associations frequently dominate, while on lower slopes, poorer drainage encourages Gleysolic soils, and where soils are almost permanently saturated, Organic soils develop. Where permafrost is near the surface, similar predictable soil patterns develop from ridge to depression, although some of these profiles can be badly disturbed by cryoturbation. Naturally occurring thermokarst features, already noted as having importance in the shrub tundra, also occur here in Cryosolic areas. In addition, human-induced disturbances can give rise to a deepened active layer, resulting in melting of the upper permafrost and thermokarst subsidence (Mackay, 1970).

3.5 Forest-Tundra in Canada

3.5.1 Ecoclimatic Sub-Provinces

The distribution of forest-tundra in Canada is shown on Figure 1.2 as ZE VIII–IX and on Figure 1.3 as the boreal-tundra ecotone. Using terrestrial ecozone terminology, we are dealing with the northern zones of the Taiga Plains and Taiga Shield (ecozones 6 and 8 in Figure 1.5). Ecoclimatic subzoning of the forest-tundra is necessary to help distinguish between areas which are more tundra-like (the shrub subzone, or northern forest-tundra) and areas where the tree component gives a more boreal-like appearance (the forest subzone, or southern forest-tundra). The boundary between the shrub subzone and the forest subzone is designated as the physiognomic forest limit. In Quebec the gradual rise in elevation towards the south allows for a broad forest-tundra zone which can be conveniently subdivided into the High Subarctic (shrub zone of the forest-tundra) and the Mid-Subarctic (forest zone). To the west of James Bay, the more abrupt climatic gradient produces a much narrower zone, creating a rapid transition from High to Low Subarctic. Because of this, west of James Bay, both the shrub subzone and the forest subzone fit within the High Subarctic Sub-Province (Figure 1.4 and Table 3.2). It is important to note that the Low Subarctic Sub-Province is not considered part of the forest-tundra region. While the lichen-moss-shrub ground cover of the Low Subarctic possesses many similarities to the forest-tundra, it is more suitably described as "open lichen woodland" or "open boreal forest."

The climate of the High Subarctic is typically humid, with a strong positive moisture index, and precipitation maxima occur in summer and result from cyclonic activity along the polar front. In the High Subarctic, such as at Churchill, Manito-

Table 3.2
Zoning within the Subarctic Ecoclimatic Province

Sub-Province	Ecoclimatic Region	Vegetation Zone
High Subarctic	High Subarctic (HS)	Forest-tundra shrub subzone in Quebec, and shrub plus forest subzones west of James Bay
Mid-Subarctic	Mid-Subarctic (MS) Maritime Mid-Subarctic (MSm)	Forest subzone in Quebec
Low Subarctic	Low Subarctic (LS) Atlantic Low Subarctic (LSa)	Northern boreal forest (i.e., open lichen woodland)

ba (Figure 3.3A), almost five months average above 0°C. Here melting ground ice combines with a mean annual precipitation of 402 mm to ensure ready access for well-rooted vascular species to soil moisture during the growing season. At Kuujjuarapik, a Mid-Subarctic station on the east shore of Hudson Bay, the extra month above 0°C, combined with a 2–3 degree higher average annual temperature, appears to be sufficient to promote tree growth, except on the more exposed uplands or slopes.

Typical ecosystems in the forest-tundra would include (1) rock outcrops and upper slopes dominated by lichen–heath–dwarf shrub communities, (2) lower slopes with scattered low trees and lichen–moss–dwarf shrub ground cover, (3) sheltered but drained depressions and lower slopes where open *Picea* forest develops, and (4) wetlands, which often take the form of peatscapes or muskeg (often peat plateau bogs and ribbed fens) that cover large, level areas, both within the forest-tundra and in the boreal forest to the south. Although wetlands are not discussed in detail in this chapter (see chapter 8), Figure 3.5 is provided to give some background for references to wetlands in the discussion of both forest-tundra and boreal forest. It is noted from Figure 3.5 that large percentages of the forest-tundra, particularly around the southern shores of Hudson Bay, are classified as having in excess of 75% wetlands.

In the forest-tundra, a combination of high latitude (i.e., relatively low sun angles), wind, and a high moisture index often promotes greater arboreal growth on southeastern and eastern slopes than at similar topographic positions facing west. This dependency by forest-tundra trees on energy and wind contrasts with other ecosystems such as the mixed-grass prairie, where on undulating topography, negative moisture indices and greater absolute warmth sometimes encourage a moister, tree-promoting microclimate only on more northerly facing slopes, leaving herbaceous communities to dominate on the effectively more xeric southern slopes.

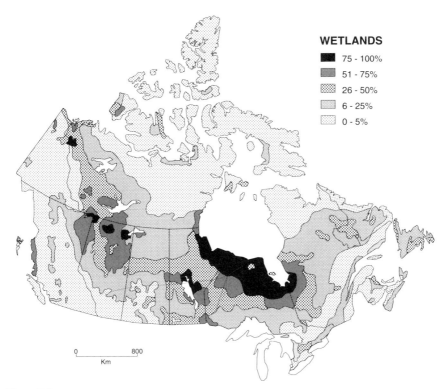

WETLANDS

- 75 - 100%
- 51 - 75%
- 26 - 50%
- 6 - 25%
- 0 - 5%

Figure 3.5
Wetlands of Canada (after National Wetlands Working Group, 1986). Categories refer to per cent
of land area occupied by wetlands. See chapter 8 for more detail on Canadian and world wetlands.

3.5.2 The Shrub Subzone (Northern Forest-Tundra)

Figure 3.6 shows the distribution of the shrub and forest subzones of the High and
Mid-Subarctic of northern Quebec and Labrador (after Payette, 1983), while
Figure 3.7 shows the same subzones in the High Subarctic west of James Bay (after
Timoney at al., 1992). In the shrub subzone the major change from typical tundra
involves a marked increase in boreal species without any significant reduction in
those more characteristic of the Arctic (Morisset et al., 1983). The distribution of
some of the forest-tundra species is strictly correlated with the tree line. Certain species, such as *Chamaedaphne calyculata* (leather leaf), *Drosera rotundifolia* (round-
leaved sundew), and *Scirpus hudsonianus* (bulrush), are found throughout the
forest-tundra right up to the tree line, but not within the tundra (Morisset et al.,
1983). This phytogeographic barrier also holds true for true tree life form representatives of the *Picea* and *Larix* genera. It is also noted that the rate of change in forest

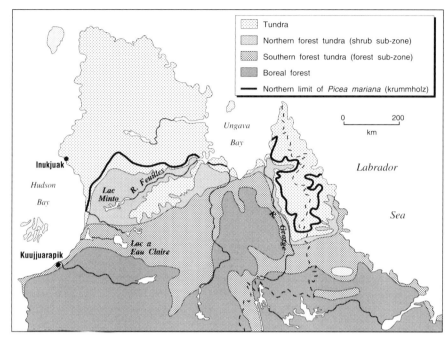

Figure 3.6
Vegetation zonation of northern Quebec-Labrador (after Payette, 1983)

cover percentages is not simply a progressive one from the shrub tundra to the boreal forest. Rather, as one moves south from the tree line, tree cover initially increases very slowly, increases rapidly in the middle of the forest-tundra transition, and then decelerates as the Low Subarctic boundary is reached (Timoney et al., 1993b)

In the Lac Minto region (Figure 3.6), *Picea mariana* is the dominant tree species, although it is frequently accompanied by *Larix laricina*. In terms of cover the tundra-like lichen–heath–dwarf birch community dominates. Small forest stands are located in protected sites facing north, east, and southeast, whereas krummholz forms occur on the more exposed sites facing the predominantly westerly winds (Payette, 1983). Associated with krummholz forms of *Picea mariana* in northern Quebec are also forms which spread laterally close to the ground, that is, layering to form cushion or mat krummholz forms (Payette and Gagnon, 1979). The loss of the true phanerophytic form does not necessarily prevent these individuals from surviving low winter temperatures and desiccating winds by growing infranivally, that is, remaining below snow cover (Lavoie and Payette, 1992). Some *P. mariana* individuals that exhibited supranival skirted and whorled forms before the Little Ice Age (1570–1880) were only able to survive this colder period by reverting to infranival mat krummholz forms. That these same individuals have not since reverted to supranival forms during the warmer conditions of the last century may well reflect the fact they can not trap enough drifting snow to allow for vertical growth (Lavoie

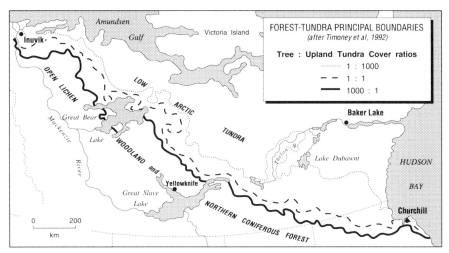

Figure 3.7
Forest-tundra ecotone zonation in the Northwest Territories (after Timoney et al., 1992) and north-ern Manitoba (after Scott et al., 1987b)

and Payette, 1992). Some areas of old-growth lichen-spruce krummholz burned off during the Little Ice Age only support lichen-tundra cover today because of the lower snow-trapping potential of this new cover and associated soil thermal regime effects (Arseneault and Payette, 1992).

To the southwest of Lac Minto, near the shores of Richmond Gulf, lies the boundary between the shrub and forest subzones (Figure 3.6). Here *Picea glauca* becomes an important component in forest stand cover. In the shrub subzone, trees generally not only have to tolerate the harsh climate but also have to contend with unstable soil conditions resulting from solifluction and the growth of ice lenses in the root zone (Jacoby, 1983). Along the coast itself, littoral *Salix* spp. shrub populations reflect continuing isostatic rebound. Here shoreline shrub vitality shows healthy individuals, while backshore communities are in decline (Mörs and Bégin, 1993).

West of Cape Churchill in northeastern Manitoba, the broad belt of recently emerged coastline has calcareous soils which have generally had less than 2,000 years for profile development. As a consequence, the high carbonate content of eskers, raised beaches, kames, and ground moraine provides lime-rich drainage waters to enter wetlands, ensuring they are primarily fens. Responding to these eutrophic soil conditions, *Picea glauca* "tree islands" are found in sheltered, better-drained locations close to tundra margins. To a lesser extent, some scattered *Larix laricina* are also present in the wetlands, but *Picea mariana* are uncommon here, preferring the somewhat thicker, and therefore more acid, wetland soils some distance inland. Typical white spruce sites would include the windward and leeward lower slopes of eskers and the lee of pressure ridges formed by ice around lake shorelines. Here white spruce often form infranival skirts or aprons (thickly layered branches) with upper flag forms which vary in shape depending on age, height, relative mutual protection, and the severity of ice-crystal wind blasting (for a classification of flag-related forms, see Scott et al., 1987a). Some very exposed spruce even develop into clones when additional vertical stems arise from the basal branches of the "parent" tree. Within the protection of larger white spruce islands, trees can reach heights of 10 m, and supranival skirt forms can be encountered. These supranival skirts often have as much as a metre of dead branches, or branchless trunk, between them and the ground. Above the skirts, only several metres of the trunk may reflect the more classical flag form or be devoid of branches, while above that, a typical spruce shape develops, with branches on all sides. Normally, the most severely impacted abrasion zone is within 80 cm of the snow surface (Scott et al., 1993).

Ecotone characteristics northwest of Hudson Bay possess some notable differences, as well as similarities, to the northern Quebec situation. Climatically, it is noted that both summer heat and net radiation increase significantly between areas northwest of Great Bear Lake and areas between Great Bear Lake and Hudson Bay (Timoney et al., 1992), and this may contribute to the previously mentioned greater importance of *Picea glauca* over *P. mariana* to the northwest. It is also possible that this separation in dominancy among spruce species may be related to the aforementioned soil nutrient conditions of the region. Such differing mineral soil conditions do influence species composition and general community diversity. On the eutric mineral Cryosols of the northwest, white spruce are associated with *Dryas*, legume, and bryophyte communities, with high species diversity, while on nutrient-poor Brunisols to the southeast, more acidophilous species such as ericaceous shrubs and lichens join black spruce to form low species diversity communities (Timoney et al., 1993a).

Latitudinal effects on growing conditions may also account for the general broadening of the ecotone to the southeast which is apparent both in Figure 3.7 and when Figures 3.6 and 3.7 are compared. Timoney et al. (1992) report that ecotone breadth west of Hudson Bay varies from a mean of 112 km in the western half of this region to 179 km in the eastern half. In northern Quebec, ecotone width sometimes exceeds 300 km due to the gradual decrease in elevation north from Schefferville, which lessens the apparent latitudinal decline in temperature (Elliott-Fisk, 1988). Note also the similar impacts of microclimate both west and east of Hudson Bay, which allow the shrub subzone to extend north along the Thelon and Feuilles rivers, while being forced almost due south inland of Hudson Bay by cold advective winds off the pack ice (Figures 3.6 and 3.7).

3.5.3 The Forest Subzone (Southern Forest-Tundra)

The forest subzone of the forest-tundra is the dominant cover of the Mid-Subarctic Ecoclimatic Sub-Province in Quebec and the southern portions of the High Subarctic west of James Bay. Typical cover in the Lac à l'Eau Claire region of northern Quebec (Figure 3.6) includes sizeable forested areas on the lowlands and slopes with less area covered by lichen–heath–dwarf shrub openings on exposed and water-free sites. Some of these exposed sites lack tree cover due to a combination of harsh growing conditions and low levels of tree regeneration after fire. Occasionally, non-forested lowlands are found on the downwind sides of large lakes, where cold winds off the ice create a more tundra-like microclimate. On fire-free islands in Lac à l'Eau Claire, Payette (1988) found both cyclic-autogenic sphagnum–*Picea mariana* communities and allogenic succession communities characterized by sphagnum–*P. mariana* replacement by lichen-heath. The autogenic succession results primarily from the constancy of the water supply and the absence of fire, while the allogenic succession is a consequence of the groundwater becoming more acidic (oligotrophic) as the peat accretes upwards over time. In this allogenic succession, the community shifts from *Sphagnum girgensohnii–Picea* to *S. russowii–Picea*, then to *S. fuscum–Picea,* and finally to lichen-heath (Payette, 1988).

While the dominant tree species in the Mid-Subarctic of Quebec are *Picea mariana* and *Larix laricina,* a general improvement in tree-growing habitat in the forest subzone is reflected by the presence of *Picea glauca, Betula papyrifera* (white birch), and *Populus tremuloides* (trembling aspen) along some of the rivers, such as the George River (Payette, 1983). A general increase in temperature in the twentieth century has also led to conditions conducive to permafrost degradation in parts of this critical climatic transition zone. On the eastern shores of Hudson Bay near Kuujjuarapik, Laprise and Payette (1988) found a 49% decrease in the total cover of palsas and collapse scars between 1957 and 1983. During the same time period, thermokarst lake area had increased by 44%. On the extensive coastal sandy terraces of this region, post-fire succession begins with moss-lichen dominants, then *Cladina* and *Stereocaulon* lichens take over and are joined by white spruce seedlings, indicating a slow return to forest-tundra. One interesting feature associated with this succession is the development of "lichen polygons." These are not related to ice-wedge polygons, but result from forest-floor lichen desiccation. As spruce forest develops, *Cladina stellaris* (formerly *Cladonia alpestris*) ultimately becomes the dominant lichen ground cover (Filion and Payette, 1989). In upland hilly areas, not only does aspect influence the distributional tendencies of some of the more narrowly distributed lichen species as well as krummholz spruce, but the spruce-lichen woodland is found at higher elevations on southwest-facing slopes than on those with a north-northeast aspect (Petzold and Mulhern, 1987). While *C. stellaris* is the dominant lichen throughout the forest-tundra area shown in Figure 3.6, other lichen species also characterize the region. *Alectoria ochroleuca* and *Cornicularia divergens* are more typical of the north, while *Cladina mitis* characterizes the south (Crete et al., 1990).

In areas such as the Twin Lakes kame terrace region, 35 km southeast of Churchill, *Cladina stellaris* forms broad, thick, almost-monodominant lichen carpets between tall, widely spaced *Picea glauca.* Here lichen polygons similar to those reported from Quebec seem common, particularly on sandy Eutric Brunisols which are prone to summer drought. Below the Twin Lakes kame, to the northeast, is open eutrophic fen with some stunted and krummholz forms of *P. glauca* and *Larix laricina.* Areas to the south of the kame are not only at somewhat higher elevation, but are better protected from winter winds from Hudson Bay. Here fen has begun to give way to treed proto-bog, with its large mounds of red *Sphagnum warnstorfii* and *S. fuscum* accreting outwards over fen-type bryophytes and *Ledum groenlandicum*. These Organic Cryosol profiles range in pH from 6.8 in fen peat just at the permafrost contact to 3.6 at the top of sphagnum mounds 60 cm above. Not surprisingly, this open forest is quite tall and comprised of a combination of the more acidophilous *Picea mariana* and, to a lesser extent, the more basophilous *Larix laricina*.

Along the east side of the Churchill River estuary just south of Churchill, forest-tundra is squeezed into a narrow 10–20 km wide strip between coastal tundra and the open lichen woodland to the south (Figure 3.7) (Scott et al., 1987b). Here can be found not only established forest-tundra but areas of emerging sediments resulting from recent isostatic uplift. *Picea glauca* is an important colonizing species of this new land, although studies show its success at becoming established closely parallels both air temperatures and low soil temperatures that result from the insulative properties of the dry lichen-heath, which slows heat influx (Scott et al., 1987b). In some well-developed white spruce communities in the same region, unequal thickening of permafrost in peat mounds sometimes leads to the formation of "drunken forest," a condition where many of the more mature spruce are leaning at different angles and in different directions to each other. It would appear that the slow upward expansion of a peat mound on one side of a tree can force tree roots as much as 30–40 cm above the level at which these same roots enter the trunk base, causing the tree gradually to lean away from the mound. In addition to peat mound expansion, the forest floor in general accretes upwards due to bryophyte accumulation, raising the forest floor as well as the permafrost level below. To help negate the effects of permafrost rising into the lower rooting zone, long-established white spruce can sometimes develop a few higher-level adventitious roots, which radiate out into the newer surface bryophyte peat. Rouse (1984a, 1984b) found that forest soils near Churchill are substantially warmer in the active layer than were open tundra soils and that the thaw period in the rooting zone under open forest is six months, two months longer that for the same depths under tundra. Surprisingly, he also found that the role of differing vegetation covers in evapotranspiration was greatly modified by regional advective wind patterns, both offshore and onshore from Hudson Bay pack ice (Rouse 1984c).

Fire has played an important role in helping develop the mosaic characteristics of the boreal-tundra ecotone. The upland tundra–lowland forest mosaic typical of this ecotone region to the northwest of Hudson Bay is significantly modified by fire. Fires are less frequent here than in the boreal forest to the south, with most fires spreading in from the boreal and often penetrating 25–75 km into the forest-tundra (Timoney and Wein, 1991). In the Northwest Territories, fires are both more numerous and cover larger areas northwest of Great Slave Lake, while to the southeast the opposite is the case. This may reflect the greater abundance of fire-impeding lakes, together with different weather patterns (Timoney and Wein, 1991). In the forest-tundra uplands of Ungava, the sustained reduction of tree population density following repeated destructive fires appears to be one of the primary deforestation processes of the subarctic (Sirois and Payette, 1991).

3.6 Eurasian Forest-Tundra

The Eurasian forest-tundra ecotone (Zonoecotone VIII-IX) stretches along the southern periphery of the southern tundra from northern Finland to northeastern Siberia (Figure 2.8). Because of its significance to wintering caribou, this region is considered a separate vegetation zone by the Russians and not an ecotone. Like its North American counterpart, it consists of a mosaic of tundra and boreal forest communities, but unlike the North American forest-tundra, both extreme maritime and extreme continental conditions dictate differences in the tree species types which dominate.

In Fennoscandia a birch zone dominated by *Betula pubescens* var. *tortuosa*, together with some *Pinus sylvestris* (Scots pine), corresponds most closely with the concept of the Canadian subarctic forest-tundra (Hustich, 1966). Oceanicity and the presence of triploid varieties influence the ability of this broadleaf deciduous species to do well here (Valanne, 1983). In addition, intensive logging and catastrophic outbreaks of herbivore insects have added to fragmentation (Sirois, 1992). As continental conditions strengthen to the east, add to these factors the ability for *Betula* to propagate vegetatively below the snow. In western Siberia, birch is replaced by *Larix russica* (Siberian larch), which can also develop non-phanerophytic forms to help with survival (Valanne, 1983). *Abies sibirica*, *Pinus sibirica,* and *Picea obovata* are also present in the southern portions of the western Siberia transition. In northeastern Siberia, *L. gmelinii* is the only phanerophyte throughout the northern forest-tundra, although *Pinus pumila* (dwarf Siberian pine) is encountered towards the south. In addition, there are three species of *Larix* in mountains between Sikkin and western China and an additional species in peninsular Korea and Japan (Gower and Richards, 1990). For additional information on the typical southern tundra and boreal forest margin species see sections 2.5, 4.5.1, and 4.5.2.

3.7 Primary Production and Phytomass in Forest-Tundra

Because of the mosaic nature of the forest-tundra transition, biomass and net primary productivity values vary considerably. For values typical of warmer tundra communities, see section 2.7, and for northern boreal forest, see section 4.6. A typical K factor value (comparing litter biomass with NPP) for spruce-lichen woodland in northern Quebec is 0.11 (Moore, 1978). Of interest is the degree to which grazing influences the dominant lichen cover throughout this region. Studies in northern Quebec by Crete et al. (1990) show that lichen biomass averages 1,223 kg ha^{-1} and that a positive correlation exists between lichen biomass and mean annual precipitation, with the larger values found in the wetter west. They also determined that the caribou consume approximately 0.5–0.9% of the available lichen biomass annually.

Boreal Forest (Taiga) and Mixed Forest Transition

4.1 Distribution

As with the forest-tundra, the boreal forest is a northern hemisphere vegetation formation (Zonobiome VIII). Because these forests are dominated by conifers from the *Picea, Pinus, Larix,* and *Abies* genera, this zone can also be called the boreal coniferous forest (Greller, 1989). The term "taiga" is frequently used interchangeably with that of boreal forest, and it is derived from the Russian word for coniferous forest. Some authors give a narrower definition for "taiga," considering it to include only those portions of the boreal forest between the southern boundary of the forest-tundra and the closed crown coniferous forest to the south (Sirois, 1992); essentially, the open lichen woodland. "Taiga" is the term more commonly used to describe boreal forest in Eurasia, although it is rapidly gaining acceptance in the North American literature. For convenience, the mixed forest transitions from boreal to prairie (ZE VIII–VII) and boreal to broadleaf deciduous forest (ZE VIII–VI) are also discussed in this chapter.

The boreal forest is a vast, almost continuous coniferous forest belt, covering some 6.7 million km^2 (Figure 4.1). It includes about one-third of the world's forested land and 14% of the world forest biomass (Kauppi and Posch, 1985). In North America it occurs widely in Alaska and is the dominant forest cover in Canada, stretching in a vast swath from the Yukon though south-central Canada and the Great Lakes states of the USA to Newfoundland (Figures 1.3 and 1.4). It lies between the forest-tundra to the north and usually the prairie grasslands or temperate deciduous forests to the south. While many boreal-like forests are also found in the Canadian cordillera, these will be treated separately. In Eurasia the boreal forest is found between the forest-tundra and either the temperate deciduous forests or steppe grasslands to the south. In Scandinavia it is found as far south as 62°N, while it reaches as far south as 53°N in the Urals and 43°N in Japan. It covers vast areas of northern Asia (Siberia). Being associated with recently deglaciated areas, humid climates, low evaporation rates, and relatively low elevations ensures the presence of many

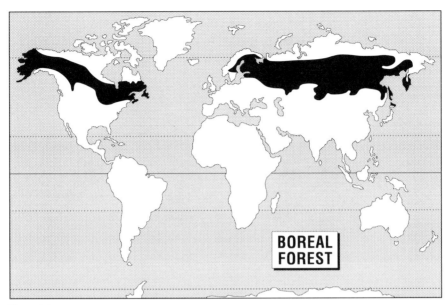

Figure 4.1
The distribution of boreal forest

wetland areas (Figure 3.5). It is also noted that coniferous forests can be found in such places as the Mediterranean, Florida, Central America, and the Philippines, but these forests are not the product of a continental climate, rather, of other combinations of environmental conditions (e.g., physiological drought) that favour the evergreen needleleaf type over the broadleaf.

4.2   Climate

Frequently described as having a cold continental climate, the boreal zone really consists of a variety of contiguous ecoclimates which support coniferous and mixed conifer-hardwood forests. Mean annual temperatures vary widely from those typical of the *Larix* (larch) wetland forests of Verkhoyansk (Figure 3.3C) in central Siberia, which average below −16°C, to values well above freezing, such as at Yarmouth, Nova Scotia (Figure 4.3C), which has a maritime continental climate. Due to the availability of moisture and a combination of latitude and growing season energy and length, forest growing conditions are present even if the forests vary considerably in dominancy from one ecoclimatic region to the next. The unifying criterion is that climate promotes conditions more favourable for the success of coniferous species over broadleaf deciduous or broadleaf evergreen types. Tuhkanen (1984) points out that a similar climatic zone can be found in the southern hemisphere. He refers to this zone as the "antiboreal" but although this region does not support a typical coniferous type cover, the *Nothofagus* (southern hemisphere beech) forests of

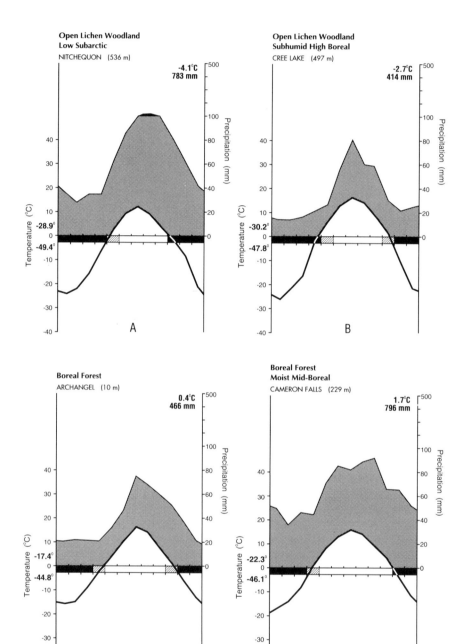

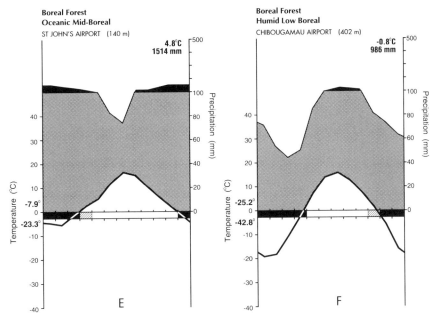

Figure 4.2
Boreal forest climate diagrams. A = Nitchequon, Quebec; B = Cree Lake, Saskatchewan; C = Archangel, Russia; D = Cameron Falls, northwest Ontario; E = St John's Airport, Newfoundland; F = Chibougamau Airport, Quebec.

southern Chile might well fill a similar ecological role to that of the *Betula* forests of northwestern Europe.

The Canadian boreal forest is associated with the Low Subarctic (LS) Ecoclimatic Sub-Province and the High, Mid-, and Low Boreal (HB, MB, LB) Ecoclimatic Sub-Provinces, while the mixed forest transition (to temperate deciduous forest) is associated with the High Cool Temperate (HCT) and Mid-Cool Temperate (MCT) Ecoclimatic Sub-Provinces (Figure 1.4). The Low Subarctic and northern portions of the High Boreal Ecoclimatic Sub-Provinces are associated with open lichen woodlands such as at Nitchequon, Quebec, where the mean annual temperature is −4.1°C and mean annual precipitation is 783 mm (Figure 4.2A), and Cree Lake, Saskatchewan (Figure 4.2B), where the mean annual temperature is −2.7°C and mean annual precipitation exceeds 400 mm yr^{-1}. The High Boreal and Mid-Boreal Sub-Provinces are each subdivided into seven ecoclimatic regions ranging from the Subhumid Boreal of the interior to the very wet Oceanic Boreal on the east coast. In Subhumid Mid-Boreal locations such as The Pas, Manitoba (Figure 1.10B), mean annual temperatures are around or above zero, while mean annual precipitation normally exceeds 450 mm. In the moist Mid-Boreal at Cameron Falls, Ontario (Figure 4.2D), mean annual precipitation exceeds 700 mm, while at St John's

Airport, Newfoundland, in the Oceanic Mid-Boreal, the mean annual temperature is 4.8°C and mean annual precipitation is 1,514 mm (Figure 4.2E).

The Low Boreal Ecoclimatic Sub-Province is subdivided into ten ecoclimatic regions, again ranging from the Subhumid of the Prairie provinces to the Very Wet of the Maritimes, but also including several transitional regions bordering the mixed forest transition to deciduous forests. In the Humid Mid-Boreal, such as at Chibougamau, Quebec, mean annual precipitation approaches 1,000 mm (Figure 4.2F), while in oceanic areas such as Yarmouth, Nova Scotia, conditions are even wetter (Figure 4.3C). It should be noted that southern Subhumid Boreal regions can have mean annual temperatures around zero, while coastal Mid-Boreal regions normally (Maritime and Oceanic) average 3–5 degrees above zero. This apparent anomaly merely emphasizes the effect of cold continental winters in depressing mean annual temperatures and is not a measure of the length or available energy of the growing season. Warmer conditions in the High Cool Temperate and Mid-Cool Temperate Ecoclimatic Sub-Provinces provide sufficient extra growing-season energy to permit the growth of the mixed forests of conifers and broadleaf deciduous species, known as the Great Lakes–St Lawrence–Acadian forests.

The great arc-shaped distribution pattern to the North American boreal is not repeated in Eurasia, primarily because weather patterns move generally from west to east there. In North America, the cordilleran topographic barrier, the Arctic Ocean extension into Hudson Bay, and the potential for moist tropical air to be drawn up from the south combine to produce a contact zone between polar and tropical air masses known as the polar front. Frontal systems travelling along this contact frequently follow a southeasterly track from the Yukon south to the Great Lakes, where they swing east and then northeast, crossing the Maritimes. So pronounced is the influence of this polar front that many researchers consider the southern boundary of the northern boreal forest to coincide with its mean winter position (Bryson and Wendland, 1967). The Siberian boreal experiences a more severe winter in both human and tree-species terms than does the North American, primarily because of Eurasia's greater longitudinal extension and the fact that warm, moist air can not be drawn in from the south. Siberian summers, however, are sufficiently long to allow tree growth, as long as those trees can also survive the rigors of winter. Only one such species, *Larix gmelinii*, appears to be able to thrive under these conditions, and it does so because of its deciduous nature (Gower and Richards, 1990).

Not all Eurasian boreal regions are like central Siberia. Archangel (Figure 4.2C), a port on the Arctic coast of northwestern Russia, is more typical of the European boreal and has a temperature regime similar to that of Cameron Falls in northwest Ontario, although it is a little drier. A noticeable decline in precipitation values from western Europe east to Siberia is to some extent compensated for by a parallel decline in evapotranspiration resulting from its higher latitude and longer winters. Moscow (Figure 4.3D), a mixed forest station located at 56°N in eastern Europe, is at a similar latitude to mixed forest in Saskatchewan but is at the same latitude as the forest-tundra in northern Quebec. It is also noteworthy that in central and eastern

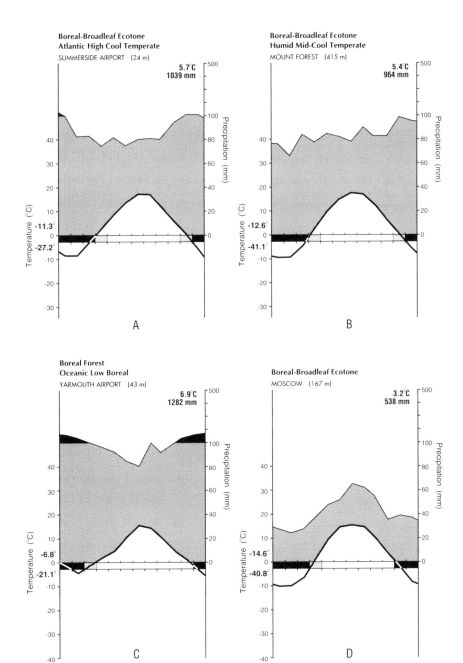

Figure 4.3
Climate diagrams for the mixed-forest transition (boreal-broadleaf ecotone). A = Summerside Airport, PEI; B = Mount Forest, Ontario; C = Yarmouth, Nova Scotia; D = Moscow, Russia.

North America, climates remain moist (and therefore potentially forested) as one moves south. In eastern Europe and Siberia, drier conditions to the south generally promote steppe grasslands.

One important general observation as to the significance of boreal climatic on vegetation cover is worth making. The moist, cool conditions of the boreal climate seem ideally suited to promote direct resource competition between two important physiognomic types in habitats where decomposition processes are slow. On the one hand, phanerophytes as water pumpers (phreatophytes) are constantly transpiring soil moisture and are even capable of lowering water tables, while on the other hand, bryophytes can be viewed more as water retainers and in fact often encourage water accumulation. Much of the dynamics of the moister portions of the boreal mosaic hinges upon the relative success of one or the other of these two groups in dominating local soil moisture balances. It would appear that in no other major tree-dominating ecoclimatic province, with perhaps the exception of the Pacific Cordilleran, do phanerophyte dominants so often reflect the success of other plant groups.

4.3 Soils

Soils associated with much of the boreal forest are those typical of humid continental climates where permafrost is either absent or only occasionally found. Only in the northern parts of the High Boreal and in the open lichen woodland of the Low Subarctic does permafrost influence pedogenesis. In the absence of permafrost, major variations in soil types result from (1) differences in acidity of parent material, (2) climate, and (3) drainage.

In terms of parent material, Podzols are generally associated with the more acidic regolith of the Canadian Shield, where strongly positive moisture indices encourage leaching and acidification (see Figures 1.9 and 3.4). Podzols have strong acidity and low nutrient status characteristics, and they comprise the dominant soil order of the boreal region. In a typical boreal forest, such as the southern Laurentian Highlands, Ferro-Humic Podzols would be the most frequently encountered great group (Wang and McKeague, 1986). In central Ontario, Humo-Ferric Podzols exhibit tonguing of the Bf horizon to depths greater than 2.5 m. Such tonguing may have been initiated by root casts formed by *Pinus strobus* windthrow or *in situ* tap-root decay and may well modify local nutrient cycling by acting as conduits for both water and downward-moving solutes (Watters and Price, 1988). Because of the significance of forestry within the boreal region, variations in the productivity potential of Podzolic soils will be addressed later. Where parent materials are more alkaline, such as on the carbonate-rich glacial deposits of the central Prairie provinces, Brunisols and Luvisols are frequently encountered (see Figures 3.4 and 4.4).

In response to variations in boreal climate, Cryosols are often encountered over permafrost in the Low Subarctic and High Boreal (see Figures 2.3 and 2.4). Eutric Brunisols dominate the calcareous deposits of the cooler High and Mid-Boreal

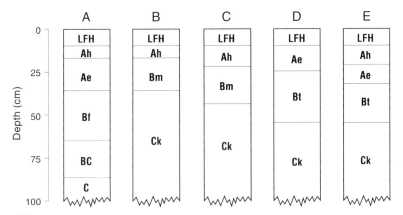

Figure 4.4
Typical boreal forest soil profiles (after Canada Soil Survey Committee, 1978). A = Humo-Ferric Podzol, B = Eutric Brunisol, C = Melanic Brunisol, D = Gray Luvisol, E = Gray-Brown Luvisol.

Ecoclimatic Sub-Provinces, while in the milder moist climate of the Low Boreal of southern Ontario, Melanic Brunisols can be found, particularly if the cover includes broadleaf deciduous trees. In the Low Boreal of the Prairie provinces, positive moisture indices and high base saturations encourage the development of Gray Luvisols instead of Podzols, while in the warmer equivalent of southern Ontario, Gray-Brown Luvisols, with their thicker Ah horizons and mull litter, are found.

In terms of drainage, areas with only modest drainage but with high water tables are generally associated with Gleysolic soils (especially Humic Gleysols, often with peaty phases), while organic soils (primarily Fibrisols and Mesisols) cover vast expanses of the more poorly drained lowlands, flood plains, and lake margins. While these conditions will be elaborated on more fully in the chapter on wetlands, it is important to stress that within the boreal there are constant changes in wetland soil profiles and wetland area. The process of forest paludification, where moist-site forest floor organic matter accumulates and eutrophic-mesotrophic organic layers are replaced upwards by more oligotrophic, is active in many areas. The opposite process, or forest depaludification, is also occurring where conditions are favourable for a reversal of water-table peat-accumulation driving forces. The importance of these two process is not just that boreal wetland soils may be undergoing constant change and that their forest cover dominants are also reflecting these changes, but that wetlands are either accreting outwards, are in equilibrium, or are retreating, depending upon which process may be the dominant one in any particular place (Glebov and Korzukhin, 1992).

Another organic soil type, the Folisol, also forms on rock outcrops throughout the Canadian Shield region because it is here that the extremely acid nature of the parent material (acid igneous and metamorphic rock) creates such slow decomposition rates that organic matter decomposes more gradually than it accumulates. While most other organic soil types depend on high water tables to discourage decomposition, it is the acidity of the parent material and the acidophilous nature of those few species which can tolerate this environment (lichens, mosses, etc.) that combine to discourage decomposition. Often less than 15 or 20 cm thick, and formed because of quite different reasons than the thick Fibrisols and Mesisols, these soils are capable of supporting tree growth, especially *Pinus banksiana* and *P. strobus* (white pine). They are, however, very vulnerable to fire, due to the fact they desiccate easily in dry years and because of their exposure on ridges.

4.4 Boreal Forest in North America

The North American boreal forest is commonly subdivided into three forest regions: the open lichen woodland, the northern coniferous forest, and the mixed forest (boreal-broadleaf ecotone). On Figure 1.3, "open lichen woodland" is synonymous with the northern margins of the boreal forest, as well as the large areas of bog at the southern end of Hudson Bay. In terms of terrestrial ecozones, this area includes the southern portions of the Taiga Plains and Taiga Shield, together with the Hudson Plains and Boreal Shield (Figure 1.5). The term "northern coniferous forest" is usually used to describe the typical *Picea-, Pinus-,* and *Abies*-dominated coniferous forests of the Mid- and Low Boreal Ecoclimatic Sub-Provinces. Large portions of central Alaska are covered by similar boreal forests, although broadleaf deciduous species play a more important role (Van Cleve et al., 1986).

The "mixed forest" region can be more accurately described as an ecotone which includes both conifers (especially pines) and broadleaf deciduous species (ZE VIII–VI on Figure 1.2). It is for this reason that not all authors consider mixed forest as part of the boreal forest proper (Rowe, 1972). Mixed forests are found in the Mixed Wood Plains of southern Ontario and Quebec and Atlantic Maritime terrestrial ecozones (Figure 1.5), as well as in a band through the northern parts of Minnesota, Wisconsin, Michigan, and New York in the northern USA and the New England states in the northeast. As a group they are frequently named the Great Lakes–St Lawrence–Acadian mixed forest. In addition, a narrow band of mixed forest separates the southern boreal of the Prairie provinces from the aspen parkland of the prairie grasslands, but it is quite unlike the Great Lakes–St Lawrence–Acadian forests. It will, however, be discussed in some detail since it represents an example of a southern transition from boreal forest dictated more by moisture stress than by conditions favouring broadleaf trees.

Conifers seem well adapted to boreal soil conditions in that they variously possess adaptations to help seek out moisture, tolerate near-saturated soils, or cope with strongly acid, low nutrient supply conditions. Retaining photosynthetically viable leaves for several decades implies a conservative demand for soil nutrients by evergreen conifers (Gower and Richards, 1990). In addition, it would appear that on extremely low base-saturated soils, such as Podzols, many of the conifers have the ability to live almost "saprophytically" off their own thick mor litter. They achieve this by having the great majority of their fibrous rooting systems confined to the lower litter layer and thin Ah horizon (if present). While some of the larger roots may penetrate to depth to provide stability and seek out moisture, it is doubtful that they derive much in the way of nutrients except perhaps from humus exchange sites in the illuvial (B) horizon. While acid mor provides an inhospitable environment for most decomposers, fungal mycelia and symbiotic mycorrhizae do well. All conifers possess these symbiotic ectotrophic mycorrhizae, which live partly between cells within the tree root while sending their mycelia out into the surrounding litter/soil. These mycorrhizae not only greatly increase the effective nutrient and water-absorbing surface of the host, but actively partake in the extraction of nutrients from litter. Studies in the acid lowland forests of Amazonian Brazil led Went and Stark (1968) to refer to this phenomenon as "direct mineral cycling." While the boreal situation may not be as extreme as that of the humid tropics, due to climatically induced slower growth rates and a proportionally larger nutrient supply via rainfall, evidence certainly supports the conclusion that conifers seem quite indisposed to letting soils with poor nutrient-supplying and poor nutrient-retaining capabilities squander nutrients released through decomposition. Despite the high leaching potential of Podzols, mycorrhizae clearly help prevent this potential from being realized (Mahendrappa et al., 1986). As with tropical forest ecosystems (Scott, 1978), the major nutrient pool for typical well-drained boreal forest sites is in the combined phytomass of the living tree and litter. Unlike in tropical forests, however, the role of natural fires both in releasing nutrients locked into boreal phytomass and in preventing climax boreal forest from developing must be stressed.

The same frontal systems which guarantee the humid climate upon which boreal forest depends also brings lightning. So adapted are conifers to fire frequency that some researchers consider the relationship more than coincidental. Mutch (1970) goes as far as stating that some species actually encourage fire and that fire works to their advantage if they have high flammability and quick recuperation. In essence, fire encourages regeneration, a form of competitive struggle (Pyne, 1984). Whether or not there is an exact cause-effect relationship is difficult to determine, but a large body of research clearly demonstrates that fire and boreal forests have a long association and that fire is one of the most important forcing functions in boreal forest dynamics (Payette, 1992; Pyne, 1984). Because of the importance of fire reoccurrence (fire regimes) to boreal ecosystems in general and to forest succession chronosequences in particular, its role will be noted in the discussion below. For a more detailed discussion of fire regimes and fire as a controlling process in Canadian boreal forest, see Payette (1992) and Bergeron and Dansereau (1993).

The role of herbivores within boreal forests must also be acknowledged. Not only does there appear to be an important association between open lichen woodland and caribou, but other species which selectively graze on tree seedlings can impact on succession. Selective foraging by moose on hardwood seedlings and their avoidance of conifers may play an important part in mixed forests (Pastor et al., 1993). In addition, browsing can prevent preferred species from becoming part of the canopy, thereby promoting a more open canopy with associated increase in understorey shrubs and herbs (McInnes, 1992). Beaver can also dramatically alter local drainage systems through dam construction, flooding, and selective herbivory (Naiman et al., 1986).

4.4.1 Open Lichen Woodland

The open lichen woodland, or "open spruce forest," as it is sometimes called, is located between the forest-tundra to the north and the closed-crown forests of the northern coniferous forest to the south and is essentially coincident with the moist Low Subarctic Ecoclimatic Region. Open lichen woodland is a suitable descriptive term for this vegetation zone because a blanket of light-coloured lichens dominates on the surface of better-drained sites, while trees do not form a closed-crown canopy, but are generally well spaced. This vegetation type is considered different from the forest-tundra zone (where forest stands can be found, but are discontinuous) because here open forest is continuous throughout the landscape, occurring in depressions and on ridges. Only on bare rock outcrops and in seasonally inundated wetlands are trees absent. Phanerophytes are more numerous here and account for approximately 4% of the spermatophyte species (Scoggan, 1978).

An explanation as to why lichens dominate large areas together with scattered conifers is still somewhat elusive. It would appear that following fire, lichens are quick to recolonize open sites, particularly where soil moisture is limited (Kershaw, 1977). Thick lichen mats develop, which in turn inhibit successful tree establishment, partly because of the mat itself and partly because of the lower summer soil temperatures produced under this high-albedo lichen cover. To this must be added the significance, certainly in drier years, of the thick lichen-moss layer in reducing water input to the spruce rooting zone in the mineral soil below (Jessee, 1990). A 250-year post-fire chronosequence in the northern limits of lichen woodland in the Grande rivière de la Baleine in Quebec suggests that well-drained lichen woodlands are self-perpetuating in the absence of fire (Morneau and Payette, 1989). Studies within 3,000-year-old relict spruce-lichen communities isolated from fire in the northern forest-tundra of northwestern Quebec show little tendency towards closure and therefore also lend support to the hypothesis that open lichen woodland is self-perpetuating (Payette and Morneau, 1993). However initially formed, then, open lichen woodland "can persist as a stable form over long periods even in the absence of fire" (Rowe, 1984, p. 234).

On moister sites less prone to this seral cover sequence, other community types develop. Successfully established conifers can spread by the slow mechanism of layering, which shades out lichen in favour of mosses (*Pleurozium schreberi*), and if fire is not again to intercede for at least 200 years, a closed-canopy *Picea-Pleurozium* climax forest can develop, particularly in the west (Kershaw, 1977). However, as lightning-induced fires are relatively frequent events throughout the region, seral open lichen woodland cover is clearly perpetuated, particularly on drier

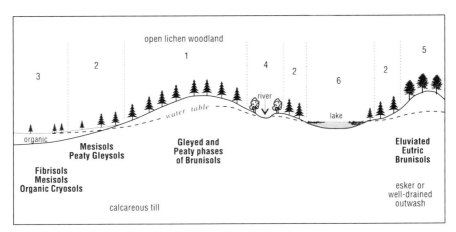

Figure 4.5
Typical transect through open lichen woodland

sites (Elliott-Fisk, 1988). Figure 4.5 represents a typical profile through the ecosystem variety expected of an open lichen woodland area in northern Saskatchewan or Manitoba, where soils have developed on calcareous tills, rock outcrops, eskers, and outwash. Community types identified by number on Figure 4.5 are described below.

1 Typical Lichen-Spruce Woodland   From the air this community type is easily seen because light-coloured lichens clearly dominate on the better-drained soils. The fruticose lichens (primarily *Cladina* spp. such as *C. stellaris, C. mitis,* and *C. rangiferina*) are very visible because the dominant tree species, *Picea glauca,* is generally not more than 10 m tall and is well spaced (Ritchie, 1962). *Cladina stellaris* tends to dominates open lichen woodland in eastern Canada, a role taken over by *Stereocaulon paschale* in the west (Kershaw, 1977). Together with the lichens are often found *Sphagnum* mosses, *Betula glandulosa,* and *Ledum groenlandicum,* along with some willow and ericoid shrubs, while under *Picea,* shading generally encourages feathermosses such as *Pleurozium schreberi.* Soils are typically Brunisolic, and moisture supply is good to the point that it often encourages the development of peaty phases, even on moderately well drained sites. Even the better-drained mineral soils (other than sands and gravels) have thick litter layers, with lichens dominating the relatively drier of these.

2 Spruce-Moss Woodland   Here the near-saturated mineral soils have a darker-coloured herbaceous cover dominated by *Sphagnum* and *Aulacomnium* moss species and scattered *Picea mariana* and *Larix laricina.* Shrubs are again common, with *Vaccinium uliginosum* joining *Ledum, Betula,* and *Salix* spp. Gleysolic soils result from the higher water tables, and peaty layers are the norm. Cryosols may be found over patches of permafrost. This zone can sometimes be considered transitional and subject to change, because if peat is accumulating, the regional water table rises. Instability results because this rising water table "forces" wetland-promoting conditions to expand laterally upslope into less moist forest (type 1 above) – in essence, exogenous paludification is taking place within the forest at the forest-wetland margin (Glebov and Korzukhin, 1992).

3 Open Tamarack Wetlands   *Sphagnum* bog or open *Larix* fen (depending on the acidity or basicity of the groundwater) dominate this nearly saturated to saturated habitat. If trees are present, they are normally *Larix laricina*. Often *Larix* is found growing in rows along the low peat ridges formed in string bogs. Moss-sedge wetlands with organic soils are also common, but here scattered and stunted *L. laricina* are still to be found (see section 8.5.3 for more detail).

4, 5, 6   Other Typical Community Types   Riverine forests (type 4) are found along the banks of some streams. Here, modest soil drainage combines with a higher nutrient status suited to stands of *Populus* spp. and *Betula papyrifera*. On sandy eskers or outwash deposits can be found relatively closed stands of *Picea glauca* and/or *Pinus banksiana*, and perhaps some *Populus balsamifera* (type 5). Of interest also is the fate of small lakes which post-glacially may well have been quite irregular in outline. Today many of these lakes appear to have very smooth edges because wetland communities on floating peat are expanding outwards from the shore (type 6).

These open lichen woodlands present harsh environments for tree growth. Conifer individuals grow slowly, and although they may not be in direct competition with their neighbours, they can die long before reaching maturity. They are also vulnerable to fire. One consequence of these harsh growing conditions is the frequent presence of dead standing trees or trees with dead branches. Although it may take decades for these tree "skeletons" to fall, they are often being decomposed as rapidly in the standing position as if they were prostrate on the moist, mossy ground. This is because they provide a very suitable habitat for epiphytic lichens. Dead branches on living trees are particularly noticeable because they are literally smothered in epiphytic lichens, while the healthy branches seem capable of limiting such invasions. The significance of this above-ground decomposition is unclear, but it may well be a major mechanism in the recycling of nutrients in wetlands where soil-surface decomposition rates are slow.

4.4.2 Northern Coniferous Forest

Closed *Picea* spp. forests characterize vast areas of the better-drained soils of the northern coniferous forests, while *Larix laricina* again dominates, along with *Picea mariana* on wetter sites and *Pinus* spp. on drier, low soil quality sites. A noticeable addition to the coniferous stands, especially in the Moist and Humid High Boreal

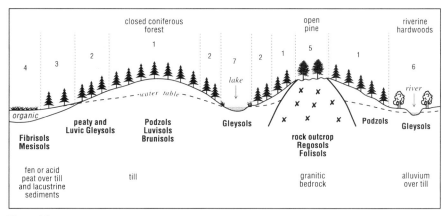

Figure 4.6
Generalized transect through the northern coniferous forest

and the Mid-Boreal of Ontario and Quebec, is *Abies balsamea*. Broadleaf deciduous tree species are also important components of many forest regions, particularly where the climate is drier or subsequent to fire, and open fen and bog remain important landscape features. Many bog sites have been so infilled by peat that they now support commercially viable stands of *Picea mariana* and *Larix laricina*. Combined with the large tracts of *Picea glauca* found on better-drained sites and *Pinus banksiana* found on excessively drained outwash, eskers, raised beaches, and outcrops, these stands support a valuable pulp and construction lumber industry.

The simple example of a profile through northern boreal forest shown in Figure 4.6 is much more complex in nature because of the influences of climate, parent material (both texture and nutrient status), aspect, logging, and fire (Mueller-Dombois, 1964; La Roi, 1992). It is also more complex than the transect through open lichen woodland (Figure 4.5) because here approximately 9% of the spermatophytes are tree and tall shrub species (phanerophytes) (Scoggan, 1978). Many successional stages are not dominated by the climax species mentioned above, and climatic variations permit the encroachment of many species associated with the mixed forests to the south, especially along river floodplains. The profile, however, makes some attempt to include community types more typical of central Canada, where deep glacial regolith covers the surface, together with the more acid regolith and igneous outcrops associated with the Canadian Shield of the eastern Keewatin District and in northern Ontario and Quebec.

1 Closed Spruce Forest   This is one of the two classic community types of the northern coniferous forest, and in the moist continental boreal of central Canada, it consists of closed *Picea glauca, P. mariana*, and *Abies balsamea* stands, together with some *Populus* spp. and *Pinus banksiana* and some *Betula papyrifera*. Soils are well drained and typically Podzolic in the Shield regions, while Luvisols and

Brunisols are more typical of the Prairie provinces, where parent materials are usually calcareous. Almost pure stands of *Picea glauca* and *P. mariana* make these stands commercially valuable, and they support large lumber and pulp mills, such as those found at Thunder Bay in northwest Ontario. In many parts of that region, fire intervals are so short that there is a preponderance of younger, non-climax stands, and it may well be that the perpetuation of *Abies*-dominated stands is in part due to the impact of spruce budworm on spruce (Zoladeski and Maycock, 1990).

The central Canadian boreal of Ontario and eastern Quebec is effectively humid (нвh and мвh). Away from this humid region, both to the west in the drier Subhumid Boreal of Alaska, Alberta, Saskatchewan, and Manitoba and to the east in the wetter Perhumid, Maritime, Atlantic, and Oceanic Boreal of eastern Quebec and Newfoundland, climatic conditions allow for variations of the classic forest community type. In interior Alaska, highly productive *Populus tremuloides, Betula papyrifera*, and *Picea glauca* stands are encountered on well-drained south-facing slopes, while on north-facing slopes, permafrost and microclimate promote moss-dominated *Picea mariana* stands (Viereck et al., 1986). To the lee of the Rockies in central Alberta, climatic conditions are somewhat drier, so that below 1,000 m, particularly on mesotrophic sites, boreal forest includes a much greater deciduous (*Populus* spp.) component (La Roi, 1992). Above 1,000 m, forests are more typically boreal, although a few conifer species associated more with the Cordillera are also present. *Pinus contorta* (lodgepole pine) replaces *Populus*, while *Picea engelmannii* (Engelmann spruce), *Pseudotsuga menziesii* (Douglas fir), and *Abies lasiocarpa* (subalpine fir) take on subdominant roles (North, 1976; Achuff and La Roi, 1977). *Pinus contorta* is not found east of the Birch Mountains of eastern Alberta (Critchfield and Little, 1966). According to Looman (1987a), *Pinus contorta* and *Pinus banksiana* are essentially two associations of the "collective" species *Pinus divaricata*. The lodgepole pine association can be considered montane-alpine, while the jack pine is boreal.

In eastern Canada, both increasing precipitation and narrower seasonal temperatures ranges influence species dominancy. For example, in the Chibougamau region of central Quebec, heavier rainfall (Figure 4.2F) greatly favours *Picea mariana* at the expense of *Picea glauca* and *Abies balsamea* on both lowland sites and on uplands. *Abies balsamea* regenerates in shade and deep litter and is a late successional species. Consequently, it dominates only in older stands. In the Laurentian Highlands, the typical upland *Picea mariana–Pleurozium schreberi* (feathermoss) forest gives way to *Pinus banksiana* in drier sites, to mixed *Abies balsamea–Populus tremuloides* forest on better soils in more humid areas, and to vast peat bogs dominated by *Sphagnum* (Cogbill, 1982). Fires are so frequent (130 years for feathermoss spruce forest and 70 years for pine and hardwoods) that no succession to climax forest in the traditional sense occurs (Elliott-Fisk, 1988; Cogbill, 1982). Throughout much of this typical boreal forest, fire is more likely to separate one tree generation from another rather than is one succession (Johnson, 1992).

The extremely heavy precipitation and maritime climate of the Avalon Peninsula, Newfoundland (see St John's Airport, Figure 4.2E), also favours *Abies balsamea* –

dominated coniferous forests interspersed with large areas of moss-heath peatscapes, which bear a striking resemblance to those of western Europe. In these per-humid forests, fires are not so characteristic, so arboreal succession can proceed (Johnson, 1992). In Gros Morne National Park in western Newfoundland, forested areas are typically dominated by *Abies balsamea, Betula papyrifera, Picea glauca*, and *P. mariana,* while along the coastal plain, the ameliorated climatic allows deciduous representatives of the Great Lakes–St Lawrence–Acadian mixed forest to be present (Bouchard et al., 1987). Studies in central Newfoundland fir forests experiencing heavy moose-grazing pressures found that while grazing favours *Picea glauca* and *P. mariana* and reduces both *Betula papyrifera* and some *Abies balsamea,* the overall effect on forest growth was positive because of thinning (Thompson and Curran, 1993).

2 Hygromorphic Forest   Because of a higher water table on these hygromesic sites, the gleying pedogenic process takes over from the podzolization or leaching regime associated with the better-drained soils upslope. Responding to these less-aerobic soil conditions are both trees and cryptogams. Here *Picea mariana* and *Betula papyrifera* become the usual dominants, along with some *Populus tremuloides* and *P. balsamifera,* with *Populus tremuloides* taking on much greater importance to the west, particularly in Alberta (La Roi, 1992). Both *Picea glauca* and *Abies balsamea* seem less tolerant of the moisture regime on Gleysolic soils and become less competitive than the shallow rooting *Picea mariana*. As noted for the open lichen woodland to the north, this zone may exhibit a tendency towards paludification, becoming a wetland if bryophytes do so well that peat accumulation causes the water table to keep rising.

3 Black Spruce–Tamarack Wetlands   Here soil moisture conditions are such that few tree species can survive other than the shallow rooting *Picea mariana* and *Larix laricina*. Bryophytes and sedges dominate these landscapes. Black spruce is relatively shade intolerant, so the open conditions of most acid-groundwater bogs favour its growth. In the most extreme sites in terms of tree growth, *L. laricina* may be the only tree species found. The survival of spruce and tamarack here seems less associated with the severity of the climate as with the severity of soil-atmospheric conditions and soil reaction.

4 Open Wetlands   Where seasonal standing water is found, trees are usually absent and a wetland vegetation cover dominated by a variety of herbaceous species and bryophytes results. *Sphagnum*-dominated bog is typical of acid groundwater, while basophilous bryophytes and sedges dominate where groundwater is mesotrophic or eutrophic.

5 Pine-Dominated Outcrops   This is the second classic landscape of the northern coniferous forests and is associated primarily with the spectacular scenery of the lichen and *Pinus* spp. dominated outcrops of the Shield. *Pinus banksiana* dominates throughout the region as a whole, although in the warmer climates of eastern Canada, particularly close to the transition to the Great Lakes–St Lawrence–Acadian forests, *P. strobus* (white pine) and *P. resinosa* (red or Norway pine) join it. In the Lac Abitibi region of western Quebec, *Pinus strobus* is exclusively associated with rock outcrops (Bergeron and Bouchard, 1983). In Newfoundland, *P. banksiana* is absent, the only indigenous pines being *P. resinosa* and *P. strobus* (Mednis, 1981). In central Manitoba, between Cedar Lake and the northern end of Lake Winnipeg, large expanses of exposed karst pavement (primarily dolomite), with only the thinnest of soils, is found. Here slow-growing scrubby *P. banksiana* almost completely dominate the post-fire succession, and the frequency of fires is so high that natural ecological succession to *Picea mariana* effectively becomes an unrealistic expectation. In western and northwest Alberta, *P. banksiana* gives up its dominancy to *P. contorta* or hybrids between the two associations (Looman, 1987a; North, 1976).

Soils on these outcrop areas are often thin organic Folisols, which may be completely mineralized during forest fires, initiating a new soil-forming sequence. Because pines are low nutrient and low water demanding species, they compete well against other conifers on Folisols. They often send roots out great distances across bedrock, through the thin layer of rotting and living lichen and moss, in search of water and nutrients. The droughty nature of some south-facing rock outcrops in southeastern Manitoba even permits the cactus *Opuntia fragilis* to survive far from its more typical prairie habitat (Frego and Staniforth, 1985). On rock outcrops near Lac Duparquet, northwestern Quebec, xeric conditions have reduced both tree-growth rates and fire frequency so much that some *Thuja occidentalis* individuals have attained ages of 900 years or more (Archambault and Bergeron, 1992)!

Outcrop ground-cover species include mosses, lichens (predominantly *Cladina mitis, C. rangiferina,* and *C. stellaris*), *Juniperus communis* (common juniper), and hardy grasses. Pure stands of pines are also associated with sandy areas of outwash and raised beaches. Although *J. communis* is frequently considered a colonizing species which takes advantage of low competition in stressed and acid environments, its distribution is somewhat restricted because of elimination by fire or by natural succession (Diotte and Bergeron, 1989). Fire may also limit distribution patterns for such species as *Pinus resinosa*, which seems to withstand burning better on islands due to local fire regimes where the larger and more intense fires of the mainland are less frequent (Bergeron and Brisson, 1990). This may well account for the presence of red pine on an island in southern Lake Winnipeg, although it is absent on the mainland until much farther to the southeast. In northwestern Quebec, forest fire histories confirm the importance of surficial material and physical environment on the delineation of fires, particularly in areas greatly fragmented by water bodies (Dansereau and Bergeron, 1993).

6 Riverine Communities   Floodplains benefit from eutrophic water supplies, alluvial sediments, and a favourable soil atmosphere during the major part of the growing season, but suffer from frequent spring flooding. Deciduous hardwoods, particularly in the Low Boreal Ecoclimatic Region, are quite suited to this regime, so *Acer, Quercus, Ulmus,* and *Fraxinus* species tend to dominate. To some extent, these are extensions of the mixed forests to the south and are found here only because floodplain habitats modify the typical boreal environment. Floodplain soils tend to be base rich, waterlogged in early spring, and well aerated in summer. Such conditions favour broadleaf deciduous species over evergreen because the former are less active during spring flooding.

7 Lake Margins   Freshwater lakes and marshes are common throughout the region. Shoreline vegetation, like the vegetation in community types 4 and 6, is greatly influenced by the nutrient content and pH of the water and by water-level fluctuations. *Fraxinus nigra* (black ash) does well on many lake floodplains within the boreal of northwestern Quebec, where it is variously associated with *Alnus rugosa* (speckled alder), *Salix* spp., and ferns such as *Onoclea sensibilis,* depending on the spring-summer flooding regime (Tardif and Bergeron, 1992). If seasonal water-level fluctuations encourage marsh development with base-rich groundwater, *Phragmites australis* (reed) dominates. Occasionally, salt-pans are found near shore, such as at Dawson Bay on Lake Winnipegosis. Here salinity discourages arboreal cover, and as salinity increases, diversity decreases and becomes more prairie salt-pan-like (Burchill and Kenkel, 1991).

Forest-Floor Species   In the above description of the seven environment types, attention is focused primarily on the arboreal species. While these phanerophytes constitute only 8–11% of the boreal species (Scoggan, 1978), the great majority of species are low shrubs, herbs, mosses, and lichen, the importance of which must not be overlooked. In addition, the distribution of understorey species with high indicator value can often be used as a better way of evaluating site quality for tree growth than simply examining the total species present (Strong et al., 1991).

Common understorey species on very moist boreal soils include *Sphagnum* mosses, *Equisetum arvense* (common horsetail), *Caltha palustris* (cowslip), and the subshrubs *Oxycoccus palustris* (swamp cranberry) and *Ledum groenlandicum* (Labrador tea). Common mesophytic forest-floor species include a variety of feathermosses, *Carex* spp., *Matteuccia struthiopteris* (ostrich fern), *Cornus stolonifera* (red-osier), *C. canadensis* (bunchberry), the ubiquitous *Lycopodium annotium* (stiff club-moss), *Petasites palmatus* (palmated-leaved colt's foot), and tall shrubs such as *Alnus incana* (grey alder) (Mueller-Dombois, 1964). An examination of three feathermoss species distribution patterns under *Picea mariana* – dominated boreal forest in northeastern Ontario showed that *Pleurozium schreberi* was most abundant in the canopy gap between trees and near tree stems and *Hylocomium splendens* dominated canopy margin areas, while *Ptilium crista-castrensis* dominated mid-way between stem and canopy margins (Kavanagh, 1987).

On dry sites, *Cladina mitis, C. rangiferina* (reindeer lichen), and the feathermoss *Pleurozium schreberi* are common, along with herbs such as *Aster laevis* (smooth aster), *Solidago* spp. (goldenrod), *Oryzopsis pungens* (northern rice grass), and *O. asperifolia* (white-grained mountain rice grass). Ericaceous sub-shrubs such as *Arctostaphylos uva-ursi* (common bearberry), *Vaccinium angustifolium* (low sweet blueberry), and *Gaultheria procumbens* (checkerberry) also do well and can be joined by *Rhus radicans* (poison ivy) and the taller shrubs *Prunus pumila* (low sand cherry) and *Rosa acicularis* (prickly rose). Blueberries do particularly well on open, well-drained, acid, sandy sites, especially in *Pinus banksiana* – dominated forest, following forest fires, or in young forestry plantations.

These few examples of northern boreal forest-floor species are provided simply to make the point that tremendous species diversity actually exists in these forests, particularly along upland-bottomland gradients. For more detailed examples of forest-floor gradient variations, see Mueller-Dombois (1964) on herbaceous species diversity in southeastern Manitoba, and Carleton (1990) on ground-cover (terricolous) bryophyte and macrolichen diversity.

Induced Forest Disturbance   The value of this region to the forestry industry should also be emphasized. Large areas of the northern boreal forest have been clearcut to produce both pulpwood and lumber, and post-logging successions often differ significantly depending on logging disturbance intensity (MacLellan, 1991). Efforts at reafforestation have been ineffective in the past (Wylynko, 1991), with regeneration being primarily by volunteering following scarification. Today, however, both private industry and provincial forestry departments are making greater efforts to promote and carry out replanting. In Manitoba, the Hadashville Provincial Nursery is currently producing 15 million seedlings each year, the majority of which are black spruce and jack pine, for use in provincial forests and by private woodlot owners and pulp-lumber companies. There is a growing realization that the ideals of sustainable development and general public opinion are coming closer, that environmental impact assessments (EIAs) for new forest operations are necessary, and that both provincial and federal governments must re-evaluate and co-ordinate their approaches to environmental reviews (Ross, 1991). In 1992, ten model forests were established nationwide under the auspices of the Green Plan in order to promote sustainable forests, not just sustainable forestry.

Additional areas have been inadvertently deforested by the mining and smelting industry, with pollutants killing vegetation both directly and indirectly. The sensitivity of forest-floor feathermosses such as *Pleurozium schreberi* and lichens to induced acid rainfall is well known (Hutchinson and Scott, 1988; Scott et al., 1989). Conifers, particularly *Pinus banksiana* and *Picea* spp., have died or been damaged around Flin Flon (Longton, 1985) and Thompson, Manitoba, by atmospheric pollution (Blauel and Hocking, 1974; Jones et al., 1984). Downwind for several kilometres from the smelter at Flin Flon, many bottomland conifers have died or adapted layering survival strategies, while conifers as far away as 20 km away have also died along rock outcrops. In contrast, deciduous hardwoods (*Betula* and *Populus* spp.) on the mineral and deep organic soils of slopes and bottomlands have generally survived. It would appear that pollutants such as SO_2 (Malhotra and Khan, 1983) and perhaps heavy metals have destroyed the living component fabric of thin organic Folisols overlying bedrock on the long rock-outcrop ridges up to 20 km downwind (south-southeast) of the smelter. They have achieved this by killing off terrestrial lichens and mosses. Ridge-top Folisols have then disintegrated and were eroded, so that the vegetation there, perhaps already weakened by pollutants, succumbed to lack of water, nutrients, and physical support in the absence of a protecting organic soil. Longton (1985) reports the total absence of a feathermoss carpet, even under lower-slope conifer stands from 10 km northwest to 17 km southeast of Flin Flon. It seems paradoxical that ridge-top conifers 10 km downwind of the pollution source are dead, while other, apparently healthy conifers cover lower slopes and bottomlands only metres away. While it would appear that these coniferous stands have survived because the slope and bottomland mineral and organic soils on which they stand have survived, much of the pre-pollution covers of epiphytic and terrestrial lichens and mosses within these stands are absent, the activity of fungi and actinomycete decomposers in the litter-topsoil appears to be curtailed (Christenson, 1990), conifer seedling development appears limited, and the forest is strangely silent. As at Thompson, Manitoba (Jones et al., 1985), there is concern that reduced regeneration will lead to future forest degeneration.

4.4.3 Mixed Forest (Boreal-Broadleaf Ecotone)

Many of the arguments used to give the forest-tundra effective zonal status (chapter 3) apply equally well to the boreal-broadleaf ecotone. Walter (1979) reflects this reality by naming it the "boreonemoral zone." This transitional forest zone possesses climatic/edaphic conditions suited to dominancy by both broadleaf deciduous species (hardwoods) and conifers (softwoods) (Pastor and Mladenoff, 1992). In Canada this region is known as the Great Lakes–St Lawrence–Acadian Forest Region, and in terms of ecoclimates it is associated with much of the Low Boreal of Ontario and southern Quebec and the High Cool and Mid-Cool Temperate Ecoclimatic Sub-Provinces (Figure 1.4). In terms of Canadian terrestrial ecozones, it is associated with the southern fringes of the Boreal Shield and the Mixed Woods Plains and Atlantic Maritime Ecozones (Figure 1.5). In the USA this boreal–broadleaf ecotone is known by such names as the Northern Hardwood Forest, the Hemlock–White Pine–Northern Hardwood forest (Nichols, 1935), or the Northern Hardwood–Spruce Forest. Authors such as Braun (1967) divide this part of the USA into the Great Lakes–St Lawrence Division to the west and the Northern Appalachian Highland Division to the east. The western forests are found in the northern parts of Minnesota, Wisconsin, and Michigan, while the Appalachian zone includes central and northern New York State, as well as most of the New England states (Figures 4.7 and 7.3). This western zone coincides with the Northern Lakes and Forests Ecoregion, while the eastern portion includes the Northeastern Highlands, the Northeastern Coastal Zone, and the Northern Appalachian Plateau

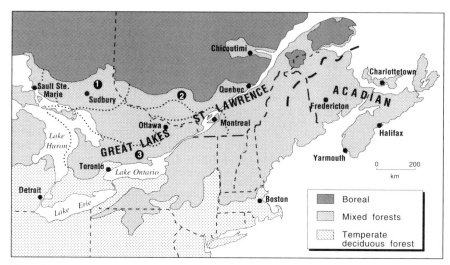

Figure 4.7
Central and eastern portions of the Canadian–USA boreal-broadleaf ecotone (mixed-forest transition). The text refers to just three specific Canadian sub-regions (after Rowe, 1972), which are: (1) Temagami and Sudbury–North Bay, (2) Algonquin-Pontiac and Laurentian, and (3) Huron and Upper St Lawrence. In the USA the mixed-forest transition is known as the Northern Appalachian Highland Division.

and Uplands Ecoregions (using the American Ecoregion Classification System of Omernik, 1987).

Podzols (Spodosols) are still the dominant soil order, but substantial areas of more agriculturally suited great groups such as Melanic Brunisols, Gray-Brown Luvisols (Alfisols), and Humic Gleysols are also found (Figure 1.9). It is these less podzolized soils, combined with a somewhat milder climate than for the true boreal, which promote the broadleaf-conifer mosaic. While in the cooler boreal, differing soil conditions allowed for tree dominancy to be fought out chiefly among the conifers, here, with a climate more conducive to broadleaf deciduous species, this latter group frequently outcompetes or co-exists with conifers on the more nutrient-rich soils. If this region has any distinguishing tree species, they are the pines *Pinus strobus* and *P. resinosa* and the yellow birch, *Betula lutea* (also known as *B. alleghaniensis*). The major botanical ranges of these three species, especially the red pine, accord closely with the mixed forest zone, where they are found primarily on Podzolic soils. *Pinus resinosa* is in fact a close relative to, and the only North American representative of, a typically Eurasian group of pines which includes *Pinus nigra* (Austrian pine) of southern Europe (Chritchfield and Little, 1966). The ranges of most other conifer species extend well north into the boreal, while the botanical ranges of the hardwoods often extend far to the south and east.

Some authors separate the Acadian forests from the rest of the Canadian transitional zone because this maritime region differs from the St Lawrence and Great

Lakes forests in that *Picea rubens* (red spruce) becomes the dominant conifer (Rowe, 1972). These maritime forests are also characterized by various amounts of *Abies balsamea, Picea glauca, Betula lutea,* and *Acer saccharum* (sugar maple), together with some *Pinus resinosa* and *P. strobus.* The Great Lakes–St Lawrence region has *Pinus strobus* and *P. resinosa* as the dominant conifers, together with the non-boreal conifer *Tsuga canadensis* (eastern hemlock). Important broadleaf dominants include *Betula lutea, Acer saccharum, Acer rubrum* (red maple), *Tilia americana* (basswood), *Quercus rubra* (red oak), and *Ulmus americana* (white elm). Of lesser importance are typical boreal species such as *Picea mariana, P. glauca, Pinus banksiana, Betula papyrifera, Populus tremuloides,* and *P. balsamifera,* together with *Picea rubens.*

4.4.3.1 The Great Lakes–St Lawrence Region

Also known as the hemlock–white pine–northern hardwood forest, this region has experienced even greater interference from human activity than the northern boreal forest to the north. This is because local soils have generally greater agricultural potential and the forests provide valuable lumber. Much of the colonizing history of this region by Europeans, in both Canada and the USA, is associated with the need to harvest the great stands of *Pinus strobus* to provide lumber for expanding settlements to the east and south (Larson, 1972). Recent conservation efforts in this region have been centred on preserving remnant pine stands such as those found in Itasca State Park in Minnesota. Because the region varies considerably in community type from Manitoba and Minnesota east to Quebec and New England, diagrams are less suited to demonstrating this variety, so a number of typical forest landscapes are described instead. These examples include (1) the Quetico and Rainy River sections west of Lake Superior (see Figure 1.3), (2) the Temagami (or Timagami) and Sudbury–North Bay sections north of Lake Huron (Figure 4.7), (3) the Algonquin-Pontiac and Laurentian sections between Georgian Bay and just north of Quebec City, (4) the Huron–Ontario–Upper St Lawrence River Sections, and (5) the Northern Appalachian Highland Division of the northeastern USA.

Quetico and Rainy River Mixed Forests   This is the more continental portion of the mixed-forest zone known as the Moist Lower Boreal Ecoclimatic Region, but unlike other mixed forests to the east, it is subject to the effects of the drier prairie climate. Consequences of the colder, seasonally drier climate encourage the dominancy of pines such as *Pinus banksiana, P. resinosa,* and *P. strobus* (Rowe, 1972). Jack pine has expanded considerably in post-settlement times since it frequently colonizes areas logged of white and red pine and because of increased incidences of fire. Itasca State Park in northern Minnesota was established in 1891 to protect its superb stands of both *Pinus strobus* and *P. resinosa.* The largest white and red pines in Minnesota are found at Itasca, and individuals have been measured with diameters at breast height in excess of one metre. Under the tall white pine canopy is normally found an understorey of *Acer saccharum, Tilia americana* and *Ostrya virginiana* (hop-hornbeam) (Peet, 1984).

Throughout this mixed-forest region, *Picea glauca, Abies balsamea,* and *Betula papyrifera* also do well on adequately drained lowland soils. Hardwoods such as *Acer rubrum, Fraxinus nigra* (black ash), and *Ulmus americana* thrive along flood-plains, while on drained lowland soils to the south are hardwood stands dominated by *Acer negundo* (Manitoba maple), *A. saccharum, Tilia americana, Quercus rubra,* and *Q. macrocarpa* (bur oak). Boggy lowlands still have the typical *Picea mariana–Larix laricina* tree components, while stands of *Thuja occidentalis* (eastern white cedar) are also common, particularly where soils are base rich. *Sphagnum* bogs and *Phragmites* fens are also common.

Temagami and Sudbury–North Bay Mixed Forest   In this upland region, remnant stands of climax forest are typically dominated by *Pinus strobus*, with *Betula papyrifera* and *Picea glauca* (Figure 4.7). Well-drained sandy soils generally have a *Pinus banksiana* cover while rocky outcrops possess more *Pinus resinosa*. Soils are primarily Podzols, and only a few hardwoods such as *Betula lutea* and *Acer saccharum* do well. An almost pristine area of old-growth white pine can still be found in the Temagami wilderness area around Lake Temagami, northwest of North Bay, Ontario. Here old-growth white pine appear to be self-replicating (Quinby, 1991). This is one of the last major areas of old-growth white pine in Canada; however, its total preservation is in doubt. In May 1988 the Ontario government approved con-struction of new logging access roads, a decision that immediately led to conflict be-tween conservationists and the Teme-augama Anishnabai Indians on the one hand and the Ministry of Natural Resources, along with loggers and hunters, on the other (Moise, 1989). Although in 1990 the provincial government gave the Teme-augama Anishnabai Indians the right to decide what logging was carried out on their land, conservationists continue the struggle to save the old-growth stands on crown land.

To the west of the Sudbury region, the forests of the Turkey Lakes Watershed just north of Sault Ste Marie can be classified into three groups: upland hardwood types, upland mixedwood types, and wetland mixedwood types (Wickware and Cowell, 1985). Typical upland hardwood stands are dominated by *Acer saccharum*, with shrub-herbaceous layers varying considerably with soil conditions. Upland mixed-wood types include *Picea glauca*–dominated forest on level, moister sites, adjacent to lakes or streams; *Acer saccharum*–dominated stands, together with *Betula lutea*, occurring on middle to lower slopes of north-northeast aspect on fresh-dry to fresh-moist soils; and *Acer rubrum–Thuja occidentalis* stands on crest positions and in frost pockets with shallow soils. Wetland mixedwood types include stands domi-nated by *Fraxinus nigra*, as well as typical boreal species such as *Picea glauca, Betula papyrifera, Abies balsamea, Thuja occidentalis,* and *Larix laricina*. Such wetland sites frequently have ground covers of sphagnum, feathermoss, and *Carex* and *Equisetum* species. Under *Fraxinus,* the ferns *Onoclea sensibilis* and *Osmunda trispera* frequently dominate ground cover (Wickware and Cowell, 1985).

To the south of the Sudbury region, across Lake Huron in the northern half of Michigan's Lower Peninsula, classic white pine forests were present in pre-settlement times on coarse-textured soils derived from outwash and ice-contact deposits (Whitney, 1986); however, fire and logging have greatly altered these forests. These sandy dry-to-mesic soils support successional forests which are eventually dominated by *Pinus strobus, P. resinosa, Acer rubrum,* and *Quercus rubra.* Where lowland sandy soils have a fluctuating water table, mature forest cover is more typically *Pinus strobus, Abies balsamea,* and the shrubs *Viburnum lentago* (sweet viburnum) and *V. cassinoides* (wild-raisin). On mesic soils derived from glacial till and lacustrine deposits, typical hardwood forests of *Acer saccharum, Fagus grandifolia* (American beech), *Tilia americana, Ostrya virginiana, Fraxinus americana,* and *Acer pensylvanicum* dominate. On these more mesic soils, *Populus* spp., *Corylus cornuta* (beaked hazelnut), and *Prunus serotina* are important early successional species, along with many of the afore-mentioned hardwoods (Roberts and Christensen, 1988). Logging and clear-cutting in this region has not always reverted succession to pioneer conifer species. Disturbance-mediated accelerated succession has been found for hardwoods such as *Acer saccharum* in logged *Thuja occidentalis* (eastern white cedar) communities, while clear-cutting and burning in *Pinus banksiana* stands encouraged rapid conversion of some sites to early successional hardwoods (Abrams and Scott, 1989).

Logging has not been the only action to impact on the conifers in this forested region. A considerable area of conifer-free broadleaf deciduous forest dominated by *Betula papyrifera, Quercus rubra,* and *Acer rubrum* is found downwind of the Falconbridge and Copper Cliff smelters at Sudbury (Amiro and Courtin, 1981). *Pinus strobus* was found to be the most sensitive local tree species to SO_2 damage and was absent or extensively damaged up to 40 km away from Sudbury towards the northeast (Hutchinson, 1973). Red oak and red maple are relatively more resistant and can be found within 2 km of the smelters. Losing their leaves in the fall and growing new ones each spring gives hardwoods an obvious competitive advantage, permitting them to survive the phytotoxic effects of sulfur fumigation periods in winter far better than evergreen conifers. Some ridge-top sites, which originally supported acid soils with pines, have been so badly acidified that all vegetation has gone, soils have been eroded, and even dead wood does not readily decompose. Now that sulfur and heavy metal emissions are being reduced, natural regeneration of hardwood communities is occurring in more favoured sites (Amiro and Courtin, 1981). Efforts at replanting *Pinus resinosa* are also being made, although in the case of the more acidified soils, liming and fertilizer applications are necessary. The effects of heavy metals on soil fungi noted near the Flin Flon smelter stack are also present here (Carter, 1978), so the rehabilitation of those mineral soils that have survived near Sudbury may take considerable time.

Algonquin-Pontiac and Laurentian Mixed Forests   This area extends east from the Sudbury forests through the upper Ottawa River region to just north of Quebec City

(Figure 4.7). Because of the dominance of *Picea mariana* in the northwestern portions of this region, it could almost be described as boreal-like were it not for the presence of typical mixed-forest hardwoods in combination with *Pinus strobus* and *Tsuga canadensis* (eastern hemlock) (Rowe, 1972). In the east, the presence of *Picea rubens* and *Tsuga canadensis* and of such hardwoods as *Fagus grandifolia* (American beech) lets this region take on a more Acadian mixed-forest appearance. On the acid Podzolic soils of the Lanoraie delta northeast of Montreal, softwood dominants include *Tsuga, Picea*, and *Abies balsamea*, while dominants on less acid Podzols, particularly where human disturbance has been great, are hardwoods such as *Fagus grandifolia, Acer saccharum, A. rubrum, Quercus rubra, Populus grandidentata* (largetooth aspen), *P. tremuloides,* and *Betula papyrifera* and the conifer *Pinus strobus* (Lamontagne et al., 1991). South of the Upper Ottawa Valley, upland forests are dominated by hardwoods such as *Acer saccharum, A. rubrum, Fagus grandifolia,* and *Betula lutea,* together with the conifers *Tsuga canadensis, Pinus strobus,* and *P. resinosa*. Mixed-forest swamps are also common here.

Huron–Ontario–Upper St Lawrence Mixed Forests   These forests stretch from the eastern shores of Lake Huron through eastern Ontario to Montreal (Figure 4.7). They are typical of the southern portion of the mixed-forest region, where winters are milder, yet summers are warm (see climatic data for Mount Forest, Figure 4.3B), and landscapes are more commonly dominated by broadleaf deciduous species than conifers. Soils reflect this milder climate and are generally base-rich Gray-Brown Luvisols derived from non-igneous tills and alluvial deposits. Here *Acer saccharum* and *Fagus grandifolia* are common, being joined by *Tilia americana, Betula lutea, Fraxinus nigra, F. americana,* and *F. profunda* (black, white, and red ash respectively), *Quercus alba,* and *Q. macrocarpa*. To the north, at higher elevations within Algonquin Provincial Park, *Picea mariana* is abundant on upland and on poorly drained sites, while hardwoods such as *Acer saccharum, A. rubrum* (red maple), *Betula lutea* (yellow birch), *Tsuga canadensis,* and *Pinus strobus* are prominently distributed. Here, hardwoods tend to segregate out on the finer-textured soils of warm slopes and hill tops (Cwynar, 1975). To the south, along the shores of Lake Ontario, hardwoods dominate, while *Tsuga* and *Pinus* are less frequently encountered except on steep slopes or swampy soils. In Presqu'ile Provincial Park, some Humic Gleysol areas have old-growth stands dominated by massive eastern hemlock often more than one metre in diameter.

Lemieux (1963) differentiates eight major hardwood associations in southern Quebec, five of which are dominated by *Acer saccharum*, two by *Fraxinus nigra,* and one by *Betula lutea*. Only on the poorer soils are more tolerant hardwoods joined by conifers such as *Pinus strobus, Abies balsamea,* and *Tsuga canadensis*. River floodplains are associated with the hardwoods *Fraxinus nigra, Fagus grandifolia, Platanus occidentalis* (silver sycamore), *Ulmus rubra,* and *U. thomasii* (slippery and rock elms respectively), while *Thuja occidentalis* thrives in base-rich swamp sites and on old fields (Rowe, 1972). To the east, in the Ottawa-Montreal region, most of the original forest has been cleared for agricultural or grazing lands, with *Ulmus americana* particularly common along fence lines and around farm buildings. Where original forest remains, dominant hardwoods are *Fagus* and *Acer* on the calcareous tills, while rock outcrops and sandy areas have mixed stands with *Pinus strobus, P. resinosa, Tsuga canadensis, Picea glauca,* and *Abies balsamea.* South of Montreal, in the Haut-Saint-Laurent region, sizeable successional associations are found on clear-cut and abandoned fields. Clear-cutting favours the sprouting of *Tilia americana* (basswood), *Ostrya virginiana* (hop-hornbeam), and *Prunus serotina* (black cherry), while abandoned fields are frequently invaded by *Ulmus americana* (American elm), *Fraxinus americana* (white ash), and *Betula populifolia* (grey birch). With both successional associations, there is convergence to an *Acer saccharum* climax forest (Brisson et al., 1988).

Clearing for agriculture, urban development, and timber production are not the only impacts human beings are having on these maple forests. A 1987 survey among Quebec's maple-syrup producers, in conjunction with a helicopter survey, revealed that only 1% of the maple forests were considered healthy, a drop from the 52% obtained in a similar 1982 survey (Norton, 1988). In 1986 the Ontario Ministry of the Environment set up some 110 forest plots across the province to monitor hardwood conditions. Results show that the most serious problems are in the Muskoka region east of Georgian Bay and that the decline is rapidly moving to the northern and eastern parts of the province. While no single factor was found to be responsible, tent caterpillars and soil acidification were tentatively identified as causal agents in many of these sites (Norton, 1988).

The Northern Appalachian Highland Division Differences between these Appalachian forests and their Canadian counterparts include the degree to which broadleaf deciduous species are associated with conifers, the conifer dominants themselves, and the response of individual species to fire. Prior to European settlement, five species dominated the extreme southern portions of these forests: *Acer saccharum, Fagus grandifolia* (American beech), *Tsuga canadensis* (eastern hemlock), *Betula alleghaniensis* (yellow birch, named *B. lutea* in Canada), and *Tilia americana* (Woods, 1984). Here, hardwoods dominated over conifers, with maples and basswood more important to the west and beech and maple to the east. To the north, *Pinus strobus* takes on increasing importance over hemlock and is joined by *Abies balsamea* (balsam fir), *Picea rubens* (red spruce), and *P. glauca.* In New England the pre-settlement forests included both hardwoods, *Tsuga,* and white pine, while today, following cutting, agricultural abandonment, and the chestnut blight, the forest again appears quite mature, but unlike anything that preceded it because it is now dominated by *Tsuga* (Foster et al., 1992).

The famous Hubbard Brook Experimental Forest in New Hampshire is located in this region and has been the object of intense nutrient-cycling and successional studies. Here *Betula alleghaniensis, Acer saccharum,* and *Fagus grandifolia* constitute some 90% of the relatively mature second-growth forest (Bormann et al., 1970;

Bormann and Likens, 1981). Among other things, researchers have concluded that clear-cutting may be an ecologically acceptable practice in northern hardwood forests provided that safeguards are taken. These safeguards include giving weight to successional species such as *Prunus pensylvanica* (pin cherry), which, although having no commercial value, is an exploitive species that reduces erosion and aids in the recovery of nutrients which might otherwise be lost to leaching, runoff, and erosion (Likens et al., 1978). After heavy cutting on fine-textured tills, succession goes through such species as *Prunus, Populus* spp., *Betula alleganensis,* and *B. papyrifera*, toward a climax of *Acer saccharum* and *Fagus grandifolia*. Where sites do not favour hardwood climax, *Acer rubrum* (red maple) is usually succeeded by *Tsuga canadensis* at lower elevations and *Picea rubens* at higher (Leak, 1991).

In the Adirondack Mountains of northern New York, sugar maple, beech, and yellow birch hardwoods dominate below approximately 700 m. Between 700 and 900 m is a transitional zone to red spruce and balsam fir that extends up to 1,100 m, above which pure subalpine balsam fir forests are found. A recent decline in the *Picea rubens* (red spruce) populations has been detected both here and in the New England states, the causes for which are speculative (Scott et al., 1984). During the last two decades, there has been an increasing rate of mortality of red spruce individuals of all size classes in such a way that decline is not logically related to successional development, while in central and northern New England, a lack of reproduction has also been observed (Siccama et al., 1982). This unexplained mortality among red spruce is creating disturbance patches in spruce-fir forests in the White Mountains of New Hampshire and is beginning to exert an influence on vegetation patterning (Foster and Reiners, 1983). Red spruce stands in southwestern Virginia, central West Virginia, and Whiteface Mountain, New York, also appear to have been undergoing a general decline during the last fifty years (Adams and Stephenson, 1989; Battles et al., 1992). In the Allegheny Mountains in Pennsylvania, old-growth stands on better-drained uplands are dominated by beech–hemlock–white pine forests, while more poorly drained flats and footslopes are dominated by hemlock and yellow birch (Whitney, 1990).

The southern part of the northern Appalachian Highlands represents a transition to true temperate deciduous forests, with their mild winters, warm moist summers, and strongly positive moisture indices. None of the hardwoods listed, or even hemlock, show resistance or reproductive adaptations to fire (Woods, 1984). The facts that hardwoods are not flammable in the way most conifers are, that hardwood litter is less acid and more readily decomposed, leaving less flammable material on the forest floor, and that wetter summer climatic conditions prevail combine to reduce the likelihood and frequency of forest fires. The presence of Indians did undoubtedly increase the frequency of fires over those caused by lightning, but there is no strong evidence that they purposely burned large areas of the forested northeastern USA (Russell, 1983). A number of mixed-forest outliers are also found in the central and southern Appalachians, and these are discussed in section 7.4.2.1 (also see Figure 7.3).

4.4.3.2 The Acadian Mixed-Forest Region

This region covers almost all of the Maritime provinces (Figure 4.7). Here *Picea rubens* is the characteristic coniferous species throughout, but it is joined by species typical of both the boreal forest to the northwest and the mixed forests to the west. As a consequence, phanerophytes constitute approximately 12–15% of the species (Scoggan, 1978). Despite marine influences reducing both winter length and severity, the broadleaf component to these forests does not always take on the dominancy that might be expected. To a large extent, this reflects the depression of summer temperatures and the increase in precipitation which result from proximity to the Atlantic and which characterize this portion of the High Cool Temperate Ecoclimatic Sub-Province. A brief comparison of climatic data between Mount Forest in southern Ontario, far to the west, and Yarmouth, Nova Scotia, shows that the more continental station has the warmer summer (Figure 4.3). Broadleaf species dominate only in areas with a milder summer climate such as that of Summerside, Prince Edward Island (Figure 4.3A).

The Eastern Lowlands section of New Brunswick and northern Nova Scotia supports predominantly coniferous forests on Podzols, with hardwoods becoming important towards the west and south. Here the climate is per-humid, with the moisture index for Saint John, New Brunswick, being +105. *Picea glauca, P. rubens, P. mariana*, and *Abies balsamea* can be found in pure stands or combined with typical mixed-forest pines and hardwoods (Warner, et al., 1991). Selective logging and fire have favoured pines and spruces, while *Thuja occidentalis* has declined. Quality-oriented logging, exacerbated by outbreaks of *Choristoneura fumiferana* (spruce budworm), has led to a decline in harvestable tree size and the urgent need for the redevelopment of the forest industry (Baskerville, 1988). It should be noted that severe insect infestations are periodically experienced throughout the boreal, mixed, and deciduous forests of Canada and the northern USA (Franklin, 1973).

Along the Atlantic Uplands of Nova Scotia, precipitation is higher and winds strong. Here the *Picea rubens*–dominated forests have experienced disturbance from fire and logging, and some of the badly disturbed areas have taken on a parkland appearance with scattered spruce, pines, and hardwoods such as *Quercus rubra* (red oak). Along the south Atlantic shores of Nova Scotia around Yarmouth, extreme maritime influences dictate a return to Low Boreal Ecoclimatic conditions, which leads almost to the return of boreal forest. Were it not for the presence of some *Acer rubrum*, the *Picea glauca, P. mariana, Abies balsamea*, and *Betula papyrifera* dominated stands found here could easily be mistaken for northern coniferous forest.

Prince Edward Island is a good example of a mixed-forest region where the suitability of soils for agriculture has contributed to major clearing of native forest. Soils are usually quite acidic but suited to such crops as potatoes. These soils originally supported a predominantly broadleaf forest of *Fagus grandifolia, Acer saccharum,* and *Betula lutea*. Typical boreal conifers are also found along the northern shore, valley bottoms, and upland flats (Rowe, 1972), while mixed-forest conifers such as

Pinus strobus, Picea rubens, Thuja occidentalis, and *Tsuga canadensis* can also be found, together with such hardwoods as *Acer rubrum*.

4.4.4 Mixed-Forest Transition to Grassland (Northern Mixedwoods)

In the Prairie provinces, the transition zone on the southern side of the Subhumid Low Boreal Ecoclimatic Region is dictated more by moisture stress than by improving growing-season temperatures. Unlike the mixed forests of southern Ontario and Quebec, where broadleaf species take over from conifers as one moves south into deciduous forests, in the Prairie provinces, the boreal forest gives way to aspen or aspen-oak parkland, communities typical of the moister edges of the prairie grasslands. Between typical northern boreal forest and the aspen parkland is a generally narrow transition zone dominated by a mixed forest where the only common broadleaf species are *Populus* and *Betula*. None of the typical Great Lakes–St Lawrence–Acadian maples (except the occasional *Acer negundo*), beeches, and ashes are found here, as are neither of the two important mixed-forest species of pine, *Pinus resinosa* and *P. strobus*. The somewhat drier conditions are also reflected in less soil leaching. Luvisols, Brunisols, and some Dark Gray Chernozems are therefore typical of the transition zone, and some areas have agricultural potential.

While most authors agree that the southern boundary of this transition zone is reached at the southern limit of conifer distribution, some difficulty is encountered in defining where the northern boundary should be drawn. To simplify this decision, Zoltai (1975) suggests that the transitional zone is where at least one, but not more than three, of the four typical Prairie province boreal conifers is encountered. The northern margin of this transition zone (Figure 5.7) therefore represents the southern limit of at least one of *Picea glauca, P. mariana, Pinus banksiana*, and *Larix laricina*, while the southern limit of the zone is where conifers no longer volunteer. Rowe (1972), on the other hand, gives a broader distribution to this mixed forest, basing his northern margin purely on the presence of large stands of *Populus* spp., which extend well north of the southern limit of these four conifers. While Rowe's northern margin is therefore not based on species distribution boundaries, there is agreement that this mixed forest is quite unlike that in the Great Lakes–St Lawrence–Acadian regions, and like most other authors, he includes it within the boreal forest region proper (forest sections B.15, B.18a, and B.19a). For convenience, these mixed forests of the Prairie provinces are locally known as either the "mixedwood section" of the boreal forest, the "boreal mixedwoods," or the "northern mixedwoods" (Samoil, 1988).

While the somewhat drier conditions of the dry continental boreal seem to promote the replacement of conifers by *Populus tremuloides* and *P. balsamifera*, they also favour fire, so post-fire successions are quite apparent here. In the Interlakes region of Manitoba, *Populus tremuloides* is nicknamed the "fire tree" because it comes back so quickly after burning. Smithers (1961) reports that following fire in central Alberta, *P. tremuloides* and coniferous seed often germinate together, but the aspen grows more quickly, thereby taking on dominancy. After 50 or so years, the conifers catch up with the aspen and join it in the canopy. By 80–100 years the conifers over-top the aspen, and often within as little as 120 years of the fire succession's begin-ning, the aspen has died out.

Forestry operations are also having an influence on the regeneration of stands of *Picea glauca* in such areas as Hudson Bay, Saskatchewan. Here pure aspen stands have already been cleared for the manufacture of waferboard, so aspen must now be extracted from mixed stands where white spruce is in the understorey (Froning, 1980). Unfortunately, extraction methods are such that these white spruce are dam-aged. As a result, softwood stands near the town of Hudson Bay, Saskatchewan, are on the decline, with many logged areas having degraded to brush. There is, however, potential to restore these once highly productive areas (Ball and Kolabinski, 1979). With improved pulpwood technology, there is now considerable harvesting of aspen for kraft paper production. Inevitably, associated with the utilization of both soft-woods and hardwoods in this northern mixedwoods forest comes the challenge of sustainable development and a complete rethinking about forest renewal within the forestry industry (Samoil, 1988).

4.5 Eurasian Boreal

As with the North American boreal, the Eurasian is dominated by members of the *Picea, Pinus*, and *Larix* genera. Unlike the North American boreal, however, where *Picea* dominates from west to east and *Larix* is restricted to wetlands, *Picea* dom-inates in the "dark spruce forests" of Europe, while *Larix* dominates in the "open larch forests" of eastern Siberia, and mixed coniferous stands dominate along the northwest Pacific margin. This contrast between the two major continental regions of the Holarctic is the direct influence of climate, which varies as much longitudi-nally between western Europe through Siberia to eastern Asia as it does latitudinally. Another noticeable difference is that in Europe, *Pinus sylvestris* is the dominant pine from the northern boreal to the mixed-forest transition to the south, while in North America, pine dominancy changes from *P. banksiana* in the north to *P. strobus* and *P. resinosa* in the mixed-forest transition to the southeast.

In terms of the relationships between climate and vegetation types, it is convenient to discuss Europe separately from Siberia and the Pacific fringe. Figure 4.2C gives climatic data for Archangel, a typical dark spruce forest station, while Figure 4.3D illustrates climatic data for Moscow, which lies in the mixed-forest transition to the south. When these stations are compared with an open larch forest station east of the Urals, such as Verkhoyansk (Figure 3.3C), the reason seems obvious. It should be pointed out that the boreal of western Europe typically has a maritime climate even milder in winter than, say, for Yarmouth, Nova Scotia (Figure 4.3C).

4.5.1 The European Boreal

The most notable difference between the forests of the European Boreal and their North American counterparts is the reduced number of dominant conifer species.

The cool maritime climate and Podzolic soils produce excellent conifer-growing conditions so that climax spruce forests are closed canopy. In Scandinavia (including Finland) these dark spruce forests are dominated almost exclusively by *Picea abies* (Norway spruce), while drier or recently burned sites may be dominated by almost pure stands of *Pinus sylvestris* (Scots pine) and wetter by *Larix decidua* (European larch). Hardwoods such as *Betula pendula* and *B. pubescens* are also encountered (Arnborg, 1990). Contrary to what the name "dark spruce forest" might suggest, natural forests are rarely too dense to eliminate ground cover completely.

Following the destruction of climax spruce forest by fire, *Betula* is often the primary successional tree dominant, succeeded later by *Pinus sylvestris* and ultimately, *Picea abies* (Walter, 1979). Unlike the relatively short length of seral stages of *Populus* and/or *Pinus* in North America, northern European post-fire succession takes much longer. As a result, *Pinus* can be found dominating large areas that should otherwise end as *Picea* forest climax. According to Walter (1979), the *Betula* stage in northern Sweden can last 150 years, while the *Pinus* dominancy may last 500 years before *Picea abies* climax forest results. As fire often reoccurs before this spruce stage has been reached, it is not surprising Scots pine is so common. In northern Sweden, climax old-growth *Picea abies*–forest nurse logs help in maintaining spruce dominance much as they do for hemlock and Douglas-fir on the west coast of Canada. Hofgaard (1993) notes that in old-growth spruce stands, some 40% of the spruce seedlings were found on old logs, which could serve as important regenerating substrates for up to 150 years.

Dark spruce forests are valuable as sources of construction material, pulp, and fuel and have long been integrated into the economies of northern European nations. *Picea* plantations are now a common sight through the southern boreal and mixed-forest areas of continental Europe and the British Isles. While *Pinus sylvestris* is the only native conifer to the British Isles, its commercial value there has been greatly surpassed by the introduced *Picea* spp. Figure 1.11 shows the relationship between the ecological and physiological optima for *Picea glauca* and *Pinus banksiana* for North America, and these relationships should equally well apply to Norway spruce and Scots pine. In fact, Figure 1.11 is based on ideas coming from field studies performed on *Pinus sylvestris* by the famous Finnish forester Cajander. It was Cajander who first clearly explained these ecological versus physiological relationships, when his studies showed that plantation *Pinus* did best on moderately well drained, moist, mildly acid soils, while under natural competition from *Picea*, it was restricted as the climax species to either drier, sandy sites or wet, boggy sites. Cajander is also credited with pioneering the use of ground vegetation in forest site classification (Mikkola, 1982), an approach which has met with great success when applied in the Canadian boreal in southeastern Manitoba (Mueller-Dombois, 1964), in northern Sweden (Arnborg, 1990), in refining Cajander's own Finnish forest site types (Tonteri et al., 1990), and in classifying Scots pine successional stages (Nieppola, 1992).

European boreal herbaceous cover bears many resemblances to the boreal forest floor cover in North America. *Lycopodium* spp., *Pteridium aquilinum* (bracken), and sub-shrubs of the *Vaccinium, Erica,* and *Calluna* genera give the open pine forests and treeless uplands their characteristic rusty-brown and purple summer colours. Where dark spruce forests do live up to their name by reducing forest-floor light conditions, herbaceous cover is often impoverished or absent. Contributing to this is the thick acid mor litter layer that results from the slow decomposition of the acidophilous litter. Here too, contributing to the reduced light are epiphytic lichens which hang from branches. As already noted for Canada, lichen are particularly sensitive to certain air pollutants, and it has been reported that over the last twenty-five years, there has been a decline of epiphytic lichen species over more than 100,000 km^2 in Finland alone (Kauppi et al., 1992).

Typical of the European boreal and mixed forests to the south are large areas of treeless bog and grassy and true heathland (Marcuzzi, 1979). Many of these areas were no doubt forested in post-glacial times, but a return to cooler, more humid conditions, combined with increased incidents of fire, the introduction of domesticated grazing animals, and clearing for fuel and agricultural/grazing land, has no doubt contributed to their disappearance. Both pine stumps and bog oak are frequently discovered in peat cuttings and attest to earlier forest cover (Bellamy, 1986). Layers of petrified pine and birch are common in upland bog in northwestern Ireland, each recording an earlier forested phase, interspersed with a more humid peaty phase. These "peatscapes" have served, certainly for hundreds of years, as valuable sources of fuel. In 1980, 60 million metric tons of peat were used to generate electricity, primarily in Ireland and the Moscow-Leningrad region of Russia, and another 30 million tons were used for domestic fuel (Bord na Móna, 1985). As in North America, considerable amounts of *Sphagnum* peat moss are also extracted for horticultural use. In 1980, 130 million tons of horticultural peat were used worldwide, and of this, 120 million tons were used in the former USSR alone (Bord na Móna, 1985).

In hilly maritime areas, such as in the mixed-forest region of northwest Ireland, it is not uncommon to see *Sphagnum*-dominated Atlantic lowland blanket bog on the poorly drained lowlands and farmland on the lower and mid-slopes, while thin ericaceous blanket bog covers the uplands and hill crests (Scott, 1967). Comparatively speaking, the *Sphagnum* bogs are moderately thick, while the uplands have thin, poorly drained, oligotrophic ericaceous peat developing because of heavy precipitation, acidic parent materials, and the loss of nutrients downslope through runoff and throughflow. Hillcrest conditions are such that decomposition rates under *Calluna* (ling) and *Erica* (heather) dominated heath are slow, thin peats develop, and trees do not readily become established. The exact role of climate/soil conditions in this landscape mosaic is difficult to evaluate because of the long history of landscape "domestication" by humankind. So intense was the disturbance that forests in western Ireland were declining rapidly by 5,000 BP, and *Pinus sylvestris* was eliminated by 1,385 BP, although it was subsequently reintroduced (Hannon and Bradshaw, 1989). Wood from an earlier post-glacial forest cover has often been so well preserved in bogs that it was the source of the extremely strong "bog oak" often used as roof beams in Irish cottages (Evans, 1957). Strips of peat-preserved *Pinus sylvestris* excavated from upland bogs burn like candles. Even today it is not uncommon to see ericaceous peat as shallow as 20 cm being removed for fuel. Another characteristic of western Europe is the virtual absence of trees along coastlines in areas where high winds and salt spray combine to limit woody shoot growth. Frequently, these areas are heathlands, while under less severe conditions, *Ulex europaeus* (gorse) makes a continuous scrubland.

Compared to the hemlock, Sitka spruce, and lodgepole pine mesothermal forests along Canada's west coast (see chapter 6), this region looks remarkably impoverished in terms of tree cover. The bleak, blanket bog landscape of western Ireland at 52–55°N is at approximately the same latitude as the forested Queen Charlotte Islands, yet they seem to share little in common. It is possible that human interference over the last 5,000 years has indeed contributed to this dramatic contrast. With suitable drainage, fertilizer, and lime, many of these Irish peatscapes are again supporting a forest cover, and by 1982 some 140,000 ha of Ireland's blanket bog had been planted to conifers (Bellamy, 1986). Interestingly, 45% of today's plantings in Ireland are *Picea sitchensis* (Sitka spruce), and 40% are *Pinus contorta* (lodgepole pine) (Goodwillie, 1987). The balance is primarily made up of Douglas fir and noble fir, both introduced from North America, and Norway spruce and larch, both introduced from the European mainland.

4.5.2 The Siberian Boreal

This forested zone can be readily classified into two subzones on the basis of the response of the dominant tree species to climate (Tuhkanen, 1984). Unlike the North American boreal, however, this zonation is as much longitudinal as it is latitudinal. To the west, where the climate is less severe, closed-canopy evergreen coniferous forest dominates, while in the east, due to the severity of the continental climate, an almost unique forest landscape exists where only the deciduous larch thrives. Locally, these zones are known as the "dark taiga" and "light taiga" respectively (Walter, 1979), and locally they also exhibit dominant species variations from north to south.

The dark taiga zone shows many similarities to the European boreal, except that the increasing severity of the winter causes a change in dominant tree species. *Abies sibirica, Picea obovata, Pinus sylvestris,* and *P. sibirica* (cedar pine) constitute the dark boreal dominants, while *Larix russica* (Siberian larch) does well on poorly drained sites. The West Siberian Plain, a huge "depression" drained to the north by the Ob and Irtysh rivers, dominates this western region. A humid climate, drainage problems inherent in the gentle gradients, and summer flooding create ideal conditions for wetlands (Walter, 1979). Here a mosaic of forest types can be found, where dominants reflect the nutrient status of the moist forest soils. On eutrophic soils the more nutrient-demanding tree species, such as spruce and fir, dominate. Where peat thickening (paludification) reduces nutrient content, birch take on importance, and on oligotrophic wetland soils, pines (primarily *Pinus sylvestris* but also some *P. sibirica*) dominate (Glebov and Korzukhin, 1992). Some southern outliers of this forest can be found in the Karkaralinsk Mountains in eastern Kazakhstan (Gorcha-kovskii, 1987).

In the light taiga to the east, climate/soil conditions permit *Larix gmelinii* (Asian larch; synonymous with *L. dahurica* and *L. cajanderi*) to be the "unrivalled domi-nant species" (Tuhkanen, 1984; Nikolov and Helmisaari, 1992). Summers are suf-ficiently long for tree growth, but given permafrost and mean annual temperatures of −15°C or lower, it is surprising that any trees do well. It would appear that summer heat is sufficient to melt surface layers, particularly in well-drained soils, to depths of one metre or more. While soil microclimates seem best suited for *L. gmelinii*, *Picea obovata* grows on floodplains and in damp depressions towards the south, and *Pinus pumila* (dwarf Siberian pine) bushes are scattered in with *Larix* in the north (Tuhkanen, 1984). In Siberia, larch forests cover some 2.5 million km^2. Because forests are open, sufficient light reaches the ground for a ground cover of dwarf shrubs and lichen to develop. In wetter sites, dwarf shrubs such as *Ledum palustre* and *Vaccinium* spp. dominate. On somewhat drier sites, the dwarf shrubs *Dryas integrifolia* ssp. *crenulata* and *Vaccinium* spp. are common, while on the dri-est of forest floors, lichens dominate (Walter, 1979).

4.5.3 Northwest Pacific Fringe Boreal

Along the northwestern Pacific margin, true boreal forests are found primarily in the Russian Far East between Amur Oblast and the Pacific Ocean, on the southern Kamchatka Peninsula, Sakhalin Island, and the northern islands of Japan (Fig-ure 4.1). As well, mixed forests of broadleaf deciduous and needleleaf evergreen trees cover large portions of central Japan, northeastern China, and the north coast of the Korean Peninsula (Figure 7.5). As in Canada, members of the *Larix, Picea, Abies,* and *Pinus* genera dominate boreal forest. Larch is the dominant conifer in commercial forests, particularly in continental regions. In Amur Oblast, larch com-prises 77.2% of commercial forests, while spruce and pine combined only constitute 6.7%. As can be expected in areas with an oceanic climate, such as on Sakhalin Island, the role played by larch is reduced. Here the dominant conifers are *Picea obovata, P. ajanensis* (yezo spruce; synonymous with *P. jezoensis*), and *Larix gmelinii*, with spruce and larch representing 38.6% and 33% of the cover respec-tively in commercial forests. North of Korea, in the mixed forests of the Russian Khabarovsk and maritime regions are found valuable stands of *Fraxinus* (ash) and *Ulmus* (elm) (Barr, 1989).

As with northwestern Europe, it is possible to delimit a corresponding heath for-mation which comprises the majority of the area of the Kuril Islands and parts of Kamchatka and the Aleutians (the amphi-Beringia heaths of Tuhkanen, 1984). The only conifer in these northern parts is *Pinus pumila*. Deciduous forests dominated by *Betula ermanii* (Russian rock birch) and *Alnus kamtschatica* are common on Kamchatka and resemble the birch forests of Fennoscandia. Over 55% of the cover in Kamchatka Oblast's commercial forests are hardwoods (Barr, 1989). The Kurils stretch from the northern boreal zone to the southern, with *Abies, Picea,* and *Populus* being joined in the south by *Quercus, Acer,* and *Ulmus* species. *Pinus pumila* and *Alnus* thickets are also common here.

Farther to the west, Mongolia experiences greater continentality, and like the forests of eastern Siberia, the dominant conifer here is *Larix gmelinii*, together with some *Betula dahurica, B. platyphylla,* and *Pinus sylvestris* var. *mongolica*. Mongolia differs from Siberia, however, in also having an important broadleaf deciduous component dominated by *Tilia amurensis* and *Quercus mongolica* (Tuhkanen, 1984). While most of the native forests of China have been destroyed, sizeable areas of coniferous and mixed forests are found in the Da Hinggan, Changbai, and Xiao Hinggan Mountains in northeastern China (*Population Atlas of China*, 1987). As with boreal forests elsewhere, this region is susceptible to fire (Wang, 1961), with the worst fire in modern China's history destroying 1.01 million ha of forest in the Da Hinggan Mountains in 1987 (Peart, 1988). Areas of mixed forests are also found at subalpine elevations (above 2,800 m) in the mountains of central Sichuan. Here *Abies faxoniana* dominates the canopy, with *Betula utilis* (Himalayan birch) and *B. albosienensis* forming a subcanopy layer, while *Sinarundinaria fangiana* (bamboo) dominates the understorey (Taylor and Zisheng, 1988). *Abies faxoniana* individuals often exceeds 1 m in diameter (dbh) and 45 m in height.

Although Japan is small in area, its unique position on the eastern edge of the Asian continent creates dramatic differences in climate over short latitudinal distances. Four major vegetation types are encountered: boreal (including subalpine), mixed-forest, temperate deciduous, and subtropical evergreen. Found primarily in northern and eastern Hokkaido and in small areas of northern Honshu, where the mean annual temperature is below 7°C, coniferous forest covers 3.65 million ha (14.6% of the country) (Trewartha, 1965). *Abies sachalinensis* and *Picea jezoensis* dominate, along with the classic *Pinus* spp., *Tsuga diversiflora, Larix leptolepis,* and *Betula ermanii* (Geographical Survey Institute, 1977). Subalpine forests in northern Yatsugatake Mountains in central Honshu include *Abies mariesii* (Marie's fir) and *A. veitchii* (Veitch's silver fir) stands, as well as *Abies*–hardwood (mainly *Betula ermanii*) mixed forests (Kohyama, 1984). On Mount Fuji, subalpine forests can be differentiated into four primary types, and in order of descending elevation these are *Alnus maximowiezii, Betula ermanii, Abies veitchii,* and *Tsuga diversifolia* (northern Japanese hemlock). *Larix leptolepis* and *Sorbus americana* var. *japonica* (mountain ash) are also important pioneer species and members of ecotonal communities (Ohsawa, 1984). Lowland mixed-forest woody genera are typically Holarctic and include *Abies, Fagus, Betula, Alnus, Populus, Quercus mongolica* var. *grosseserrata,* and *Tilia japonica*. Important mixed-forest species in western Japan include *Abies homolepis* (Nikko fir) and *Fagus crenata* (Nakashizuka, 1991). Soil types associated with these coniferous and mixed forests include Podzols and Lithosols (young and immature mountain soils).

Table 4.1
Biomass and NPP values for natural stands of boreal and mixed forest (from a summary table of biomass and NPP values in Art and Marks, 1971, with one value (*) from Sprugel, 1984)

Boreal Forest	Location	Biomass $t\ ha^{-1}$	NPP $g\ m^{-2}\ yr^{-1}$
Abies balsamea	Canada	133.01	935
Abies balsamea	Canada	200.09	1,258
Abies balsamea*	USA	118.00	960
Picea mariana	Canada	93.50	156
Picea-Abies	USA	340.96	1,024
Pinus sylvestris	UK	118.72	na
Picea abies	Sweden	108.60	na
Picea	Russia	260.00	700
Picea	Russia	330.00	850
Pinus	Russia	80.70	na
Pinus	Russia	280.00	610
Pinus densiflora	Japan	63.98	1,578
MIXED FORESTS			
Pinus-Quercus	USA	181.52	867
Pseudotsuga menziesii	USA	235.00	na
Tsuga-Fagus	USA	193.00	1,333
Tsuga canadensis	USA	610.06	1,183

na = not available.

4.6 Primary Production and Phytomass in Boreal Forest

Biomass values for typical coniferous boreal forest are given in Table 4.1. It should be noted that at the ecosystem level, biomass values can be broken down into three important subcategories: soil organic matter, litter, and living phytomass. In addition, the living phytomass can be subdivided into living roots, trunk plus branches, and foliage. It should also be remembered that mosses often contribute significantly to living phytomass values. Larson (1980) notes that feathermosses often form the second largest biomass component in boreal forest, while Kavanagh (1987) reports they can contribute up to 85 per cent of total ecosystem biomass.

The stage of forest succession and the softwood-hardwood mix can give rise to great ranges in above-ground biomass, as compared to more stable values for soil organic-matter storage. One study comparing typical northern Michigan boreal softwood forest with northern hardwood forest showed above-ground living biomasses of 108 and 267 t ha^{-1} respectively (Rutkowski and Stottlemyer, 1993). On freely drained Podzolic soils, generalized values for climax boreal forest biomass would be in the order of 100 (approx.), 120, and 200 t ha^{-1} respectively for mineral soil organic matter, litter, and living phytomass respectively. On poorly drained soils and on Organic Cryosols, the values for soil/litter increase dramatically. Schlesinger (1984) reports 412 t ha^{-1} (based on carbon being 50% of the organic matter) for boreal forest soil/litter in general, but some of the data used in this average were determined for Organic Cryosols. Two values from northern Manitoba were 610 and 2,540 t ha^{-1} (Tarnocai, 1972)! By contrast, the relatively small value for organic matter in the A horizon of well-aerated, freely drained mineral soils reflects the influence of the acidic soil/litter environment and its inhibiting effect on soil mesofaunal activity. Here, fungi are the primary decomposers, and although they are slow in decomposing acidophilous litter, most of the organic matter is eventually volatilized back into the atmosphere rather than being incorporated into the soil by mesofaunal activity. Research in subarctic Quebec forests concludes that these slow litter decomposition rates are due in large measure to low soil temperatures (Moore, 1981). A direct consequence of this is that litter biomass values are among the highest for any freely drained ecosystem worldwide. Another consequence is that the Ah horizon is rarely of significance in Podzols, with sizeable organic accumulations in mineral horizons occurring only in the illuvial zone (Bh, Bhf, or Bfh horizons).

When root phytomass values are subtracted from living phytomass totals, it can be seen that typical Canadian boreal forests frequently have as much biomass in the mor litter layer as in the above-ground living phytomass. In the Alaskan boreal, it was found that only 20% of the total ecosystem organic matter was actually in the above-ground tree component (Cole and Rapp, 1981). Comparative values for temperate deciduous forest would be approximately 40%. It also takes much longer for litter values to reach equilibrium in the boreal than in other forested ecosystems. In Alaska, mean residence time for coniferous forest litter is 353 years! This was determined by dividing litter return into total mature forest litter accumulation, and assumes that litter production and forest-floor litter are both in steady state, which may not be the case (Cole and Rapp, 1981). Comparative values for boreal broadleaf deciduous forest and temperate deciduous forest are 26 and 4 years respectively.

An additional consequence of this large litter biomass is seen when fire spreads through boreal forest. If litter is dry, fire can be extremely destructive, although it serves the purpose of releasing both the nutrients locked into litter and perhaps some in the older conifer trunks as well. Fires in boreal forest are usually more effective in consuming the litter layer than the standing heartwood, and the post-burning landscape following most boreal conflagrations is one of a ground covered in wood ash, while the great majority of trunks still stand. Although these trunks are dead, it may take decades before they succumb to windthrow and decomposition. In this community type, where the presence of heterotrophic detritovores and grazers is minimal, fire can be considered a necessity to ensure the recycling of mineral elements. It has been shown that the more rapid the turnover of nutrients in an ecosystem, the greater the role of, and relative biomass of, heterotrophs (O'Neill and DeAngelis, 1980). With the limited presence and effectiveness of heterotrophs in the boreal, it is clear that other mechanisms such as fire are needed to aid in nutrient recycling.

In terms of NPP, boreal forests are moderately productive by forest standards, and much more productive that either the tundra or forest-tundra communities. One advantage the evergreen conifers have over deciduous tree species is that photosynthesis can begin early in the growing season and continue later into the fall. In spite of this advantage, the deciduous conifer genera *Larix* can, surprisingly, support an annual net carbon gain similar to the evergreen conifers (Gower and Richards, 1990). Typical NPP values for boreal stands are $9-12$ t ha^{-1}yr^{-1}). For mature *Abies balsamea* stands on Whiteface Mountain, New York State, NPP values of 960 g m^{-2}yr^{-1} were determined (Sprugel, 1984). A typical K factor value (comparing litter biomass with NPP) for northern Quebec spruce forest would be 0.14, while for Minnesota pine forest the range would be $0.1-0.2$ (Moore, 1978).

Biomass and NPP values have considerable importance when the impacts of wildfires, logging, and pollution are being considered. Since 1977 the extent of North American boreal wildfires has increased six- to ninefold over long-term trends (Auclair and Carter, 1993). Increased incidence of forest fires, combined with logging, contributes to a declining average age for boreal stands, with related declines in litter biomass values. Although it is important to account for all relevant carbon fluxes in the evaluation of the impact of wildfires on CO_2 release (Kurz et al, 1991), one significance of this is that lowering biomass values means lower carbon storage. Along the southern fringes of the northern coniferous forest, where forest on peaty Gleysols is being converted to agricultural land, the release of CO_2 is even greater. This is because burning quickly volatilizes much of the felled trees and the litter/peaty layer, while ditching allows oxygen, and therefore fungi/bacteria, access to mineral soil organic matter. Loss of both nutrients and possible long-term productivity may also be associated with whole-tree harvesting of sugar maple in mixed-forest stands on the Canadian Shield (Bird and Chatarpaul, 1988).

Pollutants such as sulfur and nitrous oxides have been demonstrated to cause the demise of conifers close to smelters in Ontario and Quebec (Amiro and Courtin, 1981). Less quantifiable is the effect of acid rain on boreal forests in general. Research clearly points to the reduction in growth rates which results from acid fallout, and it must be assumed that this is the result of lower NPP values. One consequence is a lowering in the production of O_2 from affected stands and a lowering in the ecosystems demand for atmospheric CO_2. The excess release of CO_2 over consumption as a direct consequence of clear-cutting, induced fires, wetland-forest drainage to promote better tree growth (Hillman, 1988), and fen clearing and drainage for agriculture has yet to be accurately quantified. Nevertheless, this net release must be viewed as a contribution to increasing atmospheric CO_2 and must detract from the role of both wetland organic soil growth and forest biomass increases as carbon sinks. Any efforts at improving regeneration following extractive forestry activities, the reafforestation on abandoned lands, or a reduction in forest consumption (say, by greater recycling of wood and paper products) should therefore lead to the sequestering of CO_2 and to slow atmospheric buildup rates (Dixon et al., 1993; Sedjo,

1989). Preserving large areas of the remaining boreal forest in the Russian Far East and in Siberia would be important in this regard as well, because these areas contain 82% of the boreal carbon sink of the region of the former Soviet Union (Kolchugina and Vinson, 1993).

CHAPTER FIVE

Prairie (Steppe)

5.1 Distribution

The prairie, or steppe, environment consists of grasslands and parklands in middle latitudes where climate is semi-arid to subhumid (Figure 1.1). This is the equivalent of the arid-temperate climate of Walter (Zonobiome VII in Figure 1.2). "Prairie" is the most frequently used term for describing this North American environment, although some authors prefer "steppe" (Daubenmire, 1978). The term "steppe" is most commonly used in eastern Europe and western Asia. In addition to these better known Holarctic grasslands, this vegetation type is often extended to include grasslands in Mongolia, northern China, the Middle East, the High Veldt of South Africa, the pampas of Argentina, and the campos of Uruguay (Figure 5.1). While the pampas/campos region can be considered almost subtropical, the prairie vegetation type is not extended to incorporate tropical grasslands, which are more properly included under the vegetation type called "savanna." Figure 5.1 also includes small regional grasslands in the Great Valley of California, and the Palouse prairie of the intermontane region straddling the Canada–USA border. These Palouse grasslands are discussed in chapter 6.

North American prairie grasslands are found between the Rocky Mountains and the Mississippi River valley, and from southern Saskatchewan to the Gulf of Mexico. Unlike the boreal and tundra of North America, this vegetation type has a greater latitudinal than longitudinal extension because of its association more with the aridity induced eastward of the Rocky Mountains than with variations in temperature regime. Because it extends through a variety of climates, it is variously described as "tall-grass" (true), "mixed-grass" and "short-grass" prairie depending on growth potential (Coupland, 1992a). In southeastern Europe, distribution is more typically longitudinal, with the steppe extending in a relatively narrow belt east from the Ukraine to the Kirgiz steppe of Siberia. This Eurasian distribution is due to the latitudinal moisture gradient found between the humid taiga to the northwest and the central Asian deserts to the south. Were it not for the mountain belts of central Asia,

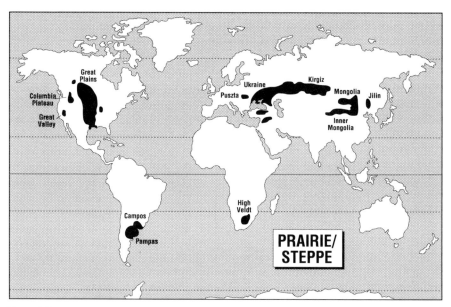

Figure 5.1
Distribution of prairie and steppe

the steppes of Mongolia and those of Inner Mongolia and Jilin in northern China would be continuous with those to the west. A small region of moist, tall grassland is found in the Puszta region of Hungary-Romania, while short-grass steppe is also found in the Anatolia-northern Iraq region.

5.2 Climate

5.2.1 North America

In the northern hemisphere, grassland climates are associated with extreme continentality. In North America, winters are normally very cold and summers hot, annual temperature range is large, and a negative moisture index with summer moisture stress is typical. Climate diagrams for four Canadian prairie locations are given in Figure 5.2. All are found in the Grassland Ecoclimatic Province (Figure 1.4), which includes three ecoclimatic regions: the Arid, Subhumid, and Transitional Grassland (Ga, Gs, Gt). Using formation terminology the Arid Grassland Ecoclimatic Region has a short-grass (xeric mixed-grass) and mixed-grass prairie cover, the Subhumid is associated with the fescue mixed-grass prairies of southwestern Alberta, and the Grassland Transition is associated primarily with both the aspen parkland and the tall-grass (true) prairie.

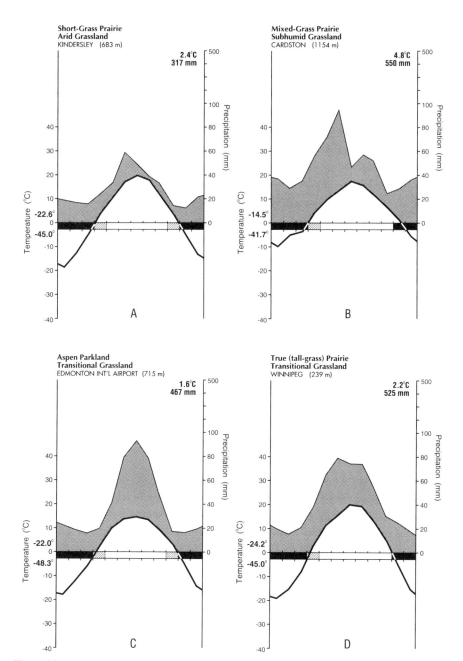

Figure 5.2
Climate diagrams for Canadian prairie stations. A = Kindersley, Saskatchewan; B = Cardston, Alberta (close to the border with Montana); C = Edmonton, Alberta; D = Winnipeg, Manitoba.

Mean annual temperatures for the Canadian prairies are usually between 1.5° and 5°C, while monthly averages often range from below −15°C to above +20°C. Summer evapotranspiration rates exceed the ecosystem's ability to supply soil moisture, so moisture indices are negative, and any tree cover is stressed. Stands of trees are found throughout the prairies, but they owe their origins to groundwater supply, effects of elevation, or protection from fire. Forests in the Cypress Hills of Alberta and Saskatchewan owe their origin to the effects of relief, while Spruce Woods in Manitoba has a tree cover that results from groundwater and an undulating topography which reduced any major impact from prairie fires. Aspen parkland is also encountered on the moister sides of the Canadian prairies, but the exact relationship between this cover type and climate is difficult to quantify since the parkland appears to be spreading into drier regions now that prairie fires are less frequent. Tall-grass prairie in the Winnipeg region is found on moist soils and must be maintained from invasion by *Populus tremuloides* (trembling aspen) by regular burning.

In the United States, grasslands extend from the Canadian border south to the warm climates of Texas. It would appear that negative moisture indices are primarily responsible for their distribution in North America, rather than growing-season temperatures. An analysis of some 9,500 prairie sites in the USA confirmed the overwhelming importance of water availability to primary production (Sala et al., 1989). Figure 5.3 includes three stations from a north-south transect on the eastern side of the prairies from Fargo, North Dakota, to Abilene, Texas. It is seen that while average annual precipitation rises as one moves south, increasing temperatures ensure that negative summer moisture indices continue to influence vegetation cover.

5.2.2 Climate in Eurasia and Elsewhere

It should be noted that, unlike its North American counterpart, where distribution of grassland types appears more longitudinal, the distribution of grasslands in Eurasia is essentially latitudinal. Having few topographic barriers between the Atlantic Ocean and eastern Europe, cyclonic disturbances travel more west to east through the centre of the continent. As one moves south-southeast from the boreal forest of northeastern Europe to the Caspian Desert, the latitudinal effects of both declining precipitation and increasing potential evapotranspiration combine to produce a marked increasing moisture-deficit gradient. So pronounced is the latitudinal distribution of vegetation, soils, and climate in this region, that it provided convincing evidence to the nineteenth-century soil scientists Sibertzef and Dokuchaev as to the interrelationships between these three variables.

Climatic data for Odessa on the northern shores of the Black Sea are given in Figure 5.3D. Odessa has a dry *Stipa* steppe (short-grass prairie) cover on Southern Chernozems and is almost the east European equivalent of the Kindersley, Saskatchewan, data given in Figure 5.2A. Due to Odessa's longer, warmer summer, moisture stress is slightly greater than in southern Canada, so is more correctly the equivalent of short-grass prairie in western Nebraska. Grassland regions elsewhere

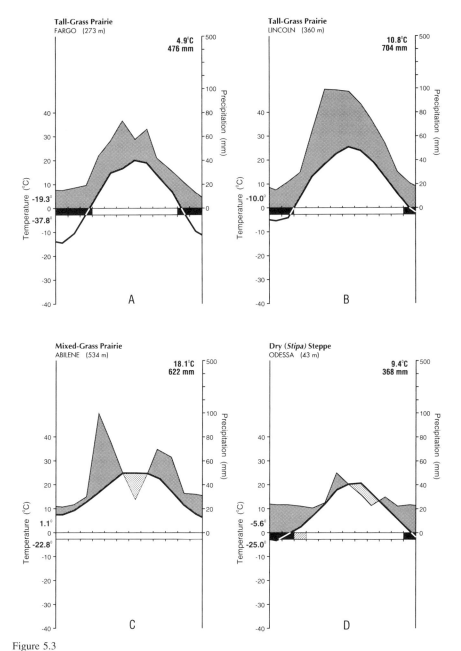

Figure 5.3
Climate diagrams for prairie stations in the USA and the Ukraine. A = Fargo, North Dakota; B = Lincoln, Nebraska; C = Abilene, Texas; D = Odessa, Ukraine.

in Eurasia also owe their origins to moisture stresses. In Turkey and northern Iraq, aridity results primarily from a combination of modest precipitation with high potential evapotranspiration, while in Mongolia and northern China, moisture stress results from the cooler, semi-arid summers and dry, cold winters. In the pampas of Argentina and Uruguay, conditions are by no means dry continental. Here, mean annual precipitation values normally exceed 500 mm; however, mild winters and warm summers create moderate negative moisture indices and give rise to tall-grass prairie. In the High Veldt of South Africa, elevation combines with modest precipitation and relatively warm temperatures to create the same grassland-promoting conditions.

5.3 Soils

Soil types found within the prairie environment possess surprising variety. As expected, there are intrazonal soils associated with poor drainage (e.g., Gleysols) and localized peculiarities of parent material (e.g., Solonetzic soils), as well as zonal types resulting from the more direct influences of different climates (e.g., Chernozems).

Prairie soils reflect a combination of soil-forming factors which normally include cold winters, hot summers, and the influences of varying herbaceous vegetation covers, along a moisture gradient with negative moisture indices. In terms of traditional soil classification terminology (see Figure 1.7 and Table 1.3), these zonal soils are described as Brown, Reddish- Brown, Reddish Chestnut, and Chernozemic soils, as one moves from semi-arid to subhumid grasslands. In terms of soil colour, the lighter or brighter prairie soils colours reflect the warmer and/or drier regions where less organic matter darkens the well-developed Ah. Using Comprehensive System terminology, most prairie soils would be known as Mollisols (Figure 1.8). In Canada the dominant zonal great groups are the Brown, Dark-Brown, Black, and Dark-Gray Chernozems (Figure 5.4), a sequence which reflects an increasing moisture–phytomass–soil organism activity gradient. At the order level, the term "Chernozem" is expanded to include most prairie soil types, other than the intrazonal Gleysolic and Solonetzic soils.

In eastern Europe the equivalent Chernozemic gradient goes through Southern, Normal, Thick, Northern, and Degraded Chernozems along the increasing moisture gradient (Walter, 1979). The use of the name "Chernozem" is not an attempt to Anglicize steppe soil terminology; rather, the name is derived from the Russian words *chernyi*, meaning black, and *zemlya,* meaning soil (Acton, 1992) and was a term later adopted by Canadian soil scientists. The Degraded Chernozem of Russia and the Ukraine is equivalent to the Canadian Dark-Gray Chernozem (previously also known in Canada as a Degraded Chernozem, or Degraded Black-earth). Both these soils owe their great group descriptions to the effects of eluviation on the Ah (Figure 5.4), which results from a higher moisture input than in Black Chernozem regions, combined with the influence of a broadleaf deciduous tree cover on leaching.

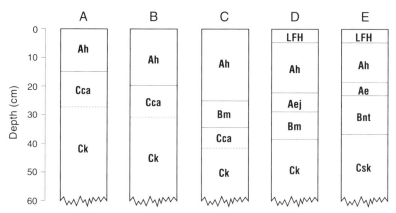

Figure 5.4
Typical soil profiles in the Canadian prairies (after Canada Soil Survey Committee, 1978). A =
Brown Chernozem; B =Dark-Brown Chernozem; C = Black Chernozem; D = Dark-Gray Chernozem;
E = Solodized-Solonetz. In the profiles, k = carbonate present, ca = carbonate enrichment, and
n = columnar structure resulting from a high sodium (s) content.

Because of the pronounced negative moisture indices, little excess soil moisture
is available for leaching within Chernozems. Carbonate enrichment in the lower horizons (the "ca" in Figure 5.4) indicates the degree to which downward movement
of weathered surface materials occurs, but below which leaching is not significant.
Characteristics of these Chernozems include high organic contents, thick Ah horizons, good structural properties, alkaline reactions, high base saturations, and low
available moisture contents. Following sod turning, this combination of attributes
provides excellent agricultural soils for cereal crops and other relatively low moisture demanding annuals (Figure 5.6). It is to be expected that the moisture stress conditions which promote the development of prairie grasses, but not trees, still exist
for agricultural crops. Agricultural exploitation of these soils must therefore take
into consideration not just the effects of moisture stress on crop productivity, but
also the effects of the dry, warm summers on both wind erosion and on promoting
greater soil salinity. A remarkable correlation between the distribution of soil types
and the Canada Land Inventory (CLI) classes 1–3 is seen when Figures 5.5 and 5.6
are compared. It is noted that the better agricultural lands (classes 1–3) correlate well
with the Dark-Brown and Black Chernozems, and to some extent with the wetter
margins of the Brown Chernozems and the drier, warmer margins of the Dark-Gray.

5.4 Prairie in North America

Grasslands constituted the largest of the pre–European settlement vegetation formations in North America (Küchler, 1964). In terms of the tall-, mixed-, and short-grass
prairie units, Canada had about 50 million ha and the USA some 300 million ha (figures summarized by Sims, 1988). It goes without saying that much of this area has

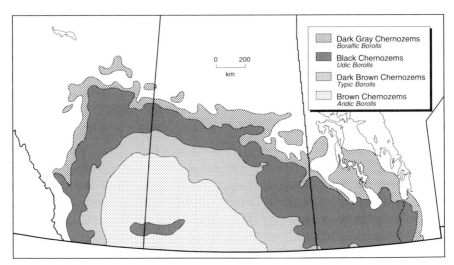

Figure 5.5
Distribution of Chernozems on the Canadian prairies (after Soils Research Institute, 1972). Note that the equivalent New Comprehensive System terminology is given and that the term "boroll" comes from *bor* for northern and *oll* from *Moll*isol.

been converted to arable lands, particularly where moisture indices are not strongly negative and where soils are Dark Brown and Black Chernozems (Mollisols). The tall-grass prairie, in particular, has provided some of the most productive agricultural lands in Canada, unfortunately at the expense of the almost total demise of this community type. Even in the USA, less than 5% of the original tall-grass prairie remains today.

It also goes without saying that the attraction of the remaining drier non-cultivated prairie for ranching has led to impacts on cover which vary depending on cattle grazing intensity. Under light grazing, certain desired forage climax species, such as *Koeleria cristata* (June grass) and *Schizachyrium scoparium* (little bluestem), undergo modest decline and are called decreasers, while other somewhat less desirable climax species, such as *Andropogon gerardii* (big bluestem) and *Stipa spartea* (porcupine grass), actually increase their cover, as do invaders such as *Poa pratensis* (Kentucky bluegrass). With the deterioration of range under heavy grazing, the decreasers decline rapidly, the increasers become relatively more palatable and also begin to undergo decline, while the fate of the often less palatable invaders is one of increasing dominance (Canadian Wildlife Service, 1992). Though descriptions which follow generally detail climax (little grazed) communities, some reference will be made to grazing impacts.

5.4.1 The Canadian Prairie

The Canadian prairie region coincides with the Grassland Ecoclimatic Province (Figure 1.4) and the Prairie Terrestrial Ecozone (no. 7 in Figure 1.5), and it covers

Figure 5.6
Canada Land Inventory agricultural capability classes 1–3 for the Prairie provinces (Lands Directorate, 1967)

5.03% of Canada's land area. As indicated in the climatic data in Figure 5.2, this ecoclimatic province is in turn divided into three ecoclimatic regions which correlate closely with traditional descriptive terminology for the various grasslands. The Arid Grassland Ecoclimatic Region is equivalent to the short-grass prairie (zeric mixed-grass prairie) and the majority of the mixed-grass prairie. The fescue mixed-grass prairie region in southwestern Alberta forms the Subhumid Grassland Ecoclimatic Region, while the Transitional Grassland Ecoclimatic Region corresponds with the aspen parkland and tall-grass prairie. Outliers of mixed-grass prairie are also found in the Peace River district of northwestern Alberta. Small patches of tall-grass prairie and oak-prairie parkland are also found in southern Ontario (Reznicek and Maycock, 1983), but these grasslands are essentially extensions of the so-called Prairie Peninsula of the United States. Many authors (e.g., Coupland, 1961; Gordon, 1979; Looman, 1980) point out that the term "short-grass prairie" does not strictly apply in Canada because these areas are really xeric mixed grasslands which have been overgrazed to the point that short-grass species such as *Bouteloua gracilis* (blue grama) and *Buchloë dactyloides* (buffalo grass) dominate. However, the term "short-grass prairie" is retained here simply for convenience, and its use in reference to Canadian prairie is not intended as support for the short-grass disclimax concept of Weaver and Albertson (1956).

As can be seen in Figure 5.7, tall-grass prairie (or true prairie, as it is often called) covers only a small portion of south-central Manitoba in the Red River basin. It is climatically, but not edaphically, similar to the wetter eastern side of the aspen parkland and extends south into the United States, where it lies between the mixed-grass prairie and the temperate deciduous forest. In Manitoba it is considered a separate cover type because in pre–European settlement times it thrived on the large area of

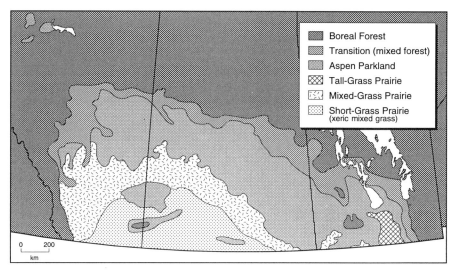

Figure 5.7
Distribution of vegetation types in the Canadian prairies (modified from Scott, 1987a). Note the inclusion of the Peace River region in the northwestern corner of the map. Riverine gallery forests are not included. The mixed forest (transition zone) is based on Zoltai, 1975; see section 4.4.4.

poorly drained lowland exposed after glacial Lake Agassiz drained. Today, few tall-grass prairie remnants survive because of agricultural development and the expansion of aspen parkland following the reduction in scale of prairie fires. The original 6,000 km^2 of tall-grass prairie of the Red River basin in Manitoba is today reduced to only about 4 km^2, or 0.07% of its former area (Canadian Wildlife Service, 1992). Likewise, much of the mixed-grass prairie has given way to the plough. While the distinction between cover types seems clearly defined in Figure 5.7, it is noted that enclaves of tall-grass prairie are found in many wetter areas within the mixed-grass prairie, just as gallery forest is found throughout the tall- and mixed-grass prairie associated with river courses.

Figure 5.8 underlines the striking correlation which exists between cover type and moisture availability as one moves north or east from the short-grass prairie region of southern Saskatchewan towards forest. Although precipitation values do not increase greatly with increasing latitude (Figure 5.8A), temperatures decrease significantly. As temperatures drop with increasing latitude, potential evapotranspiration (PE) decreases, and moisture indices (Im) change from strong moisture stress conditions to positive values. In the longitudinal transect from central Saskatchewan to eastern Manitoba (Figure 5.8B), temperature differences are minimal and PE decreases only modestly, while it is the pronounced increase in precipitation values which most demonstrably influences decreasing moisture stress. Although latitudinal vegetation zonation varies primarily because of variations in temperature, longitudinal zonation changes primarily because of variations in precipitation; both gradients show that the narrow mixed-forest zone is coincident with an im of between -10 and zero. Of interest is that im values of between zero and -10 also mark the transition zone between Dark-Gray Chernozems of the mixed forest and Luvisols and Podzols, which depend on excess moisture for eluvial-illuvial processes.

One adaptation which aids practically all prairie herbaceous species is the symbiotic association between roots and vesicular-arbuscular mycorrhizae (Reichman, 1987; Cook et al., 1988). The presence of these mycorrhizae effect significant increases in biomass production (Stall et al., 1988). Another physiological adaptation which may help certain species, particularly in the warmer, drier prairies, is the possession of a c_4 photosynthetic pathway. It would appear that under cycles of changing water potential, *Agropyron smithii* (western wheat-grass, a c_3 species) is more sensitive to water stress than is *Bouteloua gracilis* (blue grama, a c_4 species) (Brown and Trlica, 1977). Sala et al. (1988) attribute this to the fact that western wheat-grass is an earlier flowering and deeper rooting species than blue grama, while the latter, with its shallower rooting system and later flowering cycle, is ideally suited to the frequent small precipitation events typical of the short-grass prairie growing season. It has also been hypothesized that frequent fire reduces inhibiting factors for c_4 species by removing litter, warming the soil, and depressing c_3 species (Hill and Platt, 1975).

5.4.1.1 The Aspen (Prairie) Parkland

The major area of aspen parkland extends in a broad swath from just west of Edmonton to just east of Winnipeg (Figure 5.7). In addition, a number of important islands of prairie parkland surrounded by mixed forest are found in the Peace River, Grand Prairie, and Paddle River areas of Alberta and, surrounded by grassland, in the Cypress Hills and Great Sandhills areas of southeastern Alberta and southwestern Saskatchewan. In general, the parkland is coincident with the Transitional Grassland Ecoclimatic Region, moisture indices vary between -5 and -30 (see Figure 5.8), annual precipitation is 400–530 mm, and the mean annual temperature is 1.3°C (Looman, 1983a). Climatic data for Edmonton are presented in Figure 5.2C. Soil types are primarily Black and Dark-Brown Chernozems (Figures 5.4 and 5.5), although Dark-Gray Chernozems and Luvisols are also characteristic of outliers in mixed forest such as the Peace River country. Typically, parkland vegetation is associated with three important tree species: *Populus tremuloides* (trembling aspen), *P. balsamifera* (balsam poplar), and *Quercus macrocarpa* (bur oak). While *Populus* spp. are found throughout the zone, *Quercus* is only common in southeastern Saskatchewan and southern Manitoba, a possible reflection of the lower latitude and higher precipitation figures typical of the eastern parkland. Grasslands within the parkland are normally dominated by fescue grasses, with *Festuca hallii* (and sometimes *Andropogon gerardii*) dominating between central Manitoba and west-central Alberta, and *Festuca campestris* (rough fescue, or *F. scabrella*) dominating in the Rocky Mountain foothills and the Selkirk, Monashee, and Cascade mountains

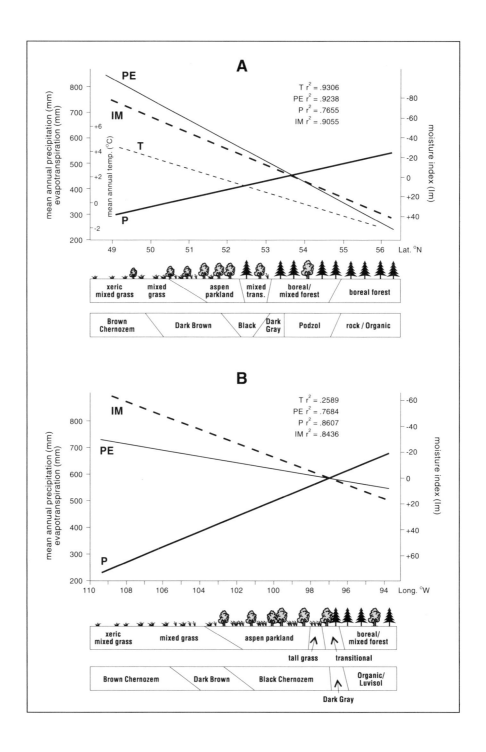

(Looman, 1982). These grasses are noticeably absent in the Peace River parkland, where *Stipa spartea* dominates, together with some *Agropyron dasystachyum*, *Koeleria cistata,* and *Carex* spp. sedges (Wallis, 1982). Effectively, the parkland proper represents a transitional zone between grassland and mixed forest (between ZB VII and ZE VII-VIII), so on its drier margins it consists of grasslands with shrub-like copses of clonal *Populus* and shrubs such as *Prunus virginiana* (choke-cherry) and *Elaeagnus commutata* (silverberry, or wolf willow) (Looman, 1983c), while on the wetter side, along the mixed-forest boundary, trees produce more or less closed stands. Other tree species are associated with river valleys and will be discussed in the sections on tall- and mixed-grass prairie.

There is support for the conclusion that since the 1880s, the aspen parkland cover type has expanded its geographic area into the grasslands. It appears that *Populus* has successfully invaded some grassland areas primarily because agricultural settlement by Europeans has greatly reduced the incidence of large-scale prairie fires (Archibold and Wilson, 1980). Studies of soil grass-opal contents in aspen forest 100 km north of Regina demonstrate that forest migration onto Black Chernozems is very recent (Fuller and Anderson, 1990). The northern boundary of the majority of this parkland zone has probably been relatively stable for some time and is well defined by the line north or east of which the presence of conifers produce mixed forest (Zoltai, 1975). On the other hand, the southern parkland boundary of the 1880s was out of phase with climatic conditions, so that following reduced burning, it has expanded its territory at the expense of mixed-grass prairie (Hildebrand and Scott, 1987).

The exact role of fire in the possible origin and/or maintenance of the more humid prairie grassland cover is unclear (Anderson, 1992; Looman, 1983b; Vogl, 1974; Nelson and England, 1971). Invasion of the wetter grassland margins by trees following a reduction in burning has, however, been documented. Ploughed fields and highways served as firebreaks, thereby reducing the spread of wildfire. With the reduction in burning, *Populus tremuloides* has had the opportunity to grow out over the herbaceous grassland dominants. Love (1959) better expresses this concept of grassland invasion by pointing out that aspen are actually growing out "under it." By this he means that the invasion of grassland is aided more by the ability of *Populus* to produce clones by suckering from roots than by spreading from seed. In

Figure 5.8
Transects from short-grass prairie to boreal forest. Transect A is based on data for the one degree of longitude between 115° and 116°W. Transect B is based on data for the one degree of latitude between 49° and 50°N.
PE = potential evapotranspiration (mm); T = mean annual temperature (°C); P = mean annual precitation (mm); IM = moisture index (see section 1.5.2). The r^2 values for T, PE, P, and IM are for regressions against latitude (A) and longitude (B). Note that in B a regression line for temperature is not given because, as expected, the r^2 indicated little correlation between longitude and temperature (after Scott and Scott, 1990).

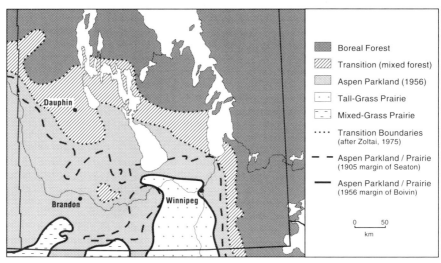

Figure 5.9
Expansion of the aspen parkland in southern Manitoba between 1905 and 1956 (after Scott, 1987a; from *Earthscapes*, by W.H. Marsh, with permission of the publisher, John Wiley & Sons, Inc. © 1987)

the past this invasion may have been slowed by grazing bison (*Bison bison*), as it still is today by the snowshoe rabbit (*Lepus americanus phaeonotus*). Examples of present-day clonal aspen spreading out into unburned remnant tall-grass prairie can be found west of Sanford, Manitoba, and the phenomenon has been documented in the Wood Buffalo National Park *Agropyron-Stipa* grasslands (Jeffrey, 1961). While annual fires effectively control aspen (Svedarsky and Buckley, 1975), infrequent fires actually encourage it because of aspen's ability to produce suckers (Moss, 1932). Looman (1987b) reports as many as 70 root suckers per square metre, following burning of aspen woods in the southern parkland. In northwestern Minnesota, Buell and Buell (1959) traced a sucker-producing root 31.7 m to its parent tree. This root was less than 1 cm in thickness for most of its length and was confined to the top 10 cm of the Ah horizon.

This expansion of *Populus* is not so obvious in the landscape only because much of the region is also suited to agriculture. One can drive through large areas of southeastern Saskatchewan's parkland and see trees only as shelterbelts, as small islands expanding in grazing lands, or where land is left undisturbed around potholes or along valley sides. Much of the pre-1880 mixed-grass prairie of southern Manitoba is now either agricultural land or parkland dominated by *Populus* spp. in drier areas or *Populus* and/or *Quercus macrocarpa* in the moister (Figure 5.9). It is important to note that parkland tree stands vary considerably in height depending on the degree of moisture stress. West of Moosomin in southeastern Saskatchewan, *Populus* stands are quite diminutive and scrubby, rarely exceeding 5 m. As one moves east into central southern Manitoba, stand height increases and in places exceeds 15 m.

Grassland communities within the aspen parkland are often nicknamed "fescue prairie" (Coupland, 1992b) because of the dominance of *Festuca hallii* (synonymous with *F. scabrella* var. *major*) and *F. campestris* (using nomenclature of Pavlick and Looman, 1984). Under closer scrutiny, however, the eastern parklands, especially in Manitoba, are also seen to include both mixed-grass and tall-grass community types (Coupland, 1961). It is therefore more appropriate, except at the local community level, to restrict the use of the term "fescue prairie" or "fescue grassland" to those in southwestern Alberta and west-central Saskatchewan. All of these parkland grasslands, however, exhibit variations due to possible disturbances from burning, overgrazing, conversion to agriculture, and invasion by woody species. While fescue grasslands are ideal for grazing, they are highly susceptible to growing-season grazing pressures (Willms and Fraser, 1992). Overgrazing leads to the decline of fescue in favour of *Agropyron, Stipa, Danthonia,* and *Carex* species (Campbell et al., 1962). This decline greatly reduces productivity, a problem that can require twenty years of greatly reduced stocking rates to ameliorate (Dormaar and Willms, 1990).

Festuca campestris is associated with the more humid side of the western parkland along the foothills of the Rocky Mountains (Pavlik and Looman, 1984). One estimate is that 13% of Alberta's native prairie is fescue dominated (Dormaar and Willms, 1990). While fescues are missing in the moister, more northerly Peace River country (North, 1976), they become even more dominant in the Cypress Hills than in the aspen parkland belt (Coupland, 1961). In the Cypress Hills and in west-central Saskatchewan parkland, *Festuca hallii* becomes the dominant fescue and is often associated with *Schizachyrium scoparium* (little bluestem, synonymous with *Andropogon scoparius*) and later flowering species such as *Andropogon gerardii* (big bluestem) (Looman, 1982). On west-central Saskatchewan Black Chernozems, *F. halii* occupies 55% of herb basal area (Coupland and Brayshaw, 1953). Although fescue as an aspen parkland component declines in Manitoba, some 2.5% of Riding Mountain National Park consists of grasslands dominated either by *Festuca hallii* along with *Koeleria cristata* (June grass) or by a combination of *Agropyron trachycaulum* (slender wheat-grass) and *Poa pratensis* (Kentucky bluegrass). It is noted that these Riding Mountain fescue grasslands are very sensitive to the effects of grazing, and under heavy to severe pressure, *Festuca* gives way to *Poa pratensis* (Trottier, 1986).

5.4.1.2 Tall-Grass (True) Prairie

A small portion of the Red River basin in southern Manitoba can be considered the only major true tall-grass prairie region in Canada, although there are small outliers, such as the Tolstoi Prairie, within the eastern aspen parkland and on sandy soils in southern Ontario. Essentially the northern extension of the true prairie, which stretches south to Texas, this ecosystem is generally dominated by *Andropogon gerardii* (big bluestem), together with *Panicum virgatum* (switchgrass), *Schizachyrium scoparium* (little bluestem), and *Spartina pectinata* (cord-grass). So significant are the big and little bluestems to this prairie type that it is sometimes nicknamed "bluestem prairie" (Kucera, 1992).

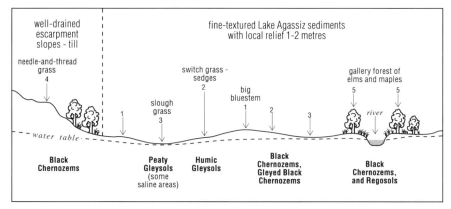

Figure 5.10
Community diversity in a typical tall-grass prairie in the Red River basin, Manitoba. Local relief within this basin is usually only a few metres, but on the heavy Agassiz clays this is sufficient to cause major variations in soil moisture availability.

Tall-grass prairie in Manitoba owes its presence to a combination of climate, poor drainage of the Lake Agassiz Humic Gleysols and Gleyed Chernozems, flooding, and a past history of frequent burning. Climatic data for Winnipeg, which has a moisture index of −12, are shown in Figure 5.2D. It is important to note that available soil moisture in the Red River valley persists later into the summer, thereby promoting the dominancy of the late-maturing tall grasses, such as *Andropogon gerardii* and *Panicum virgatum* (Looman, 1981, 1983a). Today, with reduced burning and much land taken up by agriculture, it is difficult to find any true prairie that is not at least partially invaded by aspen or by the clonal shrub *Elaeagnus commutata* (silverberry, or wolf willow) (Looman, 1983b). Although about 400 ha of true prairie remain in southern Manitoba, only a few patches, such as the 10 ha Living Prairie Museum in Winnipeg, have been officially preserved. Because of the almost total demise of this ecosystem in Canada, the Manitoba Parks Branch is attempting to rehabilitate some 242 ha of disturbed land at Beaudry Park just west of Winnipeg to its original tall-grass cover (Love, 1991).

Figure 5.10 represents a profile through the potential natural vegetation of a typical portion of the Red River basin in southern Manitoba. It shows three generalized herbaceous communities typical of the Red River lowlands, along with communities associated with better-drained uplands (no.4) and riverine "gallery forests" (no.5). These herbaceous communities vary from each other primarily because of differences in drainage, height of the water table, and slope. Spring flooding of the lower-lying areas contributed, certainly in the pre-1900s, to major ponding of water and the development of sloughs dominated by hygrophytic grasses and rushes.

1 Better-Drained Sites   This is the major community type and is dominated by *Andropogon gerardii,* which can reach heights of 2 m or more. Big bluestem represents one of the two most important species found in the true prairie of all of North America, the other being *Schizachyrium scoparium* (little bluestem). Together they comprise some 75% of the plant cover of the association (Watts, 1969). Along with the *Andropogon* and *Schizachyrium* spp. are other grasses such as *Panicum virgatum* (switchgrass), *Elymus canadensis* (Canada wild rye), *Koeleria cristata* (June grass), *Agropyron dasystachyum* (northern wheat-grass or blue joint grass), *Stipa spartea*, and occasionally *Sorghastrum nutans* (Indian grass). The dominancy of *Andropogon gerardii* over *Panicum virgatum* may well result from the ability of tall bluestem leaves to maintain high rates of carbon gain over a greater range of temperatures, and at lower leaf water potential, than does switchgrass (Knapp, 1985). Several dozen forb species are also frequently encountered, with *Aster praealtus* (willow aster), *Solidago canadensis* (tall goldenrod), *Anemone canadensis* (Canada anemone), *Rosa arkansana* (wild prairie rose), and *Lilium philadelphicum* (western red lily) being among the more common.

2 Moderately Drained Sites   On lower slopes where the potential for spring flooding is great and the water table is within 20–50 cm of the surface, soil atmospheric conditions deteriorate, reducing conditions are promoted, and the soils are Humic Gleysols. Such conditions favour a new combination of higher soil moisture – tolerant species where such grasses as *Panicum virgatum* and *Elymus canadensis* dominate. Effectively a transition zone between the better-drained *Andropogon* grasslands upslope and mesic depression bottoms, this community is joined by representatives from both.

3 Depressions or Sloughs   These poorly drained sites have Humic Gleysols often with peaty and/or saline phases, frequently have standing water late into the spring, and are dominated with hygrophytic grass, sedges, and rushes. Here, plant species variation depends on both moisture conditions and soil salinity, which is normally a function of whether the depression is a groundwater recharge (non-saline), a throughflow, or a groundwater discharge (saline) site. A cover dominated by *Spartina pectinata* (slough or cord-grass), *Eleocharis palustris* (spike rush), *Panicum virgatum, Calamagrostis inexpansa* (northern reed grass), *Juncus balticus* (Baltic rush), and *Solidago canadensis* is typical in non-saline conditions, while saline depressions are more frequently dominated by such species as *Spartina gracilis* (alkali cord-grass), *Scirpus paludosus* (prairie bulrush), *Distichlis stricta* (alkali grass), and *Puccinellia nuttalliana* (Nuttall's alkali grass).

4 Well-Drained Uplands   Here along the eastern margins of the Manitoba Escarpment, soils are well-drained, coarser-textured Black Chernozems, which reflect more closely climatic conditions and exhibit moisture stress in the latter part of the growing season. Dominant grasses include *Stipa comata* (needle-and-thread or speargrass), *Schizachyrium scoparium, Koeleria cristata, Bouteloua curtipendula* (side oats grama), and *Agropyron smithii* (western wheat-grass). Common forbs in-

clude *Solidago missouriensis* (low goldenrod), *Psoralea agrophylla* (silverleaf psoralea), *Amorpha canescens* (lead plant), and *Rosa arkansana* (Watts, 1969).

5 Riverine Gallery Forest   A distinctive feature of both the aspen parkland and
tall- and mixed-grass prairies is the riverine forests of broadleaf deciduous tree species (Looman, 1987b). Out in the tall- and mixed-grass prairie landscapes these sin-
uous forests take on true gallery forest form and are all the more obvious from a
distance because they have been less altered by human activity. One can drive across
the flat lowlands of the Lake Agassiz basin in Manitoba towards what appears to be
an abrupt forest margin, only to find that one passes through perhaps only
100–200 m of trees before once again being out in an open, almost featureless
agriculture-prairie landscape. Dominant broadleaf deciduous species include *Ulmus
americana* (American white elm), *Tilia americana* (basswood), *Fraxinus pennsylvanica* (green ash), *Acer negundo* (Manitoba maple), and *Salix* spp. (willows).
Occasionally, as in the Assiniboine Woods in Winnipeg, almost pure *Tilia* stands
can be found on Assiniboine River terraces. Some of these multi-stemmed basswood
have resprouted a number of times in response to soil aggradation on the floodplain
and may be many hundreds of years old. *Populus deltoides* (cottonwood) can also
be encountered in these gallery forests, with some individuals having diameters
(dbh) exceeding 1.25 m.

Southwestern Ontario   The presence of 1,200 km^2 of tall-grass prairie and oak-
parkland areas on sandplains in southwestern Ontario would seem quite anomalous
in light of what at first glance might appear to be intervening boreal and mixed-forest
zones between here and Manitoba. Their presence, however, is easily explained as
a northeastern extension of the Prairie Peninsula in the United States, encouraged by
the young, unstable, and droughty sandy soils of the region (see section 5.4.2.1).
Even here there are outliers as far northeast as the Rice Lake plains located just north
of central Lake Ontario. Major species here are *Andropogon gerardii, Schizachyrium scoparium, Panicum virgatum, Sorghastrum nutans,* and *Spartina pectinata*, and species composition, particularly on wetter sites and in oak–white pine
savanna, is quite similar to Wisconsin prairie to the southwest (Faber-Langendoen, 1984; Reznicek and Maycock, 1983).

   As in Manitoba, moisture gradients appear to be the major determinant in community composition variations. Here, in southwestern Ontario, *Andropogon gerardii, Schizachyrium scoparium,* and *Sorghastrum nutans* (Indian grass) normally
co-dominate on drier sites, while *Spartina pectinata* and *Calamagrostis canadensis*
dominate on wetter soils. Between the two, *Panicum virgatum* is narrowly predominant on mesic sites (Faber-Langendoen, 1984). Parkland (often locally called "oak
savanna") is also present here and includes *Quercus velutina* (black oak) on dry-to-
mesic sites, *Q. alba* (white oak) on dry-mesic sites, and *Q. macrocarpa, Q. bicolor*
(swamp white oak), and *Q. palustris* (pin-oak) on wet-mesic sites (Bakowsky,
1988). In the red oak savanna area of the Wasaga Beach Provincial Park, scattered
oak promote the nucleation of white and red pine seedlings within its shade (Kading,

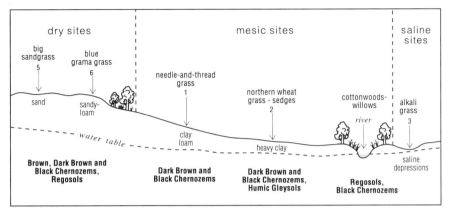

Figure 5.11
Community diversity in typical mixed-grass prairie

1989). Physiological drought promoted by low moisture retaining and well-drained regolith, together with natural fires, may well account for the absence of more mesic tree species in these oak-savanna regions (Szeicz and MacDonald, 1991). In addition to edaphic factors and natural fires, fires caused by Indians helped maintain the 172 km^2 of savanna and prairie in the Rice Lake Plains (Catling et al., 1992). In recognition of the importance of fire, it is used today as a management tool in such prairies as Ojibway Prairie near Windsor to help maintain herbaceous cover in the face of woody species encroachment (Eagles, 1993).

5.4.1.3 Mixed-Grass Prairie

Often called the mesic mixed-grass prairie to differentiate it from the drier grasslands (the short-grass or xeric mixed-grass prairie), this region is characterized best by *Stipa* and *Agropyron* grass species. Küchler (1964) uses both these genera to identify the region on his "Natural Vegetation" map of the USA and southern Canada. Climatically, the mixed-grass prairie region has moderately negative moisture indices. Cardston, located in the wetter Subhumid Grassland Ecoclimatic Region of southwestern Alberta, has an mean annual precipitation of 550 mm, but its mean annual temperature is a relatively high 4.8°C, resulting in moisture stress sufficient to limit tree growth except where soil moisture is enhanced by groundwater supplies, such as in gullies and along valley bottoms (Figure 5.2B). The majority of the mixed-grass prairie is found in the more humid parts of the Arid Grassland Ecoclimatic Region. While mean annual precipitation is somewhat less than for Cardston, lower mean annual temperatures promote significantly lower moisture stress. Soils typical of the region are the Dark-Brown and Black Chernozems, although sizeable areas of Solonetzic soil are also found. Six important plant community types can be readily identified within the region (Figure 5.11).

1 Better-Drained Mesic Sites   Perhaps typical of 50% or so of the region, this community consists of *Stipa comata* (needle-and-thread), *Agropyron smithii*, *A. trachycaulum* (slender wheat-grass), and *Festuca campestris (F. scabrella)* on moderately well to well-drained Dark-Brown, and occasionally Brown, Chernozems. In Alberta's submontane mixed grasslands, such as around Cardston, *Festuca campestris* is joined by *F. idahoensis* (Idaho fescue), *Danthonia parryii* (Parry's oat-grass), and *D. intermedia* (wild oat-grass) (Watts, 1969). Because of the dominance of *Festuca* spp., some authorities make a distinction between the fescue-dominated grasslands of the wetter prairie areas of Alberta and west-central Saskatchewan and the remaining areas of mixed-grass prairie, calling them fescue mixed-grass prairie. Unfortunately, this wetter climate, with associated Black and Dark-Brown Chernozems, is ideal for agriculture, so that today less than 5% of the original 255,00 km^2 designated as fescue grassland remains (Canadian Wildlife Service, 1992).

2 Poorly Drained Mesic Sites   On fine-textured Dark-Brown Chernozems, such as on pro-glacial lake deposits, moisture values are frequently higher, and a community dominated by *Agropyron dasystachyum*, *A. smithii*, and *Koeleria cristata* (June grass) is found. Sedges such as *Carex atherodes* (hair sedge) and *C. rostrata* are often encountered as well.

3 Saline Depressions   Common in potholes and close to river courses, where groundwater nears or reaches the surface, this community is typically associated with *Spartina gracilis* and the other species found under similar conditions in the tall-grass prairie. Salinity problems are exacerbated in agricultural soils surrounding these areas due to the practice of summer fallow, which unfortunately permits local water tables to rise slowly over time. Non-saline depressions are also dominated by species typical of moist depressions in tall-grass prairie (for more detail on these wetlands, see chapter 8).

4 Riverine Gallery Forest   Lower relative humidity levels in the mixed-prairie environment encourage *Populus deltoides* (cottonwood), *Salix interior* (sandbar willow), and *S. amygdaloides* (peach-leaved willow) to dominate the river margins and upper-bank sand bars. On lower and mid-level river terraces, other broadleaf species such as *Fraxinus pennsylvanica, Acer negundo,* and *Ulmus americana* typically dominate, while on upper terraces, *Quercus macrocarpa* is also common.

5 and 6 Xeric Sites   These sites are dominated by species typical of the short-grass prairie and are found here only where drainage is excessive and moisture-retaining properties are low. *Bouteloua gracilis* (blue grama) and *Stipa comata* dominate on sandy-loam soils, while sandy sites are dominated by *Calamovilfa longifolia* (big sandgrass). The sedges *Carex pensylvanica* and *C. eleocharis* and shrubs such as *Symphoricarpos occidentalis* (wolfberry) are frequently encountered, and in very xeric sites the diminutive cacti *Opuntia fragilis* and *Coryphantha vivipara* (pin cushion cactus) are common. Wilson and Shay (1990) report that *Bouteloua gracilis, Stipa spartea, Festuca ovina* (sheep's-fescue), and *Carex obtusata* are among the dominants in mixed-grass stands on loamy sands in the Shilo region of Manitoba. For a map showing non-cultivated sandhill prairies, see Canadian Wildlife Service (1992).

5.4.1.4 Short-Grass (Xeric Mixed-Grass) Prairie

Associated with the drier side of the Arid Grassland Ecoclimatic Region, the short-grass (or xeric mixed-grass) prairie has mean annual precipitation values generally well below 350 mm per year, mean annual temperatures are around 3°C, and moisture indices are strongly negative. In terms of plant growth, moisture availability is high only in late spring and then drops off sharply in July, to remain very low in the summer (Looman, 1983a). Figure 5.2A illustrates climatic data for Kindersley, Saskatchewan, which is located in the northern short-grass prairie just south of the boundary with mixed grassland and has an im of −51. Soils are Brown Chernozems and Solonetzic. A much greater percentage of the region is under "natural" cover compared to moister prairie regions. Agriculture is only possible under irrigation or with summer fallow rotation, but large areas have been used for cattle range. The net effect of this disturbance is that overgrazing has often deflected succession to shorter grass varieties, *Artemisia* spp. (sage brush), or cacti. It is estimated that there were originally 24 million ha of short (xeric-mixed) and mixed-grass prairie in Canada, but today only 24% remains, half of which is overgrazed (Canadian Wildlife Service, 1992). As one drives along the Trans-Canada Highway in southwestern Saskatchewan in a moist spring, it is not uncommon to see large overgrazed areas bright with the yellowpink flowers of *Opuntia polyacantha* (prickly pear) intermingled with *Artemisia frigida* (prairie sagewort) and short grasses. The scarcity of mid-sized species such as *Stipa spartea* and *Agropyron smithii* due to overgrazing is the reason many authors refer to this community type as xeric mixed-grass prairie. To see less disturbed examples of this prairie type, one must drive south of the Trans-Canada Highway to the Val Marie Prairie along the Frenchman River in Grasslands National Park. As shown in Figure 5.12, four important community types are typical of the short-grass.

1 Dry Sites   Bouteloua gracilis (blue grama) covers large areas where soils are either sandy or experience more or less continuously low soil-moisture conditions or where overgrazing has occurred, and it has been estimated that blue grama covers more than 35% of short-grass prairie in Alberta. *Bouteloua* is often joined by *Stipa comata* (needle-and-thread) and *S. spartea* (porcupine grass), as well as *Carex* spp. In sheltered depressions, small patches of tree-shrub-grass communities dominated by *Populus tremuloides, Cornus stolonifera* (red osier), *Amelanchier alnifolia* (saskatoon), *Salix* spp., and *Artemisia frigida* are encountered. Where soils are

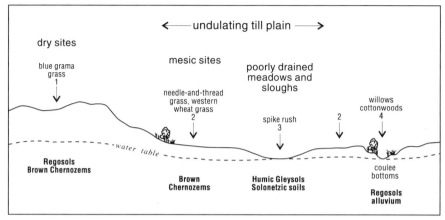

Figure 5.12
Community diversity in a typical short-grass (xeric mixed-grass) prairie

Solonetzic, such as in the area south of the Cypress Hills, *Stipa* does not seem suited to the relatively impermeable subsoil. If the Solonetzic soils lack a well-developed Ah horizon, *Agropyron smithii* and *A. dasystachyum* are found almost exclusively, while *Bouteloua gracilis* dominates together with these *Agropyron* spp. where the Ah is well developed (Coupland, 1961). While the dryer uplands and buttes of Grasslands National Park are dominated by *Bouteloua gracilis,* it is important to stress that lichens (and even some mosses) comprise much of the ground cover between the open herbaceous cover and that in wetter years, or on moister sites, *Stipa comata* and *Agropyron smithii* become common.

2 Moister Sites   *Stipa comata* dominates on deeper, moister Brown Chernozems, along with *Bouteloua gracilis, Agropyron smithii,* and sedges such as *Carex filifolia.*

3 Meadows and Sloughs   Spring snowmelt frequently fills minor depressions or potholes. Again, plant species variation depends on soil salinity, which is normally a function of whether the depression is a recharge or discharge type. Species are similar to those found under similar conditions in the more humid prairies.

4 Riverine Communities   Along seasonal stream beds (coulees) and rivers, such as the Frenchman River in the west block of Grasslands National Park, there is sufficient moisture to allow for the growth of woody vegetation. Although there are small local stands of *Acer negundo* and *Populus deltoides*, floodplains are more commonly dominated by shrubs such as *Artemisia cana* (hoary sagebrush) and *Shepherdia argentea* (buffaloberry) (Looman, 1984). As along the Frenchman River, saline seeps sometimes develop where valley sides meet the valley floor, giving rise to saltbrush-greasewood shrub communities.

5.4.2 Prairie in the USA

Seven important grassland types can be identified in the USA: tall-grass prairie, mixed-grass prairie, short-grass prairie, desert grasslands, intermontane (Palouse) grasslands, California grasslands, and eastern grasslands (Figure 5.1). Only the first three of these are part of the prairie environment (Figure 5.13). While the northern USA prairies are quite similar to those in Canada, the great latitudinal extension of USA prairie south to Texas introduces considerable climatic variety, and consequently two floristic gradients. One gradient is west-east, reflecting different soil moisture stresses, and the other is north-south, reflecting other seasonal climatic variables such as growing-season length and mean annual temperature. In addition, pre–European settlement burning due to lightning or Indian fires has also greatly influenced the prairie grasslands (Collins 1992a; Abrams, 1992; Pyne, 1983; Wells, 1970). One estimate gives pre-European burning frequencies of every 5–10 years in level regions and every 20–30 years where topography was dissected by breaks or rivers (Wright and Bailey, 1982). Axelrod (1985) estimated a burning frequency of once every 1–10 years was needed to maintain tall-grass prairie from woody encroachment.

5.4.2.1 Tall-Grass (True) Prairie

Historically, the tall-grass prairie was found in the Midwest – in parts of Iowa and Minnesota and in the eastern edges of North Dakota, South Dakota, Nebraska, and Kansas (Figure 5.13). As with the tall-grass prairie in Manitoba, however, little of this grassland type remains today. Excellent Black Chernozemic soils (Mollisols) suited to the cultivation of corn ensured that most of this region was quickly converted to agricultural land before 1900. Remnants can be found along some railway rights of way and in larger patches such as the Konza Prairie in the Flint Hills of eastern Kansas (Reichman, 1987). The boundaries of this grassland type are also problematic. In the eastern portion of its distribution, a large projection of grassland is found extending east into Iowa, Illinois, Missouri, parts of Indiana, and Ohio (Figure 5.13). Nicknamed the "Prairie Peninsula," these are really islands of grassland surrounded by broadleaf deciduous forest. An 8 ha patch of tall-grass prairie is even found on Hampstead Plains, Long Island, New York, and may well represent the last true prairie remnant east of the Appalachians (Stalter et al., 1991). The competitive abilities of grass, fires started by lightning and/or humans, previous large-scale grazing by bison, and recurrent periods of drought seem to have combined to extend the grassland-forest margin (ZE VII–VI) eastward into regions with positive moisture indices. Prairies east of the Missouri River should be considered subclimax, with the potential climax being *Quercus–Corylus americana* (oak-American hazel) forest (Axelrod, 1985; McComb and Loomis, 1944). The western boundary is a broad, more regular transition zone running almost north-south through the eastern Dakotas and Nebraska and central Kansas into central Oklahoma. Only in north-

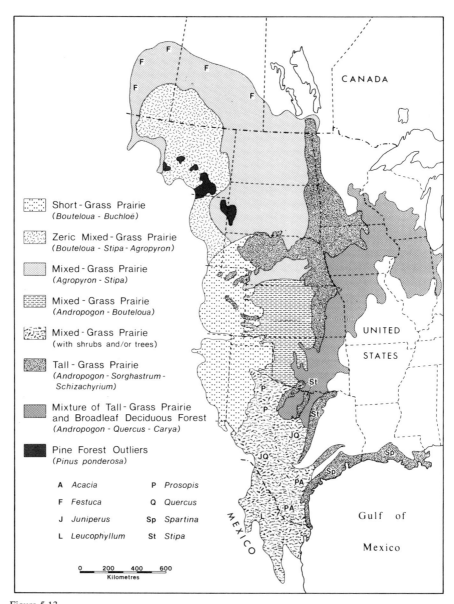

Short - Grass Prairie
(*Bouteloua - Buchloë*)

Zeric Mixed - Grass Prairie
(*Bouteloua - Stipa - Agropyron*)

Mixed - Grass Prairie
(*Agropyron - Stipa*)

Mixed - Grass Prairie
(*Andropogon - Bouteloua*)

Mixed - Grass Prairie
(with shrubs and/or trees)

Tall - Grass Prairie
(*Andropogon - Sorghastrum - Schizachyrium*)

Mixture of Tall - Grass Prairie
and Broadleaf Deciduous Forest
(*Andropogon - Quercus - Carya*)

Pine Forest Outliers
(*Pinus ponderosa*)

A	*Acacia*	P	*Prosopis*
F	*Festuca*	Q	*Quercus*
J	*Juniperus*	Sp	*Spartina*
L	*Leucophyllum*	St	*Stipa*

0   200   400   600
Kilometres

Figure 5.13
The prairie grasslands of central North America (modified from a number of sources, particularly Küchler, 1964, Risser et al., 1981, Wright and Bailey, 1982, and Diamond and Smeins, 1988).

ern Nebraska is there a tall-grass extension to the west (Figure 5.13). Even this transition zone has shown major fluctuations in location in recorded history, with documented temporary eastern retractions of up to 250 km during the drought of 1933–39 (Weaver, 1968).

Mean annual precipitation figures for tall-grass prairie can range widely. Fargo, North Dakota, which has a moisture index of −10, has a mean annual precipitation of 478 mm (Figure 5.3A), while Lincoln, Nebraska, which is 680 km due south of Fargo, has 704 mm (Figure 5.3B). In Lincoln, 80% of the precipitation falls in the growing season, yet a mean annual temperature of 10.8°C and wide annual fluctuations in precipitation ensure sufficient moisture stress to discourage easy invasion by arboreal species. The degree by which moisture indices are negative is normally the critical determinant in the regional distribution of all of the prairie grassland types, although the influence of occasional burning can not be ignored at the community level (Ewing and Engle, 1988). Only in the Prairie Peninsula do indices rise above zero (e.g., Des Moines, Iowa, with an IM of +9). Abilene, Texas, has a mean annual precipitation of 622 mm, much higher than for Fargo, yet because of the moisture stress created by an average annual temperature of 18.1°C, Abilene is in the mixed-grass prairie zone (Figure 5.3C).

When ratios of prairie growing-season precipitation to growing-season potential evapotranspiration are plotted on a map of the prairie states, they show an almost perfect longitudinal pattern, indicating that moisture stress increases due west (Risser et al., 1981). Only in the Prairie Peninsula do these ratio isolines take a bend to the east. While burning may also have played some part in the development of tall-grass prairie, particularly in the east (Pyne, 1983), it certainly encourages increased primary production and flowering due primarily to changes in surface light, soil surface temperatures, and nitrogen availability (Hulbert, 1989). In addition, frequency of disturbance by animals (Vinton et al., 1993; Gibson, 1989) or through burning influences tall-grass community heterogeneity (Collins, 1992b), although no simple relationship between burning frequency and species richness exists (Collins and Gibson, 1992). In moister sites in the Prairie Peninsula, such as in Wisconsin, trees of the *Populus, Salix, Fraxinus, Ulmus,* and *Quercus* genera invade quickly in the absence of fire (Curtis, 1971), and a program of cutting woody species and prescribed burning is recommended for preserving prairie in southeastern Ohio (Hardin, 1988). Prescribed burning also helps to ensure that the existing prairie is less prone to takeover by non-prairie shrubs and forbs (Tester, 1989) and is important in determining the frequency of certain annual species (Gibson, 1988). Tree invasion on the moister side of the tall-grass prairie is also primarily a function of burning regime (Briggs and Gibson, 1992). As in Canada, primary herbaceous vegetation patterns vary considerably in response to variations in soil moisture gradients, which are best exemplified along topographic gradients (White and Glenn-Lewin, 1984). At the local scale, however, care must be taken to consider other variables such as latitude, and soil properties such as pH, calcium, and phosphorus (Umbanhower, 1992). While Mollisols are the dominant soil order in tall-grass prairie, the influence of latitude and drainage creates variations at the suborder level. Aquoll and Boroll suborders are common in the north, while Udolls and Ustolls are more frequently encountered soils types in the south.

Dominant spring-flowering grassland species include *Koeleria cristata* (June grass) and *Stipa comata* (needle-and-thread), while the later flowering species *Andropogon gerardii* is the summer dominant in the northern true prairie. In places, big bluestem is found growing to a height of 3.5 m, but 1.25–2.5 m is more typical, and it can comprise 80% of the cover on moist, broad valley floors and lower slopes. *Panicum virgatum* (switchgrass) is another tall species often associated with big bluestem when substrate is sand. While these species are also typical of Canadian true prairie, an important difference is seen with *Sorghastrum nutans* (Indian grass), a grass found sparingly in southern Canada and the northern United States, but which can become the dominant, outcompeting *A. gerardii* on similar sites in the warmer south. Another species which increases towards the south is *Schizachyrium scoparium* (little bluestem; also called *Andropogon scoparius*), although this increase may not reflect climate but rather the inability of little bluestem to compete favourably with other more effective recolonizers in the northern prairies after the great drought of the 1930s (Diamond and Smeins, 1988). As a generalization, the tall-grass prairie can therefore be described as an *Andropogon-Panicum-Sorghastrum* community (Küchler, 1964). Other generalizations are that species with the c_3 versus c_4 pathway for CO_2 fixation increase in numbers towards the north (Diamond and Smeins, 1988), while frequent burning seems to favour the c_4 species (such as *Andropogon* and *Sorghastrum*) over the c_3 by improving summer microclimate (Hill and Platt, 1975). The general influence of mycorrhizal symbiosis can also be seen aiding warm-season grasses such as *Andropogon gerardii*, *Panicum virgatum* (switchgrass), and *Calamovilfa longifolia* (big sandgrass), but not early flowering species such as *Koeleria pyranidata* (Brejda et al., 1993; Hetrick et al., 1989).

In the northern states, tall-grass prairie is often differentiated into high, mid-, and low prairie, reflecting the moisture gradient from excessively drained high ground to poorly drained low (Godfread and Barker, 1975). Excessively drained high prairie is dominated by *Stipa comata*, *Calamovilfa longifolia*, *Koeleria cristata,* and *Bouteloua curtipendula* (sideoats grama), while mid-prairie dominants include *Andropogon gerardii*, *Schizachyrium scoparium (A. scoparius), Agropyron smithii* (western wheat-grass), *Sporobolus heterolepis* (prairie dropseed), *Stipa spartea* (porcupine grass), and *S. viridula* (green needlegrass) (Whitman and Wali, 1975). On moister, low prairie soils, *Andropogon gerardii, Calamagrostis inexpansa* (northern reedgrass), and *Spartina pectinata* (cord-grass) dominate. In saline seeps, *Spartina* and *Calamagrostis* are joined by such species as *Hordeum jubatum* (squirrel-tail grass), *Agropyron trachycaulum* (slender wheat-grass), and *Muhlenbergia asperifolia* (scratchgrass). With increasing salinity, *Distichlis stricta* (alkali grass) dominates, while *Salicornia rubra* (glasswort) becomes a monodominant in the most saline areas (Whitman and Wali, 1975).

To the south, in Kansas and through Oklahoma to the Blackland and coastal prairies of Texas, *Schizachyrium, Andropogon, Sorghastrum,* and *Panicum* generally dominate climax tall-grass prairie. In southern Oklahoma these species account for 80–93% of climax cover, but give way to shorter species in response to overgrazing (Sims, 1988). In the coastal prairie, *Schizachyrium, Sorghastrum,* and *Paspalum plicatulum* dominate remnant stands, but it is estimated less than 1% of this grassland remains in a relatively pristine state (Smeins et al., 1992). While *Bouteloua gracilis* and *B. curtipendula* become more abundant under heavy grazing, *Buchloë dactyloides* is considered the primary invader of these southern grasslands, often increasing from zero cover on excellent-condition pasture to 70% on poor-condition rangeland (Sims, 1988). For more detail on the influence of differing moisture gradients on tall-grass prairie species composition, see Kucera (1992).

Woody species are also common along rivers and frequently encountered in uplands. *Quercus macrocarpa* is particularly important as an upland species, while *Populus deltoides, Ulmus americana* (American elm), *Acer negundo* (box elder, or Manitoba maple, as it is called in Canada), *Fraxinus pennsylvanica* (green ash), and *Celtis* spp. (hackberries) dominate riverine communities in the Dakotas. It is not surprising that riverine forest near Fargo is similar to that in southern Manitoba since they are along the same Red River of the North. It is also not surprising that Dutch elm disease, which began to kill many *Ulmus americana* in the Fargo area in the 1960s, spread via rafting on the Red River to southern Manitoba in the 1970s. The shrub *Corylus americana* (American hazel) is frequently encountered along forest fringes, while *Symphoricarpos orbiculatus* (buckbrush) and *S. occidentalis* (wolfberry) commonly form low thickets out in protected prairie lowlands (Weaver, 1968). Farther to the south, in Nebraska, the lower moist slopes and ravines normally support *Tilia americana* (basswood, or linden) and *Quercus rubra* var. *borealis* (red oak), while *Q. velutina* (black oak) and *Carya ovata* (shellbark hickory) occupy intermediate sites and *Q. macrocarpa* the drier sites (Weaver, 1968). There is evidence to show that oaks (e.g., *Q. macrocarpa* and *Q. muehlenbergii*) were scarred by, but survived, repeated fire episodes in gallery forests within the Konza Prairie of northeastern Kansas (Abrams, 1985). Forest along the lower Missouri River can often reach 25 km in breadth and essentially constitutes an extension of the eastern Temperate Deciduous Forest (Zonobiome VI). So-called bluestem savannas are encountered in parts of east Texas, and *Andropogon-Spartina* grasslands border the Texas and Louisiana coastlines (Figure 5.13). In Texas some upland parklands that have been spared frequent burning have become forest that is now considered essentially non-flammable (Streng and Harcombe, 1982).

5.4.2.2 Mixed-Grass Prairie

Broader than the tall-grass zone, the mixed-grass prairie stretches south from Canada, through the Dakotas, Nebraska, and Kansas, into parts of Oklahoma and into north-central Texas. The region is also known by the names "Great Plains" and "mid-grass prairie." Climax cover is typically a blend of the species dominating tall-grass and short-grass prairie, while areas of shorter, xeric mixed-grass prairie are also found in Montana and Wyoming (Figure 5.13) and reflect overgrazing. Overgrazing, particularly in Montana, has led to grassland-rangeland climax classification based on soil and climate rather than on extant cover (Ross and Hunter, 1976). Moderately negative moisture indices characterize typical mixed-grass prairie climate, and summer moisture stresses, such as that indicated for Abilene in Figure 5.3C, are typical. *Schizachyrium scoparium* (little bluestem) dominates on moister lowlands throughout the region. This mid-grass species attains a maximum height of about 90 cm, and it is frequently found in association with other mid-grasses such as *Koeleria cristata, Stipa comata,* and *S. spartea.*

The xeric mixed-grass prairie so typical of Grasslands National Park in southern Saskatchewan extends well south into northeastern Montana. Here, in climax prairie, *Stipa comata* (needle-and-thread), *Agropyron smithii* (western wheat-grass), and *S. viridula* (green needlegrass) are particularly important, and they actively compete with *Schizachyrium.* Because of the open nature of the mixed-grass community (compared to the tall-grass prairie), shorter grass species such as *Buchloë dactyloides* (buffalo grass) and *Bouteloua gracilis* (blue grama), which are more typical of short-grass prairie, are also common here, and there is considerable intraseasonal variation among dominants (Singh et al., 1983). Where typical climax Montana prairie is subject to strong grazing pressure, *Bouteloua gracilis, Stipa comata,* and *Poa secunda* (Sandberg's bluegrass) increase in importance (Ross and Hunter, 1976). In somewhat more humid North Dakota mixed-grass prairie, the mid-grasses (*Agropyron* spp., *Koeleria, Stipa,* etc.), which constitute about 70% of the cover on ungrazed sites, reverse dominancy roles, with short grass species and sedges leaving a cover of only 16% mid-grasses (Whitman and Wali, 1975).

Experimentation with one or two years burning of mixed-prairie in west Texas showed that *Schizachyrium scoparium* increased its cover for a number of years, while species such as *Buchloë dactyloides* and *Bouteloua gracilis* showed little change, and *Bouteloua curtipendula* (sideoats grama) was reduced in importance (Wright, 1974). Experimentation with variations in fire season also showed considerable effects on the opportunistic forb species component originating within disturbed patches in mixed-grass prairie (Biondini et al., 1989). The general influences of salinity on species composition noted for Canadian tall-grass prairie potholes are similar here in the northern mixed-grass prairie. Typical zonation around semi-permanent ponds (saline groundwater discharge potholes) in North Dakota involve an outer wet meadow, a shallow marsh zone of *Scolochloa,* and a deep marsh zone of *Typha* and *Scirpus* (see chapter 8) (Richardson and Arndt, 1989). In Texas, woody species such as *Prosopis, Quercus, Acacia,* and *Juniperus* are often associated with mixed-grass prairie, although natural fires are considered to reduce significantly the importance of shrubs in southern mixed-grass prairie (Wright and Bailey, 1982).

Typical communities of broadleaf deciduous tree species are associated with rivers, while patches of shrubbery are found on sheltered slopes. Along the Missouri

River, northwest of Bismarck, North Dakota, the valley floor is dominated by *Populus deltoides* (cottonwood), *Salix* spp., and an admixture of *Acer negundo* and *Fraxinus pennsylvanica*. On the first river terrace is found a similar forest without the *Salix* spp., while on the second terrace, *Populus* declines and *Ulmus americana* joins the dominants. On the third terrace, just before opening out into mixed prairie, *Ulmus americana, Fraxinus pennsylvanica*, and *Quercus macrocarpa* dominate (Burgess et al., 1973). Gallery forest along this portion of the Missouri River can be up to 8 km across. Farther upstream, in Montana, cottonwoods generally constitute 80 per cent or more of the riverine tree cover, with randomly scattered *Ulmus* and *Fraxinus* climax stands throughout (Ross and Hunter, 1976).

5.4.2.3 Short-Grass Prairie

While the western margin of the short-grass prairie is abruptly defined along the eastern edge of the central Rocky Mountains, the eastern edge is a diffuse transition to the mixed-grass prairie (Figure 5.13). Short-grass prairie occupies southeastern Wyoming, western Nebraska, much of Colorado, western Kansas, the Oklahoma Panhandle, northwestern Texas, and much of eastern New Mexico, an area of approximately 280,000 km^2 (Lauenroth and Milchunas, 1992). Some authors also include the xeric mixed-grass prairies of Montana and eastern Wyoming in this region. It is generally agreed that short-grass communities are true climax communities (Sims, 1988), with *Bouteloua gracilis* (blue grama) and *Buchloë dactyloides* (buffalo grass) forming the two important dominants (Küchler, 1964).

Climatically, this region is much drier than in the mixed-grass community, and the landscape gives the impression of being much more open, as it is interrupted less by farmsteads or settlements. Soils are Mollisols with light brown colour and low organic matter contents. Research on the relationship between soil texture and average annual precipitation in the drier grasslands confirms the inverse texture hypothesis of Noy-Meir (1973). Sala et al. (1989) determined that when annual precipitation values are below 370 mm, sandy soils with low water-holding capacities have higher primary productivities than loamy soils with higher water-holding capacities, while the inverse is the case above 370 mm. Sandy soils perform well in relatively dry climates because they permit better moisture infiltration to depth and proportionately reduced surface evaporation, while in loams, moisture is held closer to the surface, with resultant increasing losses to evaporation. In general, all short-grass prairie moisture reserves are used up during the growing season, so strong negative moisture indices are the norm.

Warm season lower-slope cover is typically short grass dominated by the sod-forming *Bouteloua gracilis* and *Buchloë dactyloides*. Cool season mid-height grasses are typified by *Agropyron smithii, Stipa comata*, and *Koeleria cristata*, while forbs such as *Sphaeralcea coccinea* (cowboy's delight), *Oxytropis lambertii* (purple locoweed), and *Erysimum asperum* (western wallflower) are common. Heavy disturbance by mound-building pocket gophers (*Thomomys bottae*) seems to favour an increase in forb species and a decrease in grass abundance (Martinsen et al., 1990). *Opuntia polyacantha* (prickly pear, or starvation cactus) and *Aristida longiseta* are common, along with *Bouteloua gracilis*, on drier ridge tops (Schimel et al., 1985). During the drought of the 1930s, prickly pear spread so prolifically over the short-grass prairie that there was little room for animals to lie down between them. Fortunately, with the return of the rains came their natural insect predators, and populations declined rapidly (Brown, 1985). While many of the short grasses are also typical of the Canadian short-grass prairie, many of the forbs are not. Missing are some of the mid-sized grasses whose presence in the drier grasslands of southwestern Saskatchewan gave rise to the Canadian designation "xeric mixed-grass prairie." Woody species other than shrubs such as *Artemisia* spp. (sage) are not common except along rivers or in uplands, such as the Black Hills of South Dakota. In Texas, *Prosopis glandulosa* (honey mesquite) becomes an invader of highly disturbed grasslands which are low in grass productivity (Bush and Van Auken, 1989). *Populus* and *Salix* species dominate along waterways, while *Pinus ponderosa* (ponderosa pine) attempts to invade from the mountains to the west or from "forest islands" such as the Black Hills (Steinauer and Bragg, 1987).

5.5 Eurasian Steppe

The major Eurasian grassland is the "steppe," which stretches from Romania through the Ukraine to Siberia, with one major outlier in the Puszta of Hungary (Figure 5.1). As much as 1.5 million km^2 are found in the former Soviet Union alone (Sochova, 1979), while extensive grassland areas are also found in Mongolia and China. Smaller graminoid communities are found in the Paris basin, in western France on calcareous soils, and in dry or stony clearings in western European deciduous forests (Marcuzzi, 1979).

There are many similarities between Eurasian steppe and North American prairie. As previously noted, the Eurasian zonation is much more pronounced along a north-northwest to south-southeast gradient, so ascribing environmental constraints on this vegetation pattern is not difficult. Figure 5.14 summarizes many of the relationships between growing conditions and vegetation cover along such a transect through the Ukraine. As with the Canadian prairie (Figure 5.8A), critical values appear to be where potential evapotranspiration equals precipitation; these values coincide at the junction between the deciduous forest (to the north) and the forest-steppe (to the south). It should be noted that the uniformity of this gradient is enhanced by the level nature of the region and the relative uniformity of the loessic parent material. Table 5.1 summarizes the soil-vegetation correlations within this steppe while at the same time providing their North American equivalents. That the dominant turf-forming grasses in both the North American prairie and the Eurasian steppe are frequently the feathergrasses (*Stipa* spp.) and fescues (*Festuca* spp.) should not be considered surprising (Knystautas, 1987). They are, after all, both in the Holarctic floristic realm (Figure 1.6) and under almost identical "effective" climatic conditions. That these steppes have a long history of human impact through burning and grazing

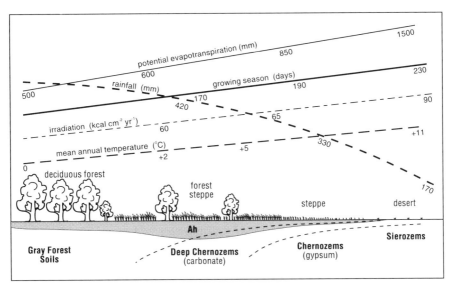

Figure 5.14
A summary of energy–moisture variables important in the differentiation of vegetation covers
and soil types in the east European lowlands (modified from Walter, 1979, with permission of the publisher, Springer-Verlag New York, Inc.). Transect is from the deciduous forest in the north-
northwest, to the Caspian desert in the south-southeast.

should also not be surprising. It has been documented that during the last 3,200
years, and particularly in the Scythian era, burning and grazing became the most
important factors in steppe evolution (Lisetskii, 1987).

   The soil and vegetation types listed in Table 5.1 would be typical of a north-south
transect in the Ukraine from Kiev to Odessa. In the Kiev region, deciduous forest
(dominated by *Carpinus* [hornbeam] and *Quercus* spp.) is separated from "damp
meadow-steppe" by the "oak-blackthorn" shrub zone, where dominants are *Quercus*
spp. and *Prunus spinosa* (blackthorn bush; the equivalent of the Canadian aspen-oak
parkland). The damp meadow-steppe and "typical meadow-steppe" have a seasonal
phenology beginning in early spring, with flowering herbs such as *Pulsatilla patens*
and sedges such as *Carex humilis* dominating (Walter, 1979). In late spring, colourful forbs such as *Ranunculus* (buttercups), *Senecio* (groundsel), and *Myosotis* species dominate, along with the maturing grasses. By early summer, grasses such as
Stipa spp. (e.g., *S. ioannis*) and *Bromus riparius* dominate, along with some flowering herbs, but by mid-July the grasslands begin to show the effects of moisture
stress, and by August they have dried out. As in Canada, long "gallery woods" penetrate south along river valleys through this typical meadow-steppe into the drier
feathergrass steppe (Keller, 1927). The feathergrass steppe is dominated by *Stipa*
spp., along with geophytes such as *Tulipa, Iris,* and *Gagea* species in spring and
Salvia, Ferula, and *Linosyris* species in summer. Spring openings within the herba-

Table 5.1
Soil and vegetation types of the Eurasian steppe (modified from Walter, 1979)

Soil Type	Vegetation Type	North American Vegetation Equivalent
Gray forest soil	Oak-hornbeam and oak forest	Mixed forest
Degraded Chernozem	Oak-blackthorn bush	Aspen-oak parkland
Northern Chernozem	Damp meadow-steppe	Tall-grass prairie
Thick Chernozem	Typical meadow-steppe	Mixed-grass prairie
Normal Chernozem	Feathergrass (*Stipa*) steppe with abundant herbs	Mixed-grass prairie
Southern Chernozem	Dry *Stipa* steppe	Short-grass prairie

ceous canopy often have a cover of moss (*Tortula ruralis*). Near Odessa a strong summer moisture deficit (Figure 5.3D) promotes "dry *Stipa* steppe," where forbs with long tap roots combine with the dominant grass species *Stipa capillata* and *Festuca sulcata* (Walter, 1979).

It is useful to compare other Eurasian grasslands with this classic Ukrainian steppe. To the west the Hungarian Puszta is, in many ways, reminiscent of the Ukrainian, but is separated from it by the Carpathians. Here the dominant grasses are *Stipa, Andropogon, Bromus,* and *Festuca.* It is considered that the herbaceous cover in the Puszta is a secondary one capable of supporting tree cover (Marcuzzi, 1979). To the east, closer to the Caspian Sea, *Artemisia larscheana* and *A. maritima* become more abundant in dry steppe (Marcuzzi, 1979) and indicate proximity to the sagebrush semi-desert that stretches east from the Caspian into Siberia.

In the forest-steppe zone in the Barabinskaya lowland in western Siberia, somewhat poorer drainage has resulted in a mosaic of steppe vegetation types on Chernozemic, Solonetzic, and saline (Solonchak) soils. Here, typical Chernozems have a forb-grass steppe dominated by *Poa angustifolia, Phleum phleoides, Festuca pseudovina, Koeleria gracilis, Carex praecox, Fragaria viridis,* and *Calamagrostis arundinacea.* On Solonetzic soils, *Festuca* and *Koeleria* are joined as dominants by *Agropyron repens* var. *glaucum* and *Carex supina.* Closer to fens, where soils become moderately saline, *Calamagrostis epigejos* and *Artemisia rupestris* are among the dominants, while in the most saline (Solonchakous) meadows, *Puccinellia distans* (alkali-grass) dominates (Titlyanova, 1982). In contrast to the wetter steppe of western Siberia, the climate of the Kurgiz steppe in eastern Siberia (east of Lake Baikal) is so continental that there is little spring moisture, and the typical spring flora associated with the wetter steppe is missing.

Large areas of steppe are also found in east Asia (Figure 7.5). In the dry steppes of Mongolia, dominants include *Agropyron cristatum* (crested wheat-grass), *Stipa krylovii* (Krylov's feathergrass), *Cleistogenes squarrosa* (cleistogenes), the sagebrush *Artemisia frigida*, and a number of *Caragana* (pea shrub) species (Borisova and Popova, 1990). Throughout northern China as a whole, the dominant grasses in dry steppes include *Koeleria, Stipa, Poa*, and *Festuca* spp. In northeastern China, in the Jilin region (Figure 5.1), the dominant grasses of the "forest steppe" are *Stipa grandis* and *Aneurolepidium chinense* (sheepgrass). The Inner Mongolian Autonomous Region of northern China has 870,000 km^2 of natural grasslands, which possess considerable variety (Ellis, 1992). In the moister "meadow grasslands" in the east, *Stipa baicalensis, Filofolium sibiricum*, and *Festuca ovina* dominate, while in mid-grass ("typical grassland"), *Stipa grandis, Aneurolopidium chinense*, and *Agropyron michnoi* dominate. In the "dry grassland" to the west of the Yinshan Mountains, *Stipa krylovii, S. bungeana*, and *Thymus serpyllum* dominate, while in the drier "desert grasslands" of the southwest of Inner Mongolia, *Stipa breviflora* is joined by woody shrubs such as *Artemisia frigida* and *Caragana* spp. (Ellis, 1992).

5.6 Southern Hemisphere Grasslands

5.6.1 The High Veldt

The High Veldt of southern Africa is located south of Pretoria, west of Durban, and north of Port Elizabeth, between 26° and 33°S latitude. Although these grasslands are frequently tall, the "high" in High Veldt refers to elevation above sea level and not to grass physiognomy. High Veldt grassland may well represent the climatic climax between 1,220 and 2,150 m, although natural fires are primarily responsible for the almost total absence of larger woody species (White, 1983). Here climate is not cold continental as in northern hemisphere prairie and steppe, but rather is warm temperate with summer rains and a dry, frosty winter (the cbw climate of Köppen). Annual precipitation varies from moderate at lower elevations to relatively high in the montane forest grasslands. The drier grasslands are dominated by tall grasses of the *Hyparrhenia* (elephant grass) and *Cymbopogon* genera, with forbs from the *Erigeron, Helichrysum*, and *Senecio* genera. Some dwarf shrubs are also present. If defoliation does not result regularly from grazing or fire, the community is invaded by some non-fire-tolerant trees. Frequently burning does occur, however, so the woody component is minimal. Another consequence of frequent burning is that *Themeda triandra* (rooigrass) becomes an important grass species (Tainton and Mentis, 1984). While fire may not be the sole determinant in the origin of the High Veldt, *Hyparrhenia* spp., *Themeda triandra,* and *Cymbopogon* spp. are perennials, with underground storage organs or rhizomes which help them survive both fire and drought. Clearly, fire is important at least in the maintenance of these grasslands (Preston-Whyte, 1985).

The grasslands of more montane areas are wetter, cooler, and consequently suited to successional change to scrub or forest (Tainton and Mentis, 1984). *Themeda triandra* is the dominant grass, while this is initially replaced by *Trachypogon spicatus* and later by shrubs such as *Athenasia acerosa* if burning becomes infrequent. If burning ceases, *Pteridium aquilinum* and other species invade. The presence of *Pteridium* as a successional fern leading to woody cover is indicative of a warmer climate than for typical northern hemisphere prairie or steppe. It is more akin to the successional sequences of the humid tropics and montane-tropics, where ferns of the *Pteridium*, *Gleichenia*, and *Dicranopteris* genera are important components in fire-free succession from herbaceous cover to forest (Scott, 1978).

5.6.2 The Pampas/Campos Grasslands

These tussock grasslands cover some 700,000 km^2 of southeastern South America in the lower Río de la Plata region. While the grasslands of west-central Argentina are known as the pampas, the campos are those in Uruguay and the southern tip of Brazil (Soriano, 1992). Lying between 30° and 38°S latitude, the region has a warm temperate climate much like that of Abilene, Texas, in the southern USA prairie (Figure 5.3C). Here the mean annual precipitation gradient is from approximately 1,000 mm in the northeast to 480 mm in the southwest. Despite the relatively high rainfall, the region has negative moisture indices throughout, approaching arid shrub-steppe-like conditions in the southwest in Patagonia.

Burning to promote grazing, the introduction and planting of exotic species more palatable to cattle, and land conversion for agriculture have all contributed to great changes within the pampas. In wetter areas, *Stipa* spp. and *Bothriochloa lagurioides* dominate, along with a large variety of indigenous grasses and herbs (Numata, 1979) growing on humus-rich Chernozems (see Figures 1.7 and 1.8). Where climate is somewhat drier, dense tussocks of *Paspalum quadrifarium* dominate. Towards the drier southwest, abutting the desert shrub–steppe of Patagonia, the original cover was dominated by *Stipa* tussock grassland (*S. trichotoma* and *S. brachychaeta*). This tussock habit, mentioned previously for South Georgia Island (section 2.6.2), is characteristic of many southern hemisphere grasslands, but almost totally lacking in the northern hemisphere (Walter, 1979). To the west, in the dry "rain-shadow" of the eastern Andes, the pampas grades into a *Prosopis calderina* (mesquite) desert shrub–steppe zone which extends all the way along the eastern Andean flanks from Bolivia to Tierra del Fuego. Dominant grasses in the Patagonian portion of this shrub steppe are *Stipa speciosa, S. humilis,* and *Poa lingularis* (Soriano et al., 1987). In turn, this steppe grades rapidly into *Nothofagus* and *Austrocedrus* (Chilean cedar) forest in the Andean foothills. The boundary between the foothills forest and the steppe is currently advancing eastward now that Indian burning to drive *guanaca* has virtually ceased (Veblen and Markgraf, 1988).

5.7 Primary Production and Biomass

Lacking any appreciable woody component, above-ground prairie phytomass (other than dead standing crop and litter) must regenerate annually and is therefore to a large extent at the mercy of available soil moisture, which in turns reflects prevailing

Table 5.2
Mean phytomass component values of typical prairie stands in the USA (from Sims et al., 1978).
Values in g m^{-2}.

	Above-ground Standing Crop (live and dead)		Litter		Live Only		Crown		Total Roots	
	G	UG	G	UG	G	UG	G	UG	G	UG
Tall-grass, Osage, Oklahoma	316	598	256	424	166	152	175	247	924	964
Mixed-grass, Cottonwood, N.Dakota	118	233	174	347	60	93	300	275	2492	1605
Short-grass, Pawnee, Colorado	127	135	128	177	62	70	315	299	1329	1149

G = grazed; UG = ungrazed. Subtract "live only" values from "above ground standing crop" to determine amount of dead standing crop. A value of 100 g m^{-2} is equivalent to 1 t ha^{-1}.

precipitation conditions. As can be expected, average above-ground phytomass values (total standing crop, both living and dead) increase from 1–2 t ha^{-1} in short-grass to 6–7 t ha^{-1} in tall-grass, with growing-season precipitation, growing-season actual evapotranspiration, and annual usable solar radiation accounting for about 80% of the variability (Sims et al., 1978). At any given place, year-to-year standing crop also varies considerably depending on soil moisture stresses. Harper (1957) found values ranged from 1.27 to 5.22 t ha^{-1} during a twenty-year study in the tall grass of Oklahoma. Walter (1979) reports similar values ranging from 4.53 to 6.25 t ha^{-1} for European feathergrass steppe during moist years and from 0.71 to 2.7 t ha^{-1} during dry years. Grazing significantly reduces all these biomass values, while biomass values are also highly coupled with burning cycles (Gibson, 1988). Generally, above- and below-ground productivity is higher on burned sites than un-burned (Dhillion and Anderson, 1993). Maximum root biomass values, on the other hand, are found in the mixed-grass prairie, with values dropping off towards both short- and tall-grass prairie (Table 5.2). Without exception, for all grassland types, the greatest reserves of organic matter are in the form of soil humus. Biomass com-partmentalization in grassland ecosystems normally declines in the order: (1) soil or-ganic matter, (2) rooting systems, (3) above-ground living phytomass (living standing crop), (4) above-ground dead phytomass (dead standing crop), (5) litter.

Of particular importance in illustrating ecosystem dynamics within prairie is the root-shoot ratio (where roots include crowns, and shoots refer to standing crop, both living and dead). In the cooler prairie regions, root-shoot ratios are generally at least 6:1, while in the warmer south the ratios are 2:1 or higher (Sims et al., 1978). In the

European meadow-steppe, some 84% of the phytomass is below ground, while in feathergrass steppe the equivalent value is 91% (Walter, 1979). The impact of grazing is to increase the ratio in cooler regions, while little change is experienced in warmer areas. Light-to-moderate grazing effects in eastern European, central Siberian, and Kazakhstan steppe actually greatly decreased living standing crop, while increasing dead phytomass at the same time (van der Maarel and Titlyanova, 1989). Standing crop in typical tall-grass prairie near Manhatten, Kansas, depends on whether or not the grassland was burned the previous fall. Here, burned prairie has typical above-ground plant phytomasses of 4–6.5 t m^{-2}, while unburned prairie has values ranging from 3.5 to 5.75 t m^{-2} (Seastedt, 1989). Litterfall was always greater on the unburned sites, however. Values typical of grazed and ungrazed short-, mixed-, and tall-grass prairie are summarized in Table 5.2. Comparable values of 1.4 and 1.2 t ha^{-1} were determined for above-ground phytomass and litter biomass respectively, for lightly grazed steppe in the Xilin River basin, northern China (Hayashi et al., 1988).

Soil organic reserves (not including litter) in North American Mollisols average 320 t ha^{-1} in undisturbed grassland, although this figure drops considerably under cultivation to 260 t ha^{-1} (Buringh, 1984). A value of 397.7 t ha^{-1} was determined for the top 60 cm of a Black Chernozem under tall-grass prairie in southern Manitoba, while a comparative figure of 304 t ha^{-1} was determined under a field converted from the same prairie some forty years earlier (Scott and Crane, 1991). Kononova (1966) gives an average value of 506 t ha^{-1} (25.3 kg m^{-2} of carbon) for litter/soil organic matter in European Chernozems. Schlesinger (1984) gives an average worldwide temperate grassland litter/soil organic-matter value of 378 t ha^{-1} (18.9 g m^{-2} of carbon). This huge Ah-horizon organic-matter storage is in part the result of breakdown and incorporation by soil macrofauna. Studies on *Diplocardia* spp. earthworms in the Konza Prairie show not only that they are the dominant macrinvertebrate biomass in the soil, but that they annually pass through their guts the equivalent of 10% of the organic matter in the top 15 cm of the Ah (James, 1991). In addition, their pedogenic role in incorporating organic matter into the A-horizon must be acknowledged.

Lieth (1975) reports an NPP range of 100–1,500 g m^{-2}yr^{-1} for temperate grasslands. This large range results from such variables as community type and the amount and distribution of annual rainfall. On the basis of examining fifty-two North American prairie sites, Lauenroth (1979) reports a strong correlation ($r^2 = 0.51$) between above-ground net primary production and mean annual precipitation. Sala et al. (1988) provide maps of above-ground net primary production for the central grasslands region of the United States under varying conditions of moisture availability. During years of average precipitation, NPP values range from just over 600 g m^{-2}yr^{-1} in tall-grass prairie to just under 150 g m^{-2}yr^{-1} in short-grass. When moisture is more limited, values drop approximately 15–20% in tall-grass prairie and 50% in short-grass. Equivalent values for favourable years show increases of 15% and 35% respectively. As with the effective moisture gradient (Risser et al., 1981), there is a definite longitudinal gradient associated with these NPP values.

Four-hundred-year-old stand of coastal mesothermal forest north of Vancouver. The three large tree in the foreground are, left to right, *Thuja plicata* (western red cedar), *Pseudotsuga menziesii* (Douglas-fir), and *Tsuga heterophylla* (western hemlock).

This windthrown hemlock is serving as a "nurse log" in an old-growth coastal mesothermal forest, southwestern British Columbia. Due to reduced competition, many conifer seedlings germinate and grow more successfully on these nurse logs than on the forest floor.

Unsalvaged blast-thrown Douglas-fir in the blast zone 25 km north-northeast of Mount St Helens. Photo taken in April 1987, seven years after destruction. Most of these 1 m plus diameter old-growth trees were simply uprooted by the blast, while deeply rooted individuals snapped like matchsticks.

A stand of *Sequoia sempervirens* (coastal redwoods) known as Cathedral Grove in Muir Woods National Monument north of San Francisco. A wedding is in progress under the shaft of light in the background.

This *Sequoiadendron giganteum* (giant sequoia) is in Sequoia National Park in the Sierra Nevada of California. Known as General Sherman, this tree is 11 m in diameter and 82.5 m tall, and contains 1,256 t of wood above ground level. It is probably the largest living individual organism in the world.

Subalpine forest–alpine meadow transition near Peyto Lake, Banff National Park. The layered branches (basal skirt) in the foreground are attached to the tall Engelmann spruce standing in their midst. These would be protected under deep snow during the winter. Fall scene with some snow patches visible.

Temperate deciduous forest along Highland Creek, Scarborough. Dark trunks are *Acer saccharum* and *Quercus rubra*, while the grey trunks in the centre are *Fagus grandifolia* and the whiter trunks to the left of centre are *Betula lutea*. Most of the forest-floor saplings are maple and beech.

The smooth, grey trunks of *Fagus grandifolia* in old-growth deciduous forest along Highland Creek, Scarborough. Note roots exposed by slope erosion. Young saplings to the left are maple, while those on the right are beech.

Temperate deciduous forest along the Niagara Escarpment above St Catharines in the Niagara Peninsula. Typical plateau and gentler-slope hardwoods include *Acer, Quercus, Fraxinus*, and *Betula*, while occasionally *Thuja occidentalis* (eastern white cedar) finds a foothold in crevices under the massive dolomite caprock.

Typical remnant hardwood forest along the Blackwater River, Caledon, Northern Ireland. These *Quercus*-dominated forests contain many introduced species from the mainland of Europe, such as *Acer pseudoplatanus* (sycamore), *Aesculus hippocastanum* (horse chestnut), and *Fagus sylvatica* (common beech). Photo by R. Ernest Scott.

Massive *Nothofagus* (southern hemisphere beech) forest on the eastern flanks of the Andes along the western shore of Lake Nahuel Huapi, Bariloche, Argentina (41°S). High humidities close to the lake encourage epiphytic lichen. Undergrowth is primarily *Chusquea quila* (bamboo), and the wagon is transporting firewood.

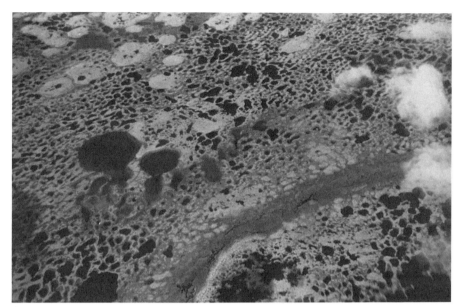

Wetlands in the Hudson Bay Lowlands, northern Manitoba. Typical thermokarst lake landscape. Note the peat polygon pattern to the right of the drainage system at the bottom of the photograph. Large light-coloured circular areas to the top left may be scars left by palaeo pingos.

Sedge fen in forest-tundra near the Twin Lakes, northern Manitoba. The low, oval mounds in the foreground and where the people are standing are massive ice features (hydrolaccoliths). Note the clean ice face below peat in the collapsed hydrolaccolith in the foreground. As the ice pushes upwards, the peat is stretched, becomes more xeric, and eventually fissures. Once fissures reach the ice core, the feature begins to melt quickly and collapses.

Sphagnum moss–*Ledum groenlandicum* (Labrador tea) floating bog growing out into a small lake in the coastal Douglas-fir forest zone just north of Vancouver. Trees in the background include western red cedar, Douglas-fir, and western hemlock.

Delta marsh at the southern end of Lake Manitoba. In the immediate foreground are the tops of *Phragmites* (reed). Behind these are several banks of *Typha* (cat-tail), between which is a narrow bank of *Scirpus* (bulrush). Ash, cottonwoods, and willows dominate the lake-margin sand dunes on the horizon.

Typical lake-margin marsh in Presqu'ile Park on the northern shore of Lake Ontario. *Typha* (cat-tail) dominates the foreground, along with some scattered *Salix* shrubs. Mixed forest on the horizon consists of *Pinus strobus, Picea glauca*, and *Acer saccharum*, with some *Quercus* spp. and *Tsuga canadensis* (eastern hemlock).

This "temporary" wetland has been produced by beaver dam flooding in the Gatineau Hills, Gatineau Park, Quebec. *Typha latifolia* and floating pondweeds dominate where only a few years earlier, *Picea glauca, Betula lutea*, and other mixed-forest tree species thrived.

Modified Atlantic lowland blanket bog, coastal County Donegal, Ireland. The *Picea sitchensis* (Sitka spruce) plantation on the left has a different ground cover to the sheep-grazed bog to the right. Taller graminoids just to the left of the fence are *Schoenus nigricans* (black bog-rush). Grazing area management includes ditching to encourage *Molinia caerulea* (purple moor grass) productivity.

Saccharum robustum and *Phragmites karka* cane grass swamp near Port Moresby, Papua New Guinea. Closely related to the domesticated sugar cane, these *Saccharum* "tassles" can reach heights of 10 m.

CHAPTER SIX

Cordilleran Environments in Western North America

Cordilleran, or mountain, systems present a major problem when vegetation is discussed on a continental or world scale, primarily because even a single mountain system presents a mosaic of vegetation types which may show little in the way of physiognomic similarity. When two mountain systems are under discussion, the differences of geographic location compound the problem. Take, for example, a transect across the Andean cordillera in central Peru or the Canadian cordillera. In Peru the transect may include desert, alpine tundra, temperate evergreen forest, cloud forest, savanna, and rain forest, none of which includes conifers; in Canada the transect includes massive hemlock–Douglas-fir forests, subalpine spruce-fir forests, alpine tundra, pine-fir forests, and pine parklands. All this variety is found within short horizontal distances, but distances which includes dramatic temperature and precipitation variations resulting from elevation, mountain barrier effects, and geographic location.

Walter (1979) takes the approach that all mountain biomes, or orobiomes, as he calls them, are unique and that they reflect to some extent characteristics of the zonobiome in which they are found. He discusses each orobiome along with its most closely related zonobiome(s) and by taking this approach, to some extent avoids the second problem, that of geographic location. "To some extent" is the operative expression because those same Andes, with one particular vegetation sequence in Peru, have a very different sequence between southern Chile and Argentina. While Walter's approach is adopted here, the problem of the local dramatic mosaic remains to remind us that if we discuss a mountain region's vegetation as a single unit in the space of only a few pages, we may need to sacrifice some of the detailed ecoclimatic descriptions to which the vegetation types in earlier chapters are so clearly suited. We would err, however, if we were therefore to conclude that mountain environments are any the less suited to ecoclimatic analysis.

The combined effects of mechanical and chemical weathering are such that even in the driest of climates, mountains are constantly being attacked by erosion and mass wasting. It is not surprising, therefore, to find tall mountain chains restricted to areas where the combined uplifting by recent orogenic folding, thrusting, and volcanic activity has, at least initially, exceeded the combined rates of erosion and mass wasting. As such regions are peripheral to continental Archean cores, it is again not surprising to find major mountain systems along the west coasts of both North and South America, in the European Alps, the backbone of Scandinavia, the Himalayas, and the spine of Indonesia and New Guinea. The following discussion will, however, confine itself only to the cordilleran system of western North America. From the point of view of physiognomic and species similarities, this can be viewed as a single region; however, for convenience it is divided into two sections, one on Canada's four cordilleran ecoclimatic provinces and the other on the cordilleran region in the western USA.

This western North American cordillera possesses unique and varying ecosystems, which have special fascination for anyone interested in vegetation. The degree to which Pacific coast conifers dominate over hardwoods is particularly impressive, a situation aided not only by a climate that favours conifers, but by their massiveness and sheer longevity, two features not shared by most hardwoods (Waring and Franklin, 1979). The rate at which these spruce, cedar, and hemlock coastal temperate rain forests are being levelled by current clear-cutting practices is also of concern. "Massiveness and sheer longevity" means that succession to classic old-growth forest, like those in the Carmanah Valley or Clayoquot Sound can take a thousand years or more. While managed clear-cut sites could be harvestable again in a few human generations, in ecosystem terms these temperate old-growth rain forests are essentially irreplaceable. In crown forests, "the B.C. rate of logging was nearly three times the annual growth rate" (FPC Research Staff, 1991, p. 46). For specific details on clear-cut areas and increases, see Forest Planning Canada (1991).

6.1 Canada's Cordilleran Ecoclimatic Provinces

6.1.1 Distribution

It is not easy to defend the argument that the cordilleran and Pacific coast vegetation types in Canada are sufficiently similar that they should automatically be grouped under one heading. It can be argued, however, that because of the repetition imposed by the mountain-valley systems over short distances, it is convenient to discuss the region as a unit. It can also be argued that a secondary unifying factor comes from the fact that forests within the region are primarily coniferous, with *Tsuga heterophylla* (western hemlock), *Pseudotsuga menziesii* (Douglas-fir), *Abies amabilis* (amabilis fir, or Pacific silver fir), and *Thuja plicata* (western red cedar) being widely distributed. In terms of terrestrial ecozones, the region can be subdivided into the Pacific Maritime and Montane Cordillera Ecozones (Figure 1.5). From an ecoclimatic standpoint, four basic ecoclimatic provinces are differentiated. The humid coastal forests of the Pacific Maritime Ecozone are included in the Pacific Cordilleran Ecoclimatic Province, while the Montane Cordillera Ecozone is differentiated into the Subarctic Cordilleran, Cordilleran, and Interior Cordilleran

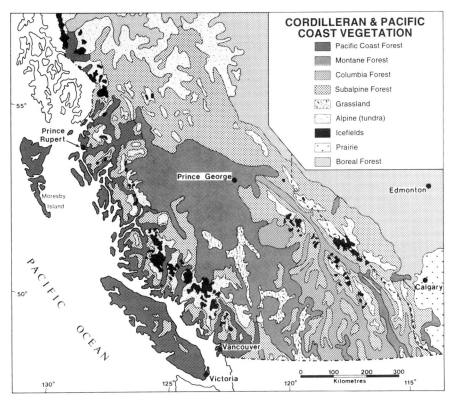

Figure 6.1
Major vegetation zones of the Pacific Cordilleran and Cordilleran bioclimatic regions in British
Columbia and Alberta (after Rowe, 1972)

Ecoclimatic Provinces (Figure 1.4). In turn, these provinces are differentiation into
seven sub-provinces and twenty-nine ccoclimatic regions on the basis of the effects
of latitude, altitude, and continentality on plant cover. Of interest is the recognition
that the Interior Cordilleran Province is stratified into numerous units which can be
recognized only on very large-scale maps or by descriptions. On Figure 1.4, ICV represents the vertically stratified Interior Cordilleran Ecoclimatic Map Unit, which
combines grassland (ICP), subhumid montane (ICM⁻), montane (ICM), subalpine
transition (ICN), subalpine (ICS), and alpine (ICA) communities. By taking this approach, the 1989 Ecoclimatic Classification System overcomes some of the problems peculiar to orobiomes discussed above.

On the basis of dominant species/life forms, vegetation cover can be generalized
into seven types: (1) Pacific coast forest, (2) montane forest, (3) Columbia forest,
(4) boreal forest, (5) grassland-parkland, (6) subalpine forest, and (7) alpine tundra
(Figure 6.1). Prairie grasslands are also found east of the Rocky Mountains

(Figure 6.1), but these are not part of the cordillera. Although there is boreal forest in the northern cordillera, boreal ecosystems (including aspen parkland) found east of the subalpine forests of Alberta are also outside the region. Walter (1979) likewise considers the region to include considerable diversity, having a mosaic of boreal forest, boreal-grassland ecotone communities, mixed forests, some Mediterranean forest elements, warm temperate coniferous forests (zB v), and orobiomes within each of these zonobiomes and zonoecotones (Figure 1.2, Table 1.1).

The complex mosaic of vegetation types in the region has been studied in detail and carefully mapped (Krajina, 1965; Rowe, 1972). While mapping units on Figure 6.1 follow the general forest types given in Rowe's 1972 map of Canadian forests, the discussion below uses the more detailed Biogeoclimatic Classification of British Columbia (Krajina, 1965) as the basis for vegetation type differentiation in both British Columbia and Alberta (Table 6.1). Some differences in terminology are seen when these two classification systems are compared (see Table 6.1 and Figure 6.1). The Canadian cordilleran forest region, listed in Table 6.1, is subdivided into Montane forest and Columbia forest on Figure 6.1. Montane forest includes Cariboo aspen–lodgepole pine–Douglas-fir to the north and interior Douglas-fir to the south, while the Columbia forest is generally synonymous with the interior western hemlock zone. The area designated as grassland on Figure 6.1 is synonymous with the cordilleran cold steppe and savanna forest region.

6.1.2 Climate

On the Pacific coast, the Northern Pacific Cordilleran Sub-Province is characterized by an extremely wet climate where precipitation in most months exceeds 100 mm, fog is frequent, and moisture indices strongly positive. The climate of Prince Rupert, with its mean annual temperature of 6.7°C and mean annual precipitation of 2,523 mm, is more typical of the northern oceanic portion of the Southern Pacific Cordilleran Ecoclimatic Sub-Province (Figure 6.2A). This moisture regime results from both the orographic effects imposed by the Queen Charlotte and Kitimat ranges and a year-round possibility of frontal activity. Snowfall can be as much as 750 cm annually and represent up to 38% of the annual precipitation (Krajina, 1969). To the south, Vancouver Island retains a humid maritime climate, although the slight summer precipitation "minimum" noted for Prince Rupert becomes quite pronounced due to a lower frequency of summer frontal episodes. Away from the extreme oceanic influence, such as around Vancouver (Figure 6.2B) and the Strait of Georgia, the climate is better described as coastal. Here precipitation is less than half that at Prince Rupert, and a pronounced summer drought permits a Mediterranean element to be identified in the flora. With decreasing latitude the probability of summer frontal activity diminishes so that at Portland, Oregon, summer drought exceeds two months. Still farther south, in California, frontal precipitation occurs only in winter, so at Yosemite National Park, California, summer drought lasts nearly four months

Table 6.1

Biogeoclimatic regions of British Columbia and Alberta (after Krajina, 1965, 1969). Ecoclimates after Ecoregions Working Group (1989).

Region	Zone	Köppen climate	Ecoclimatic sub-provinces (Figure 1.4)
Pacific coastal mesothermal forest	Coastal western hemlock	cfb	South Pacific (SP) and North Pacific Cordilleran (NP)
	Coastal Douglas-fir	csb	SP
Pacific coast subalpine forest	Mountain hemlock	Dfc	SP and NP
Canadian cordilleran forest	Interior western hemlock	Dfb	Southern Cordilleran (SC)
	Interior Douglas-fir	Dfb	Interior Cordilleran (IC)
	Cariboo aspen– Lodgepole pine– Douglas-fir	Dfb	Boreal Interior Cordilleran (ICb)
Cordilleran cold steppe and savanna forest	Ponderosa pine–bunchgrass	Dsk	Interior Cordilleran, Vertically Stratified (ICV)
Canadian cordilleran subalpine forest	Engelmann spruce– subalpine fir	Dfc	NCS, MCS, SCS, ICS
Canadian boreal forest	Sub-boreal spruce	Dfc	Boreal Southern Cordilleran (SCb)
	Boreal white and black spruce	Dfc	Boreal Mid- (MCb) Boreal Northern Cordilleran (NCb)
Alpine tundra	Alpine	ET	NCa, MCa, SCa, ICa, NPa, and SPa.

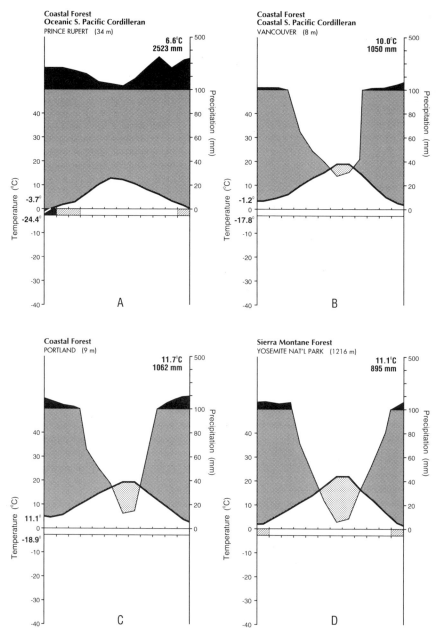

Figure 6.2
Pacific Cordilleran climatic data. A = Prince Rupert, BC (54°19'N); B = Vancouver, BC (49°16'N); C = Portland, Oregon (45°31'N); D =Yosemite National Park, California (37°40'N).

(Figure 6.2D), and at lower elevations is found a true Mediterranean climate, with its sclerophyllous vegetation.

The Cordilleran Ecoclimatic Province is sufficiently continental that oceanic influences are reduced, relative humidities are likewise reduced, and average annual precipitation values are lower (Figure 6.3). Latitudinal influences on temperature are such that this province can be subdivided into the Southern Cordilleran, Mid-Cordilleran, and Northern Cordilleran Sub-Provinces. To the north of this is found the Northern Subarctic Ecoclimatic Sub-Province. Relief also effects temperatures to the point that at higher elevations, forest is replaced by alpine tundra and even icefields. Climatic data for a typical Southern Cordilleran elevational sequence from lower subalpine forest, upslope through alpine tundra, to icefield in the Banff–Lake Louise area of Alberta are illustrated in Figure 6.3A–C. This sequence shows the typical pattern of decreasing temperature and increasing precipitation associated with increasing elevation. Orographic influences also produce pronounced rain-shadow effects in some valleys. Here moisture indices are negative, and grassland and pine parklands develop. Figure 6.3D illustrates data for the Subhumid Montane Interior Cordilleran Ecoclimatic Region (ICM⁻) within the Vertically Stratified Interior Cordilleran Ecoclimatic map unit, where a ponderosa pine–bunch grass plant community thrives in the rain-shadow along the Thompson River valley at Kamloops.

6.1.3 Soils

As can be expected in a region with such climatic diversity, the soils also show considerable variety (Figure 1.9). Along the west coast in the Pacific Cordilleran Ecoclimatic Province, freely drained soils at lower elevations are typically Humic Podzols (Spodosols), with thick mor litter horizons capping relatively infertile, but deep, strongly acid profiles. Where drainage is impeded, such as along the floodplain of the Fraser River, more fertile Gleysolic soils are found. Organic soils are also frequently encountered and include exceptionally thick Folisols, which not only overly bedrock, but in some cases overly Podzols or earlier peat deposits (Fox et al., 1987). At somewhat higher elevations in the coastal western hemlock zone, Dystric Brunisols are common, while in the subalpine mountain hemlock zone, soils are typically Regosols, Gleysols, Podzols, and Dystric Brunisols (Brooke et al., 1969).

In the Interior Cordilleran Province and Southern Cordilleran Sub-Province, soil profiles tend to reflect the rugged relief, the carbonate-rich parent materials, and reduced precipitation. Rock fields and Regosols are typical at higher elevations and on steep slopes, while alpine tundra soils are generally Gleysolic or Brunisolic. Where soil profile development has progressed, Luvisols and Brunisols dominate, while small areas of Chernozems are associated with the rain-shadow areas, such as along the Thompson River and Okanagan valleys. In the Mid- and Northern Cordilleran Sub-Provinces, soils are primarily Brunisols which exhibit limited profile development, Regosols, and rock fields. Organic soils are more common here and reflect the

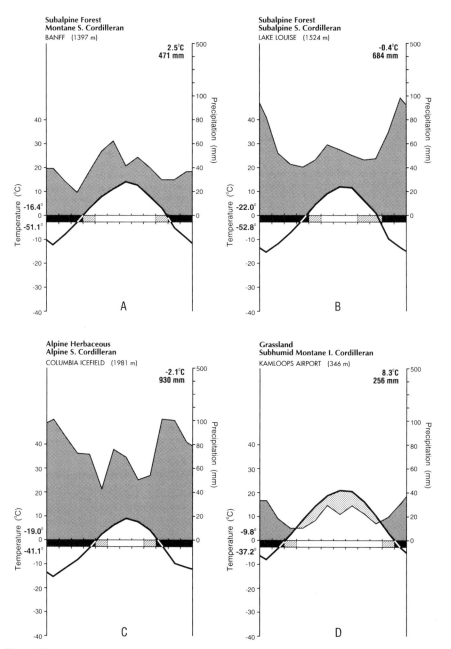

Figure 6.3
Climatic data showing the influence of elevation and rainshadows in the Southern Cordilleran
(SC) and Vertically Stratified Interior Cordilleran (ICV) Ecoclimatic Regions. A = Banff, Alberta (SCm);
B = Lake Louise, Alberta (SCs); C = Columbia Icefield, British Columbia/Alberta boundary (SCa);
D = Kamloops, British Columbia (ICp).

greater similarity between this environment and that of the Arctic Ecoclimatic Province, where Brunisols, Regosols, and Organic, Turbic, and Static Cryosols predominate.

6.1.4 Pacific Coastal Mesothermal Forest

This region includes some of the finest conifer stands and some of the tallest individual trees found anywhere in the world. Unfortunately, their high-quality lumber and huge board-feet volumes per hectare have fuelled clear-cutting on a massive scale, and a concern among many that limited numbers of pristine stands will be spared for future generations to enjoy. Fortunately, some areas, such as Strathcona Provincial Park and the Pacific Rim and South Moresby National parks, have already been set aside. In terms of vegetation differentiation, the region can be subdivided into two primary zones: the coastal western hemlock and coastal Douglas-fir zones. The western hemlock zone includes all of the Pacific coast forest shown in Figure 6.1 except for the eastern side of Vancouver Island and the mainland coast south of 50°N, which is occupied by the coastal Douglas-fir zone. It is important to note that while the region is primarily associated with its massive conifers, broadleaf deciduous trees comprise a significant portion of forest cover. Because of this conifer-broadleaf mix, Walter (1979) designates cooler portions of this region as ecotonal between temperate deciduous forest and cold temperate (boreal) forest (ZE VI–VIII) and the warmer as being ecotonal mixes between Mediterranean (sclerophyllous), cold temperate forest, and warm temperate (maritime) conifer forests (ZE VI–IV, V–IV; Figure 1.2).

6.1.4.1 Coastal Western Hemlock

This is the wetter portion of the Pacific coast forest, and it possesses a mild winter and relatively cool summer. While *Tsuga heterophylla* (western hemlock) dominates within the zone, *Pseudotsuga menziesii* (Douglas-fir), *Picea sitchensis* (Sitka spruce), *Thuja plicata* (western red cedar), *Abies amabilis* (amabilis or Pacific silver fir), *A. grandis* (grand fir), and *Pinus monticola* (western white pine) have their best growth here as well (Krajina, 1969).

Two subzones are identified within the coastal western hemlock zone. The wetter of these is characterized by a per-humid climate with an mean annual precipitation ranging between 280 and 665 cm. It is distinguished by *Tsuga heterophylla, Pseudotsuga menziesii, Thuja plicata, Abies amabilis*, and *Pinus monticola,* together with *P. contorta* (lodgepole pine). *Picea sitchensis* is also commonly found just above the shoreline in a narrow zone often only 150 m broad. Where flatlands allow gales to carry magnesium-rich spray inland, this Sitka zone also extends inland. Sitka spruce individuals, however, are rarely found more than 80 km from the coast. Except on such sites as this salt spray zone or on mesic alluvial flats, such as those in the lower Carmanah Valley on western Vancouver Island, *Picea sitchensis* is normally considered seral to *Tsuga heterophylla* in post-fire succession throughout most of its coastal range. This spruce, however, may also persist by gap-phase regeneration after small-scale disturbances without the need for intervention by fire (Taylor, 1990). The tallest tree in Canada is the 94.5 m (310') "Carmanah Giant," a Sitka spruce found in old-growth stands along the floodplain of the Carmanah Valley. Although concerted efforts have been made to protect this pristine valley from threatened logging and have it become part of the Pacific Rim National Park, approval was given (April 1990) for logging in the upper half of the valley. Another important wetter western hemlock zone species is *Chamaecyparis nootkatensis* (Alaska yellow cedar). This cedar is more common at higher elevations, where it replaces *Thuja plicata* (Orloci, 1965). Frequently encountered broadleaf deciduous species include *Populus balsamifera*, *Alnus rubra* (red alder), *Acer macrophyllum* (broadleaf maple), *A. circinatum* (vine maple), and *Prunus emarginata* (bitter cherry).

The second subzone is somewhat drier (165–280 cm mean annual precipitation). Characteristic conifers include *Tsuga heterophylla, Pseudotsuga menziesii, Thuja plicata, Abies grandis,* and *Picea sitchensis,* together with *Pinus monticola* and *P. contorta.* Douglas fir is particularly productive in this subzone. Deciduous species include *Acer macrophyllum, A. circinatum, Alnus rubra, Populus balsamifera,* and *Prunus emarginata,* and they are more abundant than in the wetter subzone. *Arbutus menziesii* (Pacific madrone), the only naturally occurring evergreen broadleaf tree species in Canada, is also found here, but only in the driest areas, particularly on calcium-rich soils – it is more characteristic of the coastal Douglas-fir zone. As the species name implies, *Acer macrophyllum* has leaves which can reach 30 cm in diameter, an adaptation that allows it to do well in shade under fir, cedar, and hemlock.

Both of these subzones have ground covers that form more or less self-evident vegetation patterns, which are closely related to soil moisture conditions. Mueller-Dombois (1965) identified six ground-cover associations in western hemlock–Douglas fir forests in the Nanaimo River valley on eastern Vancouver Island. From drier to moister sites these are (1) *Gaultheria shallon* (salal)–lichen, (2) *Gaultheria shallon,* (3) moss, (4) *Polystrichum mumitum* (western sword fern), (5) *Lysichitum americanum* (western skunk cabbage), and (6) *Adiantum pedatum* (maidenhair fern) associations. Following clear-cutting, most of the species in these associations persist, except where increased mass wasting episodes remove the soil. In northern Vancouver Island, early post-disturbance succession on burned-over clear-cut sites is dominated by *Gaultheria shallon* due to the survival of pre-disturbance salal rhizomes (Messier and Kimmins, 1991). Schwab (1983) and Rood (1984) have estimated the overall effect of logging on steepland in the Queen Charlotte Islands has been to increase mass wasting episode rates fifteen and thirty-four times respectively. On stable soil sites the original forest herbaceous species are joined by widely distributed species such as *Epilobium angustifolium* (fireweed), *Anaphalis margaritacea* (pearly everlasting), and *Pteridium aquilinum* (bracken) and a number of hardwood species such as *Acer macrophyllum, Rubus parviflorus* (thimbleberry), *Alnus rubra* (red alder), and *Populus balsamifera.* Red alder, with its nitrogen-fixing root nodules, is especially successful on the moister lower slopes, where seepage occurs well up the soil profile (Mueller-Dombois, 1965). Colonization by hardwoods poses a problem for the rapid regeneration of softwoods needed if the lumber industry has any hope of becoming self-sustaining. Despite the potential for this region's forest industry to reach the level of "sustained yield" (Bernsohn, 1983; Percy, 1986), it is still as elusive a concept here as in many other forestry regions in Canada.

Fire has always been an important factor in influencing succession in these forests. Douglas-fir, western red cedar, and western hemlock are early colonizers, but Douglas-fir is the initial dominant. As it dies out, *Tsuga* and *Thuja* take over. *Abies amabilis* (amabilis fir) is a poor seeder, so gets a slow start in post-fire succession. For this reason, amabilis is more common in the wetter climax forests, which have longer fire-free periods. It may take 1,000 years of uninterrupted competition for full expression of succession. The early successional species *Abies grandis* (grand fir) is found mixed in with Douglas-fir in the Strait of Georgia region, where it can reach heights of 80 m and diameters of 1.5 m. In Strathcona Provincial Park in the centre of Vancouver Island, Douglas-fir individuals have been recorded with heights up to 92.4 m (305') and ages of 1,500 years. In the climax forest, cedar and hemlock, which both need shade in their early growth, reproduce themselves. *Picea sitchensis* and *Tsuga heterophylla* seedlings often do better on fallen trunks or "nurse logs" than they do on the ground. These logs give rise to aligned trees with swollen, buttressed bases where the roots grow together. It would appear that recently fallen logs represent sites where competition from forest-floor mosses is low enough to favour tree seedling recruitment (Harmon and Franklin, 1989).

6.1.4.2 Coastal Douglas-Fir Zone

As with the wetter western hemlock zone, the coastal Douglas-fir zone can be readily subdivided into two subzones based on the way forest dominants respond to climate. These subzones are the wetter (102–152 cm mean annual precipitation) madrone–Douglas-fir and the drier (66–102 cm) Garry oak–Douglas-fir subzones (Krajina, 1969). Vancouver, with its mean annual precipitation of 106 cm, but marked summer dry period, is typical for the Douglas-fir zone in general (Figure 6.2B). The madrone–Douglas-fir subzone is typically characterized by *Pseudotsuga menziesii*, *Thuja plicata*, *Abies grandis*, *Picea sitchensis*, *Pinus contorta*, *P. monticola*, and *Arbutus menziesii* (Pacific madrone), with its bright orange bark. *Tsuga heterophylla* is associated with quite moist sites. Frequently encountered broadleaf deciduous species include *Alnus rubra*, *Populus balsamifera*, *Prunus emarginata*, *Acer circinatum* (vine maple), and *A. macrophyllum*.

The Garry oak–Douglas-fir subzone is dominated by the conifers *Pseudotsuga menziesii*, *Thuja plicata*, and *Abies grandis*, together with the broadleaf species *Quercus garryana* (Garry oak, or Oregon white oak) and *Arbutus menziesii*. It is important to note that the evergreen broadleaf Pacific madrone represents a new element in the Pacific Cordilleran flora, as it is more characteristic of the Mediterranean climate. It is also important to note its absence from the wetter madrone–Douglas-fir subzone. Some of the Gulf Islands are so dry that the prickly pear cacti *Opuntia fragilis* and *O. polyacantha* are encountered (Scoggan, 1978).

6.1.5 Pacific Coastal Subalpine Forest

As can be expected, the orographic effect of the Coast Mountains and the Vancouver Island ranges promotes a microthermal subalpine climate with heavy precipitation (178–432 cm yr^{-1}), which includes heavy snow cover over unfrozen ground (Krajina, 1969). As a consequence, this subalpine forest is dominated, not by *Tsuga heterophylla* as in the lowlands, but by *Tsuga mertensiana* (mountain hemlock). Geographically this region is the coastal portion of the subalpine forest identified in Figure 6.1. There are also considerable areas of subalpine forest inland in eastern British Columbia and western Alberta, but these eastern forests are dominated by *Picea engelmannii* (Engelmann spruce) and *Abies lasiocarpa* (subalpine fir). The influence of latitude is also noticeable in the distribution of this coastal subalpine forest. The southern portion is normally found between elevations of 900–1,500 m on windward slopes and 1,100–1,800 m on leeward slopes, while in northern British Columbia the elevation drops to between 300–900 m (Brooke et al., 1969).

Within the Pacific coastal subalpine region, only one major zone, the mountain hemlock zone, can be identified, although it can be readily subdivided in two subzones – the subalpine forest subzone and the subalpine parkland subzone – again on the basis of the effects of elevation on cover. In the subalpine forest subzone, *Tsuga mertensiana, Abies amabilis, A. lasiocarpa,* and *Chamaecyparis nookatensis* (Alaska yellow cedar) form closed forest and are found on soils which do not freeze in winter due to the thick insulating blanket of snow. At lower elevations within this subzone, it is normal to also find *Tsuga heterophylla* and *Thuja plicata,* while most other important lowland mesothermal tree species are found sparingly (Krajina, 1969). Gap-phase replacement in old growth at lower elevations suggests preferential replacement of all gap-makers by *Abies amabilis,* except on stumps, where almost all successful *Tsuga heterophylla* gap-fillers were located (Lertzman, 1992).

The subalpine parkland subzone is essentially the transition or ecotonal community type, separating the mountain hemlock forest below from the alpine tundra above. Within this ecotone, *Tsuga mertensiana* rarely forms closed stands but rather develops a parkland appearance, where individuals are found only in clumps or small groups and often in the krummholz form. Between these clumps are herbaceous communities which vary in dominancy primarily on the basis of the length of time they are snow free in summer. Where snowpack lasts more than nine months, *Saxifraga tolmiei* and *Luzula wahlenbergii* (woodrush) are frequently encountered on well-drained sites, while they are joined by *Juncus* and *Carex* spp. and *Luetkea pectinata* (partridge foot) on somewhat moister soils (Brooke et al., 1969). In basins where the water table is close to the surface and early growing-season flooding occurs, *Carex nigricans* dominates, with the lichen *Polytrichum norvegicum* being common. Where snowpack lasts between eight and nine months, heath-like communities dominated by *Phyllodoce empetriformis* (pink mountain heather) and *Cassiope mertensiana* (mountain heather) do well on mesic sites, while lush meadow-like communities are found on moister ground. Where snowpack lasts less than eight months, clearings between clumps of trees are frequently dominated by ericaceous shrubs such as *Vaccinium membranaceum* (mountain huckleberry) (Brooke et al., 1969).

6.1.6   Cordilleran Forest Region

This region covers the major part of the continental Canadian cordillera below subalpine forest, south of latitude 55°N. These cordilleran forests are subdivided by Rowe (1972) into Montane and Columbia forest (Figure 6.1). Krajina (1965) identifies three major zones within the region: the interior western hemlock zone, the interior Douglas-fir zone, and the cariboo aspen–lodgepole pine–Douglas-fir zone. In terms of geographic distribution, the interior western hemlock zone is essentially synonymous with the Columbia forest zone of Rowe (Figure 6.1). It is located in the Columbia River valley and other high-altitude valleys in the Columbia Mountains, eastern British Columbia. The interior Douglas-fir zone is restricted to the southern half of the Montane forests on the Fraser Plateau between the Coast Mountains to the west and the Columbia Mountains to the east, as well as to portions of the eastern valleys, such as along the Kootenay River and the upper Columbia River valley (Rowe, 1972). It is important to note that the Douglas-fir in this region is the interior Douglas-fir subspecies *Pseudotsuga menziesii* var. *glauca* (sometimes called blue Douglas-fir). The Cariboo aspen–lodgepole pine–Douglas-fir zone occupies the northern half of this montane zone on the Nachako Plateau between the Coast and Columbia Mountains. Two other important continental cordilleran montane vegetation types are identified by Krajina (1965). These are the ponderosa pine–bunch grass community found in rain-shadow areas and the sub-boreal and boreal spruce forests to the north, but both are sufficiently distinct to form vegetation regions in their own right.

6.1.6.1   Interior Western Hemlock Zone

Climatically, this zone is the wettest and most biologically productive in the interior cordillera (Southern Cordillera Ecoclimatic Sub-Province), with mean annual precipitation ranging from 56 to 170 cm (Krajina, 1969). These relatively high values result from eastward-moving Pacific air being forced over the Columbia Mountains, creating what is known as the "interior wet belt" (Rowe, 1972). While precipitation values are still much lower than for the western hemlock forests of the Pacific coast, *Tsuga heterophylla* is nonetheless still the dominant climax tree species in both regions. This is achieved with lower absolute precipitation totals simply because the growing season in the interior cordillera is shorter. The relatively high precipitation ensures soils are leached, giving rise to Ferro-Humic and Humo-Ferric Podzols (Figure 3.4). While *Tsuga* dominates in the wetter portions of the zone, *Larix occidentalis* (western larch) takes on dominancy in the drier areas. For this reason, Krajina (1969) identifies two subzones: the drier (western larch) subzone and the wetter (western hemlock) subzone.

While the broadleaf deciduous species *Betula papyrifera, Populus tremuloides,* and *P. balsamifera* are common throughout both subzones, conifers species composition differs. The drier, larch-dominated zone includes numerous conifers such as *Pseudotsuga menziesii* var. *glauca, Pinus monticola, P. contorta, Tsuga heterophylla, Thuja plicata,* and *Abies grandis. Pinus ponderosa* (ponderosa or western yellow pine) appears only in the drier southern areas, while *Picea glauca* occurs in the northern parts closer to the white spruce–dominated sub-boreal forest. Interior Douglas-fir and western larch are particularly important as early successional species following fire (Rowe, 1972). The wetter, western hemlock–dominated subzone possesses fewer conifer species. According to Krajina (1969), the only conifers found here other than the dominant stand trees, western hemlock and western red cedar, are *Pinus monticola, P. contorta, Pseudotsuga menziesii* var. *glauca, Tsuga heterophylla, Picea engelmannii* (Engelmann spruce), *P. glauca,* and *Abies lasiocarpa* (subalpine fir).

6.1.6.2 Interior Douglas-fir Zone

Climatically, this zone is drier and warmer than that of the interior western hemlock zone, and ecoclimatically, it is part of the Vertically Stratified Interior Cordilleran map unit (ICV). With a mean annual rainfall of only 36–56 cm, leaching of soils is not excessive. As a result, Eutric and Melanic Brunisols and Dark-Gray Chernozems dominate in drier areas, while Gray Luvisols are more characteristic of the wetter. The zone can be conveniently divided into two subzones: the drier (pinegrass) subzone (<48 cm) and the wetter (false boxwood) subzone (48–56 cm) (Krajina, 1969).

The drier pinegrass subzone is the most southern portion of the interior Douglas-fir zone, and only two conifers, *Pseudotsuga menziesii* var. *glauca* and *Pinus ponderosa,* attain importance. The ponderosa pine does particularly well in this zone as an early successional species where it is associated with a grass-dominated herbaceous cover. Some broadleaf deciduous species, such as *Populus* and *Betula,* are also found here; however, conditions are simply too dry for *Tsuga heterophylla* (Krajina, 1969). This subzone is essentially an ecotone between the wetter mixed forests of the false boxwood (*Acer glabrum*) subzone and the drier pine grasslands of the cordilleran cold steppe–savanna. It can be likened to the aspen parkland of the prairies, which acts as a transition zone between the mixed forest (ZE VII–VIII) and the prairie (ZB VII). It should be noted that Walter (1979) so identifies this cordilleran area in his zonobiome classification of North America (Figure 1.2). In the false boxwood subzone, *Pinus ponderosa, P. monticola, P. contorta, Larix occidentalis, Pseudotsuga menziesii* var. *glauca, Picea glauca, Abies grandis,* and *Thuja plicata* are the more abundant conifers.

6.1.6.3 Cariboo Aspen–Lodgepole Pine–Douglas-fir Zone

Geographically, this area covers most of the northern third of the Montane forest region (Figure 6.1), while ecoclimatically, it is the Boreal Interior Cordilleran Ecoclimatic Region (ICb). The southern portion of this zone is dominated by *Populus tremuloides, Pinus contorta,* and *Picea glauca*, while in the north, around Prince George, forests become more boreal-like, with *Picea* spp. dominating. The region borders subalpine forest to the west, interior western hemlock to the east, and sub-boreal spruce to the north (Krajina, 1969). Because of increasing latitude, winters are severe, yet summers are sufficiently warm to permit a distinctly non-boreal component, such as lodgepole pine, coastal Douglas-fir in the west, and interior Douglas-fir in the east, to do well. The presence of *Picea glauca* throughout and *Picea mariana* in the north, however, attests to the increasing boreal influence. While climax forest for much of the region may be spruce-fir, decimation by fire has greatly increased early succession stages dominated by trembling aspen, lodgepole pine, and birch (Rowe, 1972).

6.1.7 Cordilleran Cold Steppe and Savanna Forest

This is the region of grasslands and ponderosa pine–bunch grass parkland islands found within the interior Douglas-fir subzone (Figure 6.1). Parkland is used in this book instead of the term "savanna." Savanna is used here only in reference to subtropical and tropical formations (following Hills, 1965) or where using the term "parkland" might alter an author's meaning. Climatically, the grasslands are part of the Interior Cordilleran Ecoclimatic Region (ICp), which is the driest of all of the Pacific and Cordilleran Regions (BSk climate of Köppen), while parklands are part of the somewhat more humid Subhumid Montane Interior Cordilleran Ecoclimatic Region (ICm⁻). Both owe their distribution pattern to rain-shadow effects in the major valley systems to the lee of the Coast and Cascade ranges (Brayshaw, 1965). As shown in Figure 6.1, these grasslands are found in the middle Fraser Valley, on the Thompson River near Kamloops, on part of the southern Kootenay River, and in the Okanagan Valley. Mean annual precipitation ranges from 19 to 36 cm, winters are cool, and summers sufficiently warm that with irrigation the region supports good agricultural production and, especially in the Okanagan, valuable orchards of apricots, peaches, apples, cherries, pears, and grapes (Krajina, 1969).

As the regional name suggests, *Pinus ponderosa* is the most important tree species. In most areas it is the only tree species, although interior Douglas-fir can be found on moister slopes, and *Larix occidentalis* (western larch) and *Thuja plicata* (western red cedar) along river courses. The grasslands are to some extent the northern extensions of the "sagebrush" grasslands of the intermontane basins of the cordillera in the USA. *Agropyron spicatum* (bluebunch wheat-grass) is the dominant herbaceous species within both grassland and ponderosa pine parkland on fine- to medium-textured soils, while *Calamagrostis rubescens* (pinegrass) takes over dominancy on similar soils at higher elevations, where moisture indices are not so neg-

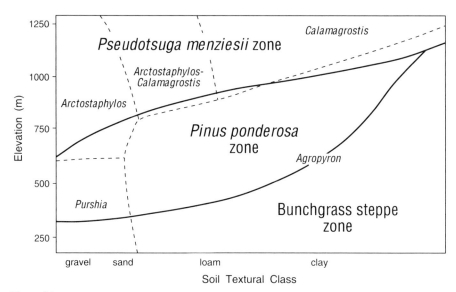

Figure 6.4
Generalized ecosystem gradients within the ponderosa pine–bunchgrass communities of the cordilleran cold steppe and savanna forest region, British Columbia (after Brayshaw, 1965)

ative and where interior Douglas-fir becomes important in the cover. Other important grass species include *Agropyron cristatum* (crested wheat-grass), *Elymus condensatus* (giant wild rye), *Festuca campestris* (rough fescue), *Koeleria cristata* (June grass), *Poa pratensis* (Kentucky bluegrass), and *Stipa comata* (needle-and-thread) (Watson and Murtha, 1978). *Purshia tridentata* (bitter brush) outcompetes *Agropyron* on sandy-gravelly soils where moisture indices are strongly negative, while *Arctostaphylos uva-ursi* (common bearberry) replaces *Purshia* at higher elevations. While the dominant sage is *Artemisia tridentata* (big sagebrush), two other sages are frequently encountered, *A. tripartita* (three-tip sagebrush) and *A. frigida* (prairie-sagewort) (Cawker, 1983). *Artemisia frigida* is palatable to cattle, but the former two are not, so they tend to increase their cover under heavy grazing. Relationships between the dominant tree, shrub, and grass species of the region, and the effects of elevation on temperature and of texture on moisture retention, are summarized in Figure 6.4 (after Brayshaw, 1965).

The generalized relationships shown in Figure 6.4 are disturbed by the effects of both burning and overgrazing. While the incidence of natural lightning fires was greatly augmented in the past by Indian-caused fires (Barrett and Arno, 1982), today the practice is more frequently associated with range management. Burning of the *Pseudotsuga-Symphoricarpos* stands leads to a succession dominated by *Pinus ponderosa* and *Populus tremuloides*, with the understorey *Symphoricarpos* being joined by Kentucky bluegrass. In areas where shrubs are in direct competition with grasses, fire favours grasses, while overgrazing favours woody species. *Artemisia tridentata* becomes the most important woody species on overgrazed fine-textured soils dominated by *Agropyron*. If these same sagebrush grasslands are then burned, big sagebrush can be eliminated, and then, with reduced grazing, *Agropyron* returns to dominancy (Brayshaw, 1965). Burning is successful against big sagebrush because the plant does not stump-sprout after fire and must reinvade by seed. The apparent grazing-related increase in sagebrush this century may well be a return to conditions existing before the grassland-promoting high burning frequency period associated with early settlement in the mid-nineteenth century (Cawker, 1983).

6.1.8 Canadian Cordilleran Subalpine Forest

This humid forested region is distinguished from other interior cordilleran regions on the basis of the response of conifer dominants to cold climates at high elevation. It is more difficult to define species continuity throughout the region, however, because dominancy changes considerably with increasing latitude. The region extends from the 49th parallel to the Yukon Territory, but its best-known forest type, the Engelmann spruce–subalpine fir forest, is only found south of 57°30'N; *Picea engelmannii* is, in fact, absent north of 57°30'N. Geographically, this region comprises the area designated as Subalpine Forest on Figures 1.3 and 6.1, other than the Coastal Zone (discussed in section 6.1.5) and portions of the interior north of 55°N (which Krajina considers to be Sub-Boreal spruce). Rowe (1972) identifies two major zones within this region: the Interior Subalpine, which lies completely within British Columbia, and the East Slope Rockies, which lies almost completely within western Alberta. This Interior Subalpine Zone is called the Engelmann Spruce–Subalpine Fir Zone in Krajina's (1969) classification.

Because the Engelmann Spruce–Subalpine Fir Zone extends the latitudinal length of British Columbia, it can be conveniently subdivided into three subzones, the southern, central, and northern. The southern subzone extends north to 52°N at elevations between approximately 1,250 and 2,200 m. While it is dominated by *Abies lasiocarpa* and *Picea engelmannii*, other commonly encountered species include *Pinus contorta, P. monticola, P. albicaulis* (whitebark pine), *P. flexilis* (limber pine), and *Larix lyallii* (alpine larch). The absence or scarceness of many of the typical montane forest species is primarily due to the greater severity of the winter and the fact that the ground often freezes before snow falls (Krajina, 1969). Despite this severity, *Picea engelmannii* and *Pinus albicaulis* individuals over seven hundred years old can occasionally be found just below tree line (Luckman et al., 1984). With increasing elevation, there is also rapid change in understorey vascular species composition, while changes in bryophyte dominants is less pronounced owing to their wider tolerance of elevation-correlated factors (Lee and La Roi, 1979). At tree line, many of the Engelmann spruce and subalpine fir individuals take on a mat krummholz form which is protected below winter snow. These mats, consisting of very high concentrations of needles, create an aerodynamic structure favouring elevated growing-season needle temperatures (Hadley and Smith, 1988). The tree line itself has frequently fluctuated in elevation during the Holocene, with the lowest recorded positions occurring only about five hundred years ago (Kearney and Luckman, 1983).

The central subzone stretches north from the southern subzone to 57°30'N and is characterized by the presence of only a few conifers, such as *Abies lasiocarpa, Picea engelmannii, Pinus contorta,* and *P. albicaulis,* and the broadleaf deciduous species *Betula glandulosa.* The northern zone has few species of importance other than *Abies lasiocarpa, Pinus contorta, Alnus crispa* (mountain-alder), and *Betula glandulosa* (scrub birch).

The east slope Rockies portion of the Cordilleran Subalpine Forest extends in a narrow belt along the eastern flank of the Rockies from the border with Montana to 55°N, at elevations between approximately 1,500 and 2,060 m. Engelmann spruce–white spruce hybrids are present at lower elevations, while Engelmann spruce takes over above. Other important lower elevation species include *Pinus contorta, P. albicaulis, Picea glauca,* and *Larix occidentalis,* while at higher elevations, *Pinus contorta* and *Abies lasiocarpa* join *Picea engelmannii. Pseudotsuga menziesii* var. *glauca* is encountered only where the subalpine forest comes in contact with patches of interior Douglas-fir forests near Jasper and Banff and at Waterton Lakes National Park (Figure 6.1).

In the last century, considerable areas of climax spruce were burned and replaced by the prolific successional lodgepole pine (Rowe, 1972). Around Banff the peak burning period, as identified by the large number of similar-aged pine stands, occurred during the prospecting–early railroad period between 1850 and 1886. Tande (1979) found that present-day forest cover in a 43,200 ha study area around Jasper originated primarily following major fires in 1758, 1847, and 1889 and that the mean fire return interval (MFRI) for all fires in the study area was 5.5 years. Since 1913, when fire suppression measures were introduced to Jasper National Park, fire periodicity and extent have both declined. In the 495 km^2 Kananaskis Watershed southeast of Banff the fire cycle appears more climatically controlled. Here the fire cycle (the time required to burn an area equal to the area of study) was 50 years before 1730, when conditions were warmer and drier, but was reduced to 90 years between 1730 and 1980, when conditions became cooler and moister (Johnson and Larsen, 1991). That fires are now understood to play a critical role in forest ecology is reflected in the 1979 Parks Canada policy, which includes an attitudinal change to park management designed to ensure minimal interference in natural processes (Finklestein, 1982). In addition, the 1989 fire management policy permits an attempt to mimic natural fire regimes (Canadian Parks Service, 1989), and the Bow River Valley in Banff National Park has already been mapped into "ignition units," with plans to burn these units at different times up to the year 2035 (Eagles, 1993).

6.1.9 Alpine Tundra and Boreal Forest

Large areas of alpine tundra are found in the Northern Cordilleran and Northern

Subarctic Cordilleran Ecoclimatic Sub-Provinces in the Yukon and Northwest Territories, as well as in small islands between subalpine forest and icefields in the Coast and Rocky Mountains of British Columbia and Alberta. See chapter 3 for fuller treatment of this vegetation type.

Areas of boreal forest are also encountered in the Northern and Mid-Cordilleran Ecoclimatic Sub-Provinces. Here vegetation is similar to that discussed for western Canada in sections 4.4.1 and 4.4.2. Krajina (1969) identifies a sub-boreal spruce zone in central British Columbia in the northern portion of the Montane forest region shown in Figure 6.1. Here an admixture of boreal species such as *Picea glauca*, *P. mariana,* and *Larix laricina* are joined by species such as *Pinus contorta, Abies lasiocarpa,* and *Pseudotsuga menziesii* var. *glauca*, species more typical of the interior Douglas-fir forest to the south. This sub-boreal spruce zone therefore represents a transition zone between boreal forest and temperate cordilleran forest.

6.2 The Cordilleran Region in the USA

6.2.1 Distribution

While the above discussion includes all of Canada's Pacific coast and cordilleran vegetation types, for the USA, discussion is restricted primarily to coastal and cordilleran vegetation types that are relatively equivalent to those found in Canada. Four major vegetation groupings are discussed: (1) the northwest coast conifer-hardwood forests, (2) the Sierra and Rocky Mountain montane pine forests, (3) the intermontane sagebrush and grassland areas, and (4) the subalpine forest and alpine tundra areas (Figure 6.5). The areas of Mediterranean (ZB IV) sclerophyllous forests and grasslands of California and the deserts and semi-deserts (ZB III) of the south are not discussed.

It should be pointed out that compared to the Canadian cordillera, a far larger proportion of the USA cordilleran region is naturally non-forested. Between the coastal ranges and Sierras of California, Oregon, and Washington and the Rocky Mountains is a broad, deeply corrugated landscape of sagebrush lowlands and small isolated forested ranges known as the Great Basin. While tectonic folding/thrusting accounts for the coastal ranges, Sierras, and Rockies, geologically recent mid-cordilleran crustal attenuation produced the ninety or so downthrown grabens, or basins, of the Great Basin region. The effect of these recent tectonic events is to create a repetitive "basin-and-range" vegetation pattern greatly dependent on rain-shadow and orographic precipitation effects. Of importance in this regard is also the fact that as latitude decreases, frontal activity episodes become fewer and more restricted to winter months. This means that in states such as Nevada and Utah, precipitation totals depend much more on orographic and convectional sources. Limiting frontal activity to winter months not only accounts for the presence of a winter-rain Mediterranean vegetation type in California, but ensures that basin-and-range country is primarily subhumid sagebrush and grassland. This basin-and-range landscape has no direct

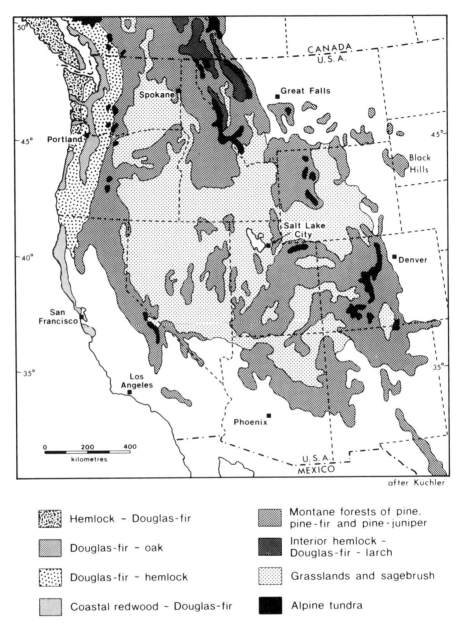

after Küchler

▦ Hemlock ~ Douglas-fir		▦ Montane forests of pine, pine-fir and pine-juniper	
▦ Douglas-fir ~ oak		▦ Interior hemlock ~ Douglas-fir ~ larch	
▦ Douglas-fir ~ hemlock		▦ Grasslands and sagebrush	
▦ Coastal redwood ~ Douglas-fir		■ Alpine tundra	

Figure 6.5
Vegetation map of the Pacific Coast–Cordilleran region of the United States, excluding
sclerophyllous forest and desert of the southwest and south (generalized from a number of sources,
especially Küchler, 1964). Areas of subalpine forest are not differentiated.

Canadian cordilleran geological equivalent, although the drier montane Douglas-fir–ponderosa pine forests and small areas of sagebrush grasslands in interior British Columbia are a reflection of the region's location to the lee of coastal ranges. The deserts of southern California and Arizona reflect an even greater reduction in frontal activity than in the sagebrush states, as well as rain-shadow effects in basins which are sometimes so deep they bottom out below sea level.

6.2.2 Northwest Coast Conifer-Hardwood Forests

Four important forest zones can readily be identified in the northwest coastal region. These are (1) the hemlock–Douglas-fir forests, (2) the Douglas-fir–oak forests, (3) the inland and montane Douglas-fir–hemlock forests, and (4) the coastal redwood–Douglas-fir forests (Figure 6.5). These forests are essentially a southern extension of the Canadian Coastal Mesothermal and Coastal Montane forests (Figure 6.1).

6.2.2.1 Hemlock–Douglas-Fir Zone

This zone borders the Pacific and extends from Vancouver Island across to the Olympic Peninsula and south to Oregon. Using ecoclimatic terminology, this zone is an extension of the Canadian South Pacific Ecoclimatic Sub-Province. Climatically, it has an extreme maritime climate with frequent fog, while summers have a short dry period. Moisture indices are strongly positive (Im of +103 for Olympia, Washington). As in Canada, a narrow coastal fringe subzone is dominated by *Picea sitchensis*, while *Tsuga heterophylla* and *Pseudotsuga menziesii* are the regional dominants, together with some *Thuja plicata*. The forest floor is normally covered with moss mats, which prove problematic for seedling germination. As noted for British Columbia, seedlings have much more success on logs than within these moss mats, due to greatly reduced competition (Harmon and Franklin, 1989). The great expanses of sand dunes found along the Oregon coast provide excellent sites to study temporal forest succession. While *Pinus contorta* is the first tree invader of long-stabilized dunes, *Picea sitchensis* generally follows, with *Tsuga heterophylla* forming the climax forest (Kumler, 1969).

Today, the three primary landscape elements in the region are clear-cut, remnant old-growth forest patches, and clear-cut-forest edges (Chen et al., 1993). The impact of logging on climax forest throughout the coastal region has led to concern for the old-growth habitat of the spotted owl and marbled murrelet (Abate, 1992). Recent studies show that in the most heavily fragmented landscapes, spotted owl social structure appeared to be abnormal, with almost all owls selecting old forests for both foraging and roosting (Carey et al., 1992). Of that portion of the old-growth Olympic National Forest present in 1940, only 24% remained in 1988 (Sampson, 1990). The Olympic Peninsula is also home to the largest known *Thuja plicata*, a 52.3 m tall and 18.6 m circumference giant, which towers alone over 26 ha of clear-cut stumps (Bronaugh, 1992a).

6.2.2.2 Douglas-Fir–Oak

Found to the lee of the coastal ranges in the lowlands which stretch south from Seattle, Washington, to Eugene, Oregon, and along part of the Oregon coast (Figure 6.5), this zone is essentially an extension of the Canadian Garry oak–Douglas-fir subzone. Reduced frontal activity and modest rain-shadow effects ensure a dry summer such as that experienced in Portland (Figure 6.2C). This summer moisture stress allows *Quercus garryana* (Garry oak), a Mediterranean-type species, to become an important part of the cover, along with Douglas-fir. In the Willamette Valley at the southern end of this zone, the pre–European settlement vegetation was a mosaic of forest, oak woodland (parkland), open oak savanna, and grassland. These Oregon white oak woodlands and savannas were maintained by aboriginal burning and have since been invaded by Douglas-fir and grand fir (Kimerling and Jackson, 1985).

6.2.2.3 Douglas-Fir–Hemlock

This forest zone is found on the lee slopes of the Coast Range and on the windward slopes of the Cascades south from Canada to northern California (Figure 6.5). In Oregon the dominant canopy species at lower elevations is *Tsuga heterophylla*. At mid-elevations (1,000–1,300 m) Douglas-fir is the canopy dominant, with western hemlock and amabilis fir dominating a lower, multi-layered canopy, which is joined by *Thuja plicata* on wetter sites. At higher elevations, *Abies amabilis* dominates (Stewart, 1986), while *Tsuga mertensiana* and *Chamaecyparis nootkatensis* (Alaska yellow cedar) become important here and in subalpine forest along the western Cascades (del Moral and Fleming, 1979). Tree-canopy composition depends to a large extent on fire and the success conifer seedlings have in becoming established in early successional stages following fire (Wright and Bailey, 1982; Turner, 1985). If *Pseudotsuga menziesii* dominates early, then regeneration of *Tsuga heterophylla* and *Abies amabilis* is rapid. When *Tsuga heterophylla* dominates the early succession, further regeneration of other species is absent or minimal, unless canopy openings result from other agencies such as windthrow (Stewart, 1986; 1988).

The logging industry is so important to the region that many clear-cut sites are replanted with *Pseudotsuga menzeisii*. Logging residues are sometimes burned, removed, or allowed to decompose in place. Prescribed burning of slash is especially valuable where competing brush species and low water-holding soils make reforestation difficult (McNabb et al., 1989). The significance of prescribed burning as a management tool, both for slash removal and within mature forest, has also lead to research on predicting the post-fire mortality of the various dominant conifers (Ryan and Reinhardt, 1988); however, recent public resistance to unwanted side-effects of prescribed burning, such as smoke, has led to a decline in this management practice. Decaying wood appears to be a factor in maintaining long-term productivity in these forests, although until decay occurs, residues can interfere with regeneration activities and may pose a fire hazard (Edmonds and Vogt, 1986). Disturbance from logging has lead to a greater regional importance for *Alnus rubra* and *Pteridium aquilinum* (Cwynar, 1987). Studies show that in Douglas fir plantations, herbaceous species increase in number for the first twenty years and then decline as canopy closes (Schoonmaker and McKee, 1988). Disturbance caused by the explosive eruption of Mount St Helens in May 1980 led to a huge salvage operation to reduce losses from the vast areas of "blastthrown" Douglas-fir. Post-eruption research in the blast zone is providing valuable information on the recovery of herbaceous species (Wood and Morris, 1990; del Moral and Wood, 1988a). Del Moral and Wood (1993) note that even after eleven growing seasons, primary succession in devastated areas has been very slow, primarily because most habitats are isolated and physically stressful, and cover is barely measurable.

6.2.2.4 Coastal Redwood–Douglas-Fir

Consisting of a narrow belt 15–30 km wide, which stretches from just south of San Francisco north to the California-Oregon border, this forest is considered a separate zone only because the massive *Sequoia sempervirens* (coastal redwood) becomes the dominant along with Douglas fir (Ornduff, 1974). Climax forest redwoods range in heights from 55 to 60 m, and in diameter from 3 to 7 m, although the tallest can reach heights of 113 m. While redwood height can exceed the tallest *Sequoiadendron giganteum* (giant sequoia) of the drier Sierras by 12 m, they do not live as long. Currently, the 110.6 m "Giant" from northern California is the tallest redwood, replacing the 112.8 m tall Dyerville Giant, which was still growing quite vigorously when toppled by other falling redwoods in 1981 (Bronaugh, 1992b). Their long life span and massive growth form give these *Sequoia* stands the distinction of having the largest living phytomass accumulations known (Franklin, 1988). The western boundary of this zone often starts several kilometres from the ocean, while inland the distributional limit of *Sequoia* is approximately that of the sea fogs. These fogs compensate for the low-precipitation period from the end of July until about the end of September by maintaining high humidities and providing the equivalent of 17–30 cm of precipitation during the "low-rain season" (Whitney, 1985). A constant supply of water is critical because coastal redwoods lack root hairs. On slopes within the coastal ranges, *Sequoia* dominates, along with *Pseudotsuga menziesii, Quercus densiflora,* and *Abies grandis.* On lower, moister ground, *Sequoia* does even better and is associated with cedar, hemlock, Douglas-fir, Sitka spruce, and oak. Coastal redwoods compete well with other species, probably because they are relatively fire tolerant (Greenlee and Langenheim, 1990) and can stand silting in floodplain situations by sending out new roots (Ornduff, 1974). Redwoods often produce basal sprouts following fires (Stuart, 1987). Muir Woods National Monument north of San Francisco and Redwood National Park along the northern California coast are just two of the areas established to protect this magnificent species and preserve examples of its old-growth ecosystem.

6.2.3 Montane Pine Forests

Montane forests that are somewhat drier than those of the Pacific Northwest are frequently dominated by pines, often in association with fir, spruce, cedar, or juniper, and they cover large areas of the cordillera (Figure 6.5). Two major montane areas can be identified: (1) the Sierra Montane forests of the eastern Cascades in Washington and Oregon, together with the Sierra Nevada in California, and (2) the Rocky Mountains Montane forests of Idaho, Montana, Wyoming, Colorado, Arizona, and western Texas, together with numerous small pine uplands within the Great Basin surrounded by sagebrush and outliers east of the Rockies surrounded by "seas" of prairie. Generally speaking, the region is characterized by im values close to zero and occasionally moderately negative. While strongly negative moisture indices result in grassland or sagebrush communities dominating, some conifers, especially ponderosa pine and juniper, are adapted to moisture stress conditions, particularly on soils which discourage water-competing herbaceous species, such as those found on rocky slopes. Low ground-cover biomass values and rugged terrain are two conditions also not conducive to the spread of fire. Where moisture indices values are positive, pines are joined by other conifers such as fir and larch, while at higher, cooler elevations, pines are often replaced by fir and spruce.

6.2.3.1 Sierra Montane Forests

Along the Pacific margin, two major montane zones can be identified, the eastern slopes of the Cascades and the Sierra Nevada. On the lower, drier slopes of the eastern Cascades, *Pinus ponderosa* dominates, while ponderosa pine is joined by Douglas-fir and grand fir at higher levels. To the south, almost pure stands of *Abies magnifica* (red fir) are found. In the Sierra Nevada, montane forest is a coniferous response to California's mild, moist winters and warm, dry summers (Whitney, 1985). Climatic data for Yosemite National Park (Figure 6.2D) on the western slopes of the central Sierra Nevada show that summer drought lasts almost four months. Sierra Nevada forests can be subdivided into three subzones: (1) mixed conifer, (2) red fir, and (3) Jeffrey pine.

The mixed-conifer forest is dominated by ponderosa pine; however, black oak dominates on the driest, rockiest places, and *Abies concolor* (white fir), *Pinus jeffreyi* (Jeffrey pine), and Douglas-fir joins ponderosa pine in higher, moister sites. *Pinus lambertiana* (sugar pine) and the majestic *Sequoiadendron giganteum* (giant sequoia) dominate cool sites that remain moist throughout the long, dry summer. The giant sequoias grow only in seventy-five groves scattered throughout the central and southern portions of the Sierra Nevada, the largest of which – the Redwood Mountain Grove in Kings Canyon National Park – is only 1,215 ha in area (Kilgore and Taylor, 1979). Found at elevations of 1,350–2,500 m, sequoias must have deep soils that remain moist all summer (Whitney, 1985). Mature sequoias are typically over 60 m tall and 1,500 years of age, with trunks over 5 m in diameter. The largest individual tree in the world, the General Sherman sequoia in Sequoia National Park, measures 82.5 m tall and 11 m in diameter and contains 1,256 t of wood above ground level.

The longevity of these sequoias is related to their lack of flammable resins, which makes them relatively fire resistant (although many have fire scars), and the presence of bark tanins, which discourage insect and fungal attack. The fact that the core areas of sequoia groves contain more small live sequoias and more dead individuals than in peripheral grove areas may be partly a response to differences in fire disturbance histories (Stohlgren, 1993). Low-intensity fires at frequent intervals characterized aboriginal burning patterns in sequoia groves up to 1875 (Kilgore and Taylor, 1979), while the occasional high-intensity fire removes surface litter and duff, opens up the canopy, and greatly facilitates sequoia germination and seedling development (Kilgore and Biswell, 1971). Since 1875, lack of light fires has allowed the buildup of fuels which could increase the likelihood of crown killing fires; however, pre-scribed burning is now being carried out in some stands. Fire exclusion and suppres-sion in the Yosemite Valley has contributed to a change in dominancy from *Pinus ponderosa* to *Calocedrus decurrens* (incense cedar) (Sherman and Warren, 1988). The policy for dealing with fires on National Forest lands was changed in 1977 to one of management rather than suppression (Nelson, 1979). This change reflects the realization that fire is an integral part of these ecosystems. The policy of logging some sequoia groves in the Sequoia National Forest, however, seems likely to coun-teract conservation efforts (Green, 1990).

Red fir forests are generally found upslope of the mixed conifer forests. Stands are dense, needle litter is thick, and understorey plants are severely limited. Following fire, lodgepole pine is an early colonizer, but red fir, which prefers shade during early growth, eventually overtops the pine and returns to dominancy (Whitney, 1985).

Pinus jeffreyi (Jeffrey pine) replaces ponderosa as the forest dominant at eleva-tions of 2,000 m or more, especially on the eastern slopes of the Sierra Nevada. This species is more tolerant of the cooler summer, the prolonged snowpack, and the greater shade found under moister forest conditions. On the western Sierra slopes, Jeffrey pine can be joined by a variety of *Pinus contorta* (lodgepole pine), red fir, and western white pine, while on eastern slopes, ponderosa pine, western juniper, and white fir are common. In parts of the central Sierra Nevada, lodgepole pine is able to replace itself successfully without crown fires by using windthrow openings as regeneration sites (Parker, 1986). In the drier southern California Sierras, Jeffery pine is joined by *Calocedrus decurrens*, *Pseudotsuga macrocarpa* (bigcone Douglas-fir), white fir, western juniper, sugar pine, singleleaf pinyon, limber pine, and black oak. *Pinus lambertiana* (sugar pine) was much more extensive when California was first settled by Europeans; however, effective fire control measures have allowed incense cedar and white fir to outcompete it through natural succession (Ornduff, 1974).

6.2.3.2 Rocky Mountain Montane Forests

Rocky Mountain montane forests are found at elevations above sagebrush or prairie and below subalpine forest. The largest area is found in the northern Rockies of Idaho, western Montana, and northwestern Wyoming (Figure 6.5). Large areas of montane pine are also found in southern and eastern Utah, central and western Colorado, northern New Mexico, and central Arizona, and some outliers are found east of the Rockies surrounded by prairie. While pine species are typically associated with these forests, they are by no means the only conifers present. Often pines are early successional species whose apparent dominancy only reflects frequent distur-bance. The rarity of true climax stands in the northern Rockies simply reflects the low probability that the hundreds of years needed for complete secondary succession to express itself are permitted, before some form of catastrophic disturbance intervenes (Turner, 1985).

Pinus ponderosa (ponderosa pine, sometimes called Rocky Mountain yellow pine) and *P. contorta* (lodgepole pine) are the most often encountered pines within this region. Ponderosa pine is frequently dominant at lower, drier elevations, just up-slope of sagebrush or grassland. In the middle Rockies, however, drier conditions prevail, except at higher, cooler elevations, and these favour lodgepole pine over ponderosa. At higher elevations to the north, or where fire suppression or overgrazing has prevented burning, pines are often associated with *Pseudotsuga menziesii*. Where ponderosa pine and Douglas fir are found together, fir normally dominates in late successional stages. Even some of the pine stands just upslope of sagebrush that have not been burned recently are experiencing takeover by Douglas-fir. Lodgepole pine does best as a colonizer of recently burned areas, so rarely dominates climax forest – it is normally found in young, even-aged stands. Mixed-aged stands occur only where the previous fire was so extensive that a source of competitor seed is too distant. Lodgepole pine cones can be serotinous or non-serotinous, with the serotinous cone type dominating in stands originating from stand-replacing burns (Muir and Lotan, 1985). They are also light intolerant and highly flammable, so lodgepole pine depends on fire much in the way jack pine does throughout the Canadian boreal. Fire suppression/control in Yellowstone National Park, northwestern Wyoming, combined with fuel buildup under lodgepole pine, set the stage for the disastrous summer fire season of 1988. Here, in the old-growth stands already decimated by twenty years of pine-bark beetle infestation, crown fires were devastating (Bolgiano, 1989).

In Montana, *Picea glauca* and *Abies grandis* (grand fir) are often associated with montane pine forests, as are some of the conifers typically found in the wetter inte-rior hemlock–Douglas-fir–larch zone (Figure 6.5). On the mid-slopes of Colorado's

Front Range east of Denver, mixed-conifer stands are the norm, with ponderosa pine dominating on the warmer, drier, south-facing slopes at lower elevations, while *Pseudotsuga menziesii* joins it on moister or higher sites and becomes the dominant on cooler, moister, north-facing slopes (Peet, 1981). At higher elevations, *Picea engelmannii* and *Abies lasiocarpa* (subalpine fir) may become important. In the prairies to the east are found forest outliers such as the Black Hills of South Dakota. Ninety-five per cent of the Black Hills is dominated by *Pinus ponderosa,* but the boreal species *Picea glauca* joins it on moist northerly slopes at high elevation (Wright and Bailey, 1982). At Scotts Bluff, Nebraska, ponderosa pine is joined only by *Juniperus scopulorum* (Rocky Mountain juniper) on rugged prairie bluffs.

In the southern Rockies, subalpine fir joins Engelmann spruce, Douglas-fir, and blue spruce on damper sites at higher elevations, while ponderosa pine prefers the lower, drier sites. In addition, numerous characteristic species of pines and oaks, probably representing a distinct flora that developed in the Sierra Madre of Mexico, are also found here and help distinguish this southern Rockies zone from areas to the north (Peet, 1988). *Abies concolor* (white fir) is important and often replaces Douglas-fir on cooler, north-facing slopes. *Populus tremuloides* is also an important species bordering streams, following fire, or ringing damp meadows, and is particularly common on the high plateaux of southern Utah. As well, *Pinus edulis* (Colorado pinyon pine) is frequently encountered here (Lanner, 1981). In the south, from central Arizona to the Guadeloupe Mountains of western Texas (Figure 6.5), forests are quite open and are dominated by *Pinus ponderosa* and juniper on south-facing slopes, while *Pinus monticola* (western white pine) prefers northern slopes. Ponderosa pine is the most important commercial timber species in the southwestern USA, and it occupies the largest area of commercial forest land in New Mexico and Arizona (Mathiasen et al., 1987). The fire-scar record revealed by ponderosa pine in the Chuska Mountains indicates a great reduction in fires since approximately 1830, a factor attributed to fine fuel reductions following the introduction of domesticated grazing animals by the Navajo (Savage and Swetnam, 1990). These park-like Chuska Mountain ponderosa stands only later underwent major structural change to become significantly denser, when warmer and wetter climatic conditions prevailed early this century (Savage, 1991).

Open pine-juniper stands are also characteristic of large areas of uplands and moister canyons scattered throughout the Great Basin (Koniak, 1985; West, 1988). *Pinus monophylla* (singleleaf or single-needle pinyon pine) and *Juniperus osteosperma* (Utah juniper) stands are characteristic of drier forest sites throughout the intermountain region, where they cover some 325,000 km^2. Additional tree dominants include *Juniperus monosperma* (one-seed juniper), *J. scopulorum* (Rocky Mountain juniper), and *J. deppeana* (alligator-bark juniper). These dry pine-juniper forests are also often associated with woody shrubs such as *Artemisia tridentata, A. arbuscula* (low sagebrush), and *Cercocarpus montanus* (birchleaf mahogany) (Austin, 1988). Pinyon-juniper forests have expanded considerably in the last 75 years due to a combination of overgrazing and reduced burning (Wright and Bailey, 1982). When fire does occur, pinyon has a much lower survival rate than the juniper, but it rapidly re-establishes itself. Although many junipers survive a burn and almost all the pinyon die, pinyon density exceeds that of juniper within 60 years and is nearly six times that of juniper within 145 years (Tausch and West, 1988).

On wetter sites at higher elevations, pinyon and juniper tend to be replaced by *Picea pungens* (Colorado spruce), *P. engelmannii,* and *Abies lasiocarpa.* Considerable local diversity is encountered in the many deep canyons characteristic of the region. In the Navajo National Monument, northeastern Arizona, Batatakin Canyon includes not only typical pinyon–juniper communities, but herbaceous, oak, Douglas-fir, and aspen communities as well (Brotherson, 1985). These communities alternate rapidly within the canyon in response to strong microclimatic, soil moisture, and shading conditions.

6.2.4 Sagebrush and Grasslands

This region includes the arid Columbia Plateau of Oregon, Washington, and Idaho, the Great Basin region of Nevada and western Utah, and numerous smaller lowland areas of Wyoming, Colorado, and northern Arizona (Figure 6.5). Because the region is located east of the coastal Sierras, precipitation values are generally very low, except on the upland blocks of the basin-and-range country. Some 563,000 km^2 from Canada to Mexico are considered sagebrush (Chadwick, 1989). Because the region is so large and vegetation cover varies somewhat from north to south, two zones can be readily identified: (1) the bunchgrass steppe of parts of the Columbia Plateau and (2) the sagebrush zone of the central Columbia Plateau, the Great Basin, and Rocky Mountain basins. Some authors partition the region more geographically into the sagebrush-steppe region and the Great Basin sagebrush region (West, 1988).

6.2.4.1 Bunchgrass Steppe

Locally known as the Palouse, these grasslands are in fact the southern extension of the Cordilleran Cold Steppe and Savanna Forest Region of the southern Canadian cordillera. Typical Palouse grassland occupies that position on the climatic gradient between *Artemisia* (sage) steppe on the drier side and *Pinus ponderosa* or Douglas-fir on the moister side (Daubenmire, 1992). While *Agropyron spicatum* (bluebunch wheat-grass), *Poa secunda* (Sandberg's bluegrass), *P. cusickii* (Cusick's bluegrass), *Koeleria cristata* (June grass), and *Festuca idahoensis* (Idaho fescue) were the original dominants, some of the periphery of the region, particularly west of Spokane in eastern Washington, has been overgrazed to the extent that *Artemisia tridentata* (big sagebrush) has invaded along with several annual species, such as *Bromus japonicus* (Japanese brome), *B. tectorum* (downy brome), and *Erodium cicutarium* (filaree) (Risser et al., 1981).

6.2.4.2 Sagebrush

Cover in this vast region is dominated by woody species which exhibit xeromorphic adaptations, together with a number of grass species. The dominant shrub is *Artemesia tridentata* (of which there are three important subspecies, big sagebrush, Wyoming big sagebrush, and mountain big sagebrush), although *A. arbuscula* (low sagebrush) is also common. The dominant native grasses are *Agropyron, Festuca, Poa,* and *Stipa* species, while the exotic *Bromus tectorum* (downy brome) has taken on considerable importance as a spring fodder. In Nevada, sagebrush dominates practically the entire landscape except for pine-juniper forest on cooler upland sites.

Considerable variation in sagebrush species cover results from both fire (Wright and Bailey, 1982) and physical and chemical properties of the soil (Jensen, 1989). None of the major sagebrush species has the capacity to resprout following burning, a factor which is important in explaining successional patterns (West, 1988). Temporal studies on *Artemisia*-grassland sites in southeastern Idaho indicated that following burning, there are early successional species, late successional species, and species common throughout the entire sequence. *Lithospermum ruderale* (puccoon) is early successional and *Artemisia tridentata, A. tripartita,* and *Gutierrezia sarothrae* (broom-snakeroot) are late successional, while base-sprouting species such as *Purshia tridentata, Symphoricarpos oreophilus* (western snowberry), *Amelanchier alnifolia, Achillea millefolium, Agropyron dasystachyum,* and *A. spicatum* are present throughout the entire sequence (Humphrey, 1984). In northeastern Nevada, *Artemisia tridentata* (ssp. *vaseyana*) is often associated with *Symphoricarpos oreophilus* (snowberry) and the grasses *Agropyron spicatum, A. trachycaulum,* and *Festuca idahoensis* (Tueller and Eckert, 1987). In internally drained depression salt-pans with saline soils, characteristic *Atriplex confertifolia* (saltbush) and *Ceratoides lanata* (greasewood) dominated communities, known as "salt-desertshrub" or "saltbrush-greasewood vegetation," often develop (West, 1988). It is noted that similar saltbrush-greasewood communities are found on salt flats farther to the north, including the Frenchman River valley in Grasslands National Park, Saskatchewan. The naturalized shrub *Tamarix ramosissima* has invaded, and now dominates, many low-elevation floodplains throughout the southwestern USA (Busch and Smith, 1993). The flammability of this species has led to a propensity for episodic burning of these communities.

One peculiar cover type in western Nevada consists of small non-upland islands of *Pinus ponderosa* surrounded by vast areas of sagebrush. Apparently these small tree-islands owe their origin to strongly acid, low available phosphorus soils developing on hydrothermally altered rock, to which typical sagebrush species are intolerant and are thereby excluded. Ponderosa pine, on the other hand, tolerates these conditions by exhibiting slow growth, high nutrient-use efficiency, and high reabsorption of nutrients before leaf abscission (Schlesinger et al., 1989).

Because of the sagebrush region's agricultural limitations, 90% of the land is publicly owned. Ranching on rented public lands is one of the few uses made of the vegetation cover, but conservationists have expressed concerns about overgrazing and low grazing fees. The Great Basin National Park in eastern Nevada (established 1986) includes large areas of little-disturbed sagebrush at lower elevations and *Pinus longaeva* (bristlecone pine) in the subalpine zone on Wheeler Peak.

6.2.5 Interior Hemlock–Douglas-Fir–Larch

In the Rockies a number of forest regions below subalpine forest, but above grand fir–Douglas-fir forest are dominated by such species as *Tsuga heterophylla, Pseudotsuga menziesii,* and *Larix occidentalis* (Figure 6.5). These forests are essentially the southern extension of the Canadian interior hemlock zone. The product of a humid climate at intermediate elevations, they derive their moisture from the eastward penetration of humid Pacific air masses. Ideal tree-growing conditions result, and this is the most productive forest region in the Rocky Mountains of the USA. In the central and southern Rockies, effectively moist climates are only found at subalpine elevations, so this montane forest type is restricted to northern Idaho and western Montana. In northern Idaho, *Thuja plicata* and *Pinus monticola* (western white pine) are also frequently encountered, while Douglas-fir is primarily a seral species which is commonly associated with *Abies grandis* at lower elevations.

6.2.6 Subalpine Forest

Numerous areas of subalpine forest are scattered throughout the cordilleran region of the USA, where they take the form of narrow closed forest at lower elevations and parkland at upper. Although not shown on Figure 6.5 these small, narrow regions are found below the identified areas of alpine tundra. In addition, subalpine forests are found along the Cascades, much of the Sierra Nevada, and on numerous small uplands in the Great Basin, where elevations do not create conditions quite cool enough for alpine tundra. In the northern Cascades and parts of the Olympic Peninsula, typical lower-elevation subalpine forest includes *Abies amabilis*, while at higher elevations, *Tsuga mertensiana* and *Abies lasiocarpa* (subalpine fir) dominate. *Pinus contorta* frequently colonizes recent burn sites or volcanic ash. In Oregon some areas of mountain hemlock have been the subject of attack by laminated root rot, which spreads radially from a central infected spot producing a "wave-regenerated sequence" of mountain hemlock death and regrowth (Boone et al., 1988).

Towards the southern Cascades, *Abies magnifica* (red fir) becomes the dominant. In the Sierra Nevada of California, *Pinus contorta* (sierra lodgepole pine variety), *P. monticola* (western white pine), *P. flexilis* (limber pine), *P. balfouriana* (foxtail pine), *P. longaeva* (bristlecone pine), and *Tsuga mertensiana* are representative. This sierra lodgepole pine is larger and longer lived than its Rocky Mountain counterpart (Whitney, 1985). At lower elevations, mountain hemlock and lodgepole pine dominate, while gnarled and windswept individuals of *Pinus albicaulis* are common near tree line (Ornduff, 1974).

In the Rockies, *Abies lasiocarpa* and *Picea engelmannii* dominate. In a sense, these cool-summer forests are a southern extension of the boreal, with the same conifer genera dominating. In the northern Rockies, subalpine fir dominates to higher elevations than Engelmann spruce and can be joined by mountain hemlock. At tree line the alpine fir is frequently joined by *Larix lyallii* (alpine larch), *Pinus albicaulis,* and mountain hemlock. In Colorado, tree-line islands of wind-trimmed Engelmann spruce and alpine fir "migrate" as much as 2 cm yr^{-1} across open alpine tundra, due to windward die-off and leeward colonization (Holtmeier and Broll, 1992).

In the southern Rockies, Engelmann spruce is found to higher elevations than subalpine fir. Here *Pinus flexilis* (limber pine) and *P. aristata* (Rocky Mountain bristlecone pine) can be found on rugged sites. Studies in Colorado bristlecone pine stands suggests that *P. aristata* primarily regenerates after fire (Baker, 1992). In the Colorado Front Range between 2,750 and 3,350 m, Engelmann spruce and subalpine fir are joined by *Pinus flexilis*, *P. contorta,* and *Populus tremuloides* (Veblen, 1986). Here *Pinus flexilis* prefers more xeric locations, and it has been noted that limber pine regeneration following fire is not occurring at some tree-line sites, possibly due to a lack of seed-caching success by Clark's nutcracker (*Nucifraga columbiana*), the bird considered the primary seed disperser for the species (Shankman and Daly, 1988). Spruce beetle outbreaks have also caused community change within *Picea engelmannii, Abies lasiocarpa*, and *Pinus contorta* dominated subalpine forest in Colorado. Unlike the shade-intolerant pine seedling succession one might expect following fire, the previously suppressed small-diameter spruce (which were not attacked by the beetle) and subalpine fir (a non-host species) take over, actually accelerating succession towards shade-tolerant species (Veblen et al., 1991).

In Arizona, subalpine fir is less common than the fir variety known as corkbark fir, with Englemann spruce extending to tree line well above both (Whitney, 1985). In the relatively dry Great Basin subalpine zone, forests are dominated by scattered *Pinus longaeva* (bristlecone pine) and *P. flexilis*, along with sagebrush. *Pinus longaeva* is best known for its ability to outlive even giant sequoias. Individuals have attained ages of approaching 5,000 years (Hiebert and Hamrick, 1984). In the Ancient Bristlecone Pine Forest of the White Mountains, eastern California, the Methuselah tree is more than 4,600 years old. One specimen cut in what is now the Great Basin National Park had over 4,900 annual rings (Whitney, 1985)! The largest known bristlecone pine, the "Patriarch," living in the White Mountains of California, has a circumference of 12 m and is 14.3 m tall, statistics which seem unremarkable until it is appreciated that it lives at 3,350 m above sea level, tolerates a mean annual precipitation of only 25 cm, and is exceeded in girth by the champions of only nine other tree species (Bronaugh, 1991)! By comparison, the oldest *P. aristata* (Rocky Mountain bristlecone pine), from farther east in Colorado, is only about 2,500 years old (Brunstein and Yamaguchi, 1992).

6.3 Primary Production and Phytomass

No other single geographic region has a larger range of biomass values for living vegetation than the Pacific Coast–Cordilleran area of North America. This is easy

Table 6.2
Biomass and NPP values for young and old-growth conifer stands in the Pacific Northwest (summarized in Waring and Franklin, 1979)

Species and Site Details	Biomas $t\,ha^{-1}$	NPP $t\,ha^{-1}yr^{-1}$
Western hemlock and Sitka spruce, 110 years old, Oregon coast	790	9.3
Noble fir and Douglas fir, 115 years old, Oregon Cascades	798	11.8
Noble fir, 400 years old, Washington Cascades	1417*	–
Douglas fir and western hemlock, >400 years old, Oregon Cascades	1442*	–
Coastal redwood "old growth," alluvial flats, California	2902	13.0
Coastal redwood >1000 years old, northern California coast	3139*	–

* Values include only stem biomass.

to appreciate because the region includes everything from alpine tundra to the largest living organisms in the world, the sequoias and giant redwoods. While a typical alpine tundra biomass value would be in the range of $5-10\,t\,ha^{-1}$, values for giant redwood stands in northwestern California can exceed $3,000\,t\,ha^{-1}$ (Waring and Franklin, 1979). By comparison, a relatively high biomass value for lowland tropical rain forest in Brazil would be in the order of $1,000\,t\,ha^{-1}$ (Fittkau and Klinge, 1973). Conifers dominate over hardwoods in this region, not only because they are somewhat better adapted to the drier summer climate, but because of their massiveness and longevity, two features not shared by most hardwoods. Perhaps the only other ecosystem which competes with the larger of these stands in terms of biomass would be the *Nothofagus* (southern hemisphere beech) hardwood forests in southern Chile and western Patagonia, which live under climatic conditions almost identical to the Pacific Northwest, but which lack competition from long-lived conifers.

The breakdown of biomass values within a 450-year-old Douglas-fir stand in the Cascades gives 11.3, 138, 640, and $195\,t\,ha^{-1}$ respectively for leaves, roots, aboveground wood, and logs plus dead standing trees (Reichle, 1969). Table 6.2 indicates biomass values in excess of $3,100\,t\,ha^{-1}$ for stems only in coastal redwood forest, a value which is the equivalent of over 3,800 t when roots and leaves are considered (using 23% as the percentage of forest biomass in living leaves and roots). If litter and dead wood values are added, the figure would rise above $4,000\,t\,ha^{-1}$! By contrast, in exposed Oregon coastal hemlock stands, wind damage reduces biomass accumulation rates due to gap formation. Here, Greene et al. (1992) report that the initial biomass peak of $600\,t\,ha^{-1}$ declines over time due to a positive feedback in which new and enlarging gaps expose more and more trees to wind damage.

Biomass values for *Abies grandis, Thuja plicata,* and *Tsuga heterophylla* stands in northern Idaho averaged $587\,t\,ha^{-1}$ (Hanley, 1976), with the largest value, $937\,t\,ha^{-1}$, being recorded in western red cedar, while net annual productivities for these stands averaged 13.4 t, of which 1.4 t is below ground (Hanley, 1976). Values are somewhat lower in the pine forests of the drier Rockies and Great Basin. Biomasses of 52–245 t ha^{-1} are reported for *Pinus contorta* (lodgepole pine) stands in Wyoming, Colorado, and southern Alberta, with net annual productivities of 2.2–8.4 t ha^{-1}yr^{-1} (as summarized by Peet, 1988), while average above-ground biomass values for Great Basin *P. monophylla* (singleleaf pinyon) are 150.5 t ha^{-1}, of which 60% is tree phytomass, the rest being litter (Everett and Thran, 1992).

Annual net primary productivity also varies considerably throughout the cordilleran region. Values as high as 36.2 t ha^{-1}yr^{-1} were reported for a 26-year-old coastal stand of western hemlock, while young Pacific Northwest stands on better sites probably accumulate 15–25 t ha^{-1}yr^{-1} (Waring and Franklin, 1979). Annual productivity rates for the cordillera as a whole are not appreciably greater than for temperate deciduous forest, so the huge biomass values reported in Table 6.2 must develop because of conifer longevity. While longevity is an obvious contributor to increasing living biomass values, less noticeable is the fact that a steady state for coarse woody debris in the Pacific Northwest Douglas-fir forests may not be reached for at least 1,000 years (Spies and Franklin, 1988)!

CHAPTER SEVEN

Temperate Deciduous Forest

7.1 Distribution

Temperate deciduous forests conjure up visions of soft green foliage, bright orange fall leaves, and scraggy grey winter branches – they in fact constitute one of the most beautiful vegetation formations worldwide and have been immortalized both in painting and on film since the dawn of either medium. With perhaps the exception of fall colours, no particular seasonal condition is remarkable in itself, but it is the variety of phenological conditions which gives this formation its attraction and holds our attention. Unlike the lush tropical rain forest of Brazil or the rugged spruce forests of the Canadian Shield, this vegetation cover keeps us anticipating its next transformation. Also, unlike the boreal or tropical rain forest, temperate deciduous forests owe much of their lushness to both soil and climatic conditions that are also ideally suited to temperate agriculture. Since the Neolithic revolution, these forests have therefore been under attack, and many deciduous forest landscapes, such as those of southern Ontario, the state of Illinois, or southeastern England, have been almost totally transformed into agricultural parklands.

The major areas of broadleaf deciduous forest are in eastern North America, Europe, and China, with minor areas in Mexico and Central America, western Asia, Korea, Japan, and Chile. Although these regions are depicted on Figure 7.1, it is appreciated that different authors give somewhat different definitions to deciduous forest, and therefore different distributions to this formation. As should be expected of a forest type characteristic of temperate conditions with cool winters (ZB VI), mixtures of temperate deciduous species occur with other types of species as one moves to different climatic regions. Recognizing that some authors consider mixed areas as deciduous forest, these too have been added to Figure 7.1. For example, in North America, typical deciduous forest becomes mixed hardwood-conifer forest to the north (ZE VIII–VI), becomes prairie parkland to the west (ZE VI–VII), and forms an ecotone with temperate evergreen forest to the south and southeast (ZE VI–V).

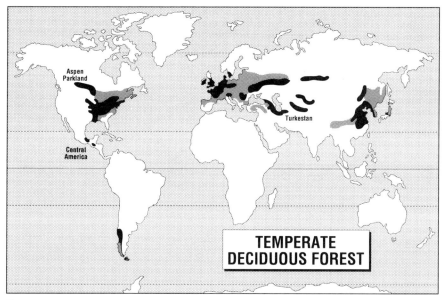

Figure 7.1
Distribution of temperate deciduous forest. True temperate deciduous forest is shown in black, while mixed forest types are stippled.

7.2 Climate

Deciduous temperate phanerophytes are remarkably in tune with their moist, temperate climate. Although deciduous trees and tall shrubs normally make up only 12% of the total vascular species count in southern Ontario deciduous forest (Scoggan, 1978) and 14% in the state of Indiana (Gleason and Cronquist, 1964), these are greater percentages than in the boreal forest to the north. While sapwood and buds experience "winter hardening," the soft leaf tissue is not adapted to prolonged frost or a winter period where temperatures often average close to zero, with at least long periods with night frost (Figure 7.2). Neither are these leaves adapted to summer moisture stress. Before obligatory abscission in the fall, leaves often turn yellow, orange, or scarlet. This yellowing results from the unmasking of carotenoid pigments by the cessation of production and the breakdown of chlorophyll. Brighter colours of red, purple, and pink are often produced by anthrocyanin pigments (Röhrig, 1991a). Before abscission, some leaf nutrients are sent for storage in other parts of the tree.

In the spring, stored nutrients aid in the rapid growth of new foliage, and provided buds are well protected, it would appear that as long as the warm summer growing season is at least four months, these trees are quite capable of producing new foliage

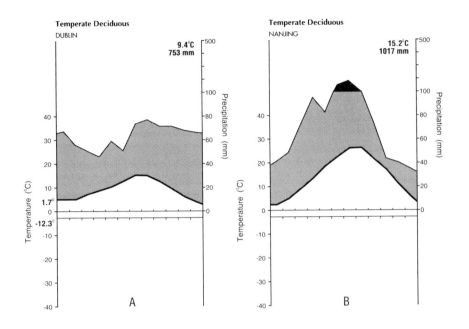

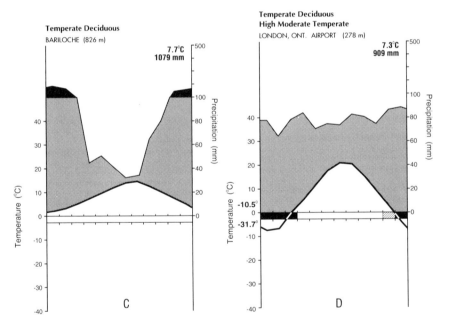

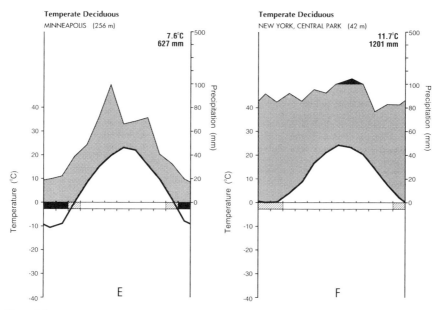

Figure 7.2
Climate diagrams for temperate broadleaf deciduous forest stations. A = Dublin, Ireland; B = Nanjing (Nanking), China; C = Bariloche, Argentina; D = London airport, Ontario; E = Minneapolis, Minnesota; F = New York, New York.

on an annual basis. The soft-tissued leaves also have large photosynthetic surfaces and high transpiration rates, and except where shaded, contribute to relatively large net primary productivities. These high transpiration rates, combined with a general lack of xeromorphic adaptations, make these soft-tissued broadleaves intolerant of moisture stress, so they must have a positive moisture supply throughout the summer. Even with sufficient moisture and little in the way of a frosty winter, deciduous species may give way to conifers if summers are not sufficiently warm. Likewise, they give way to conifers if climatic or physiologic moisture stress becomes a major factor during the growing season. Physiological moisture stress is particularly characteristic of the sandy "pine barren" soils common in the eastern and southeastern portions of the USA. It can also be seen in the aspen parkland to the northwest in such areas as Spruce Woods Provincial Park, Manitoba, an area of pro-glacial lake deltaic sands.

The broadleaf temperate forests of southern Ontario constitute the Humid High Moderate Temperate (HMTh) Ecoclimatic Region, the only ecoclimatic region within the High Moderate Temperate Ecoclimatic Province (HMT on Figure 1.4). Climatic data for London, Ontario, is typical of the region (Figure 7.2D), and mean annual temperature is above 7°C, winters are cool, and summers warm. Mean annual rainfall at London is above 900 mm, and surplus soil moisture in all months results from the frequent passage of frontal systems. New York City has a longer growing season and a higher average annual precipitation than London, Ontario, but is similar in terms of moisture surplus (Figure 7.2F). While New York demonstrates the oceanic influence on temperate forest, Minneapolis, Minnesota (Figure 7.2E), is more representative of a drier continental USA station where the forest's western margins begins to mix with prairie to form oak parkland. Minneapolis has a somewhat cooler winter, but a warmer summer, with a mean annual temperature of 7.6°C and a mean annual precipitation of 627 mm. While London, Ontario, has a strongly positive moisture index and a pronounced summer moisture surplus, Minneapolis has an im closer to zero, with periodic moisture stress.

Climatic data for Nanjing, China, and Minneapolis look quite similar except that Nanjing's more southerly and less continental geographic position makes it warmer and wetter (Figure 7.2B). Due to the influences of the North Atlantic Drift, western Europe has cooler summers than most other temperate forest regions, but these are compensated for by a longer growing season. This maritime climate also explains why these European forests are located up to 10° latitudes farther north than elsewhere. Data for Dublin, Ireland (Figure 7.2A), are typical of stations under the direct influence of moisture-bearing frontal systems coming from the west and the ameliorating influence of the North Atlantic Drift. Like Nanjing, New York has major differences between winter and summer temperatures, and like Dublin, precipitation is relatively even throughout the year.

7.3 Soils

As mentioned above, soils of the temperate deciduous forests have provided suitable agricultural conditions for farmers since the beginning of the Neolithic. By no means are all such forest soils suitable for the plough, however, because on steeper slopes, agricultural exploitation has been met with disastrous consequences. On gently rolling, level, or floodplain terrain, deciduous forests not only thrive, but soils are usually fertile and have attracted great attention. The removal of forests cover in China, western Europe, and parts of the eastern USA attest to this attraction. Unfortunately, European agricultural practices did not suit the hilly terrain of many New England states, and early colonial deforestation here has been reversed following serious erosion and land abandonment. Floodplain forest removal has also led not only to erosion episodes, but to human disasters on a vast scale. Forested areas prone to spring flooding are usually left undisturbed, so that in an agricultural landscape, such as in the London region of southern Ontario, floodplains appear as gallery forest in an otherwise agricultural parkland environment.

In Canada, typical soils types include the Melanic Brunisols, Gray-Brown Luvisols, and Humic Gleysols (Figures 1.9, 4.4). The first two of these represent soils which are moderately leached and mildly acidic, but which may well have thick Ah horizons. The development of organic rich Ah horizons under deciduous forest is to be expected. The mildly acid, relatively easily digested mull litter is an ideal food source for many types of soils organisms, in particular bacteria, fungi, and earthworms. The thick fall litter layer infrequently remains so, because upwards of 75% of one fall's residues are broken down by the next summer (Edwards et al., 1973). This rate of decomposition varies significantly with tree species and depends to a large extent on the C:N ratio. One study found that *Fagus* leaves lost 64% of their weight, oak lost 89%, and *Ulmus, Betula*, and *Fraxinus* leaves broke down completely after only one year (Heath et al., 1966). In addition to promoting breakdown, burrowing by earthworms ensures incorporation of humus into the upper mineral layers, thereby contributing to the well-aggregated, organic-rich Ah horizon, which is so attractive to farmers. The moderately high nutrient and calcium contents of deciduous foliage has also contributed to an efficient recycling of nutrients in the sense of reducing soil acidification.

In the northern deciduous forests of the USA, the dominant soils types are Alfisols (Degraded Chernozems and Brown Forest soils), while in the warmer south, Ultisols (Red-Yellow Podzolics) predominate (Figures 1.7, 1.8). In Europe, neutral to mildly acidic Brown-Earths characteristically develop under broadleaf deciduous forest (Polunin and Walters, 1985), although more acidic Brown-Earths and Gray-Brown Podzols are also encountered. In China and Japan these soils types are joined by extensive areas of alluvial (gley promoting) floodplain soils. Throughout these regions it would appear that not only can deciduous forest exist on soils with a wide range in pH and moisture content, but that many individual tree species do likewise. *Fagus sylvatica*, the European common beech, appears particularly tolerant of a wide range of pH and moisture conditions, although it appears sensitive to both the extremely dry and the very moist sites (Ellenberg, 1988).

7.4 Temperate Deciduous Forest in North America

That portion of the Nearctic Floristic Realm (Figure 1.6) which is dominated by temperate broadleaf deciduous forest constitutes the Carolinian Floral Region. Due to the presence of so many southern Carolinian elements in the hardwood forests of southern Ontario, it is considered part of this Carolinian Region (Scoggan, 1978). Based on Küchler's (1964) classification, this North American temperate broadleaf deciduous forest region can be conveniently subdivided into six characteristic forest types: (1) Beech-Maple (*Fagus-Acer*), (2) Elm-Ash (*Ulmus-Fraxinus*), (3) Appalachian Oak (*Quercus*), (4) Mixed Mesophytic (*Acer-Aesculus-Fagus-Liriodendron*), (5) Oak–Hickory (*Quercus–Carya*), and (6) Maple-Basswood (*Acer-Tilia*). The locations of these six forest types are shown in Figure 7.3, together with ecotones or transitional zones to the north, west, and south and some extensions of the Northern Floodplain forest (*Populus-Salix-Ulmus*) along western rivers. Excluded from Figure 7.1 is the aspen parkland to the northwest. For a full discussion of the aspen parkland and the transitional zone to oak-prairie parkland and Northern Floodplain forest, see chapter 5, and for detail on the mixed forests (northern hardwood forests) to the north, see chapter 4. More detailed subdivisioning of

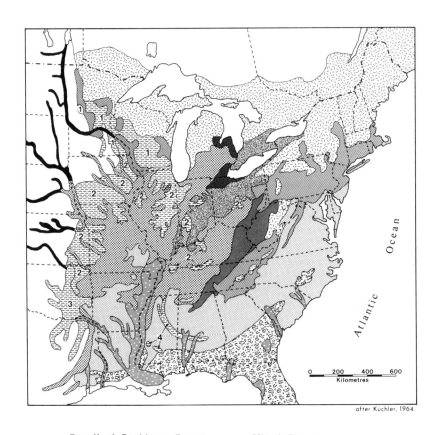

after Küchler, 1964

Broadleaf Deciduous Forest

Beech - Maple *(Fagus - Acer)*

Elm - Ash *(Ulmus - Fraxinus)*

Appalachian Oak Forest *(Quercus)*

Mixed Mesophytic Forest

Oak - Hickory *(Quercus - Carya)*

Maple - Basswood

Northern Floodplain Forest *(Populus - Salix - Ulmus)*

Mixed Forests

Southern Mixed Forest

Southern Floodplain Forest

Oak - Hickory - Pine

Great Lakes - St Lawrence
Mixed Forest (northern hardwood
and spruce)

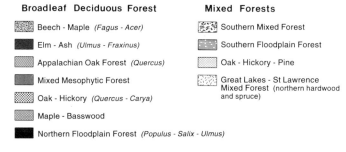 **Prairie - Broadleaf Transition**

1. Oak Parkland
(Quercus - Andropogon)

2. Mixture of Tall-Grass Prairie
and Oak - Hickory Forest

3. Cross Timbers Oak - Parkland
(Quercus - Andropogon)

4. Blackbelt Deciduous Forest with
Tall-Grass Prairie Patches

Figure 7.3
Temperate broadleaf deciduous forest in North America (summarized primarily from Küchler,
1964)

the Carolinian region is available in Barnes (1991), Braun (1967), and Küchler (1964). Small areas in southern Mexico and adjacent Central American states show affinities to this Carolinian region, but most are mixed deciduous-evergreen broad-leaf stands.

7.4.1 Canada

Although only 0.22% of Canada is included in the true Carolinian temperate broad-leaf deciduous forest (Figure 7.3 and the HMT area on Figure 1.4), it is noted that relatively large areas of deciduous hardwoods are also found in the mixed forest zone (northern hardwoods and northern hardwoods–spruce zone) of eastern Ontario, southern Quebec, and parts of the Maritime provinces (see chapter 4). What distin-guishes these two forest types from each other is (1) the degree to which conifers are present in the general region and (2) the presence or absence of a southern Carolinian hardwood element. This Carolinian forest region is therefore considered distinctive, not only because it is generally pure deciduous hardwoods, but because many of the tree species found here are at their most northerly distribution and are found nowhere else in Canada. Locating an exact transition zone between this de-ciduous broadleaf forest and the mixed hardwood-softwood forest to the north is problematic, but it generally correlates with the 7.8°C (46°F) mean annual isotherm (Thaler, 1970). In terms of geographical location, this region is therefore restricted to the southern fringes of Ontario (Figure 4.7). The northern boundary runs east from Grand Bend on the southeastern shores of Lake Huron, passes between Kitchener and London, and extends along the northern shore of Lake Ontario as far east as the Rouge River estuary just east of Metropolitan Toronto (Allen et al., 1990; White, 1977).

The whole of this region falls within the Beech-Maple forest type as delineated by Küchler (1964). It is dominated by *Fagus grandifolia* (American beech) and *Acer saccharum* (sugar maple), while *Acer rubrum* (red maple), *Tilia americana* (bass-wood), *Quercus rubra* (red oak), *Q. alba* (white oak), and *Q. macrocarpa* (bur oak) are common. In addition, many tree species found nowhere else in Canada fre-quently occur and include *Liriodendron tulipifera* (tulip-tree), *Quercus prinus* (chestnut-oak), *Q. palustris* (pin-oak), *Morus rubra* (red mulberry), *Asimina triloba* (paw-paw), *Juglans nigra* (black walnut), *Fraxinus quadrangulata* (blue ash), *Magnolia acuminata* (cucumber-tree), and *Sassafras albidum* (white sassifras). Further, the northern extension of *Platanus occidentalis* (sycamore) is almost re-stricted to this region. Also found here are some southern-element hardwoods, such as *Nyssa sylvatica* (black gum) and *Gymnocladus dioica* (Kentucky coffee-tree), which possess very narrow Canadian distributions.

Before being decimated by the bark beetle – transported fungus blight (*Endothia parasitica*), *Castanea dentata* (American sweet chestnut) was very common through-out the region (Scoggan, 1978). This blight first appeared in New York State in 1904, and as early as 1913 the chestnut, which was the single most outstanding and useful species in the hardwood region of the eastern USA, was clearly doomed (Russell, 1987; West, 1988; Cochran, 1990). While this blight causes cankers on the branches and trunk and effectively girdles them, the protective outer bark of the roots frequently remains fairly resistant and allows for sprouting. By 1950, except for some isolated individuals, all that remained of the chestnut were stumps and root and stump sprouts. Unfortunately, these sprouts do not mature to full-sized trees because individual stems become reinfected before they reach more than a few centimetres in diameter (Paillet, 1984). It is hoped that beech bark disease, which arrived in Nova Scotia from Europe around 1890, does not do the same to the stately, silvergrey barked beech. The potency of beech bark disease results from the combined activity of the beech scale insect (*Cryptococcus fagisuga*), which pierces the bark to feed on the living cells below, and a fungus of the *Nectria* genus, which kills the cambium already weakened by the scale (Cammermeyer, 1993). Fortunately, research shows that unlike the chestnut, a small percentage of beech trees in infected areas are resistant to this disease.

Sizeable areas of remnant deciduous forest are found along creeks such as Bronte Creek near Burlington and Highland Creek in Scarborough. Here, wooded slopes are dominated by *Quercus rubra, Q. macrocarpa, Acer saccharum, A. negundo, Fagus grandifolia, Ulmus americana, Fraxinus pennsylvanica, Betula lutea,* and *Tilia americana*. Beech is particularly common in the understorey, while oaks and maples normally comprise the canopy. Ground cover is sparse under dense beech cover, while more open forest is often associated with *Pteridium aquilinum* (bracken), *Rhus vernix* (poison sumac), and shrubs. Towards the north and northeast, *Pinus strobus* (white pine) and *Tsuga canadensis* (eastern hemlock) are also occasionally encountered on slopes, and *Thuja occidentalis* (eastern white cedar) along floodplains, particularly along the lower Rouge River, Scarborough. All but three of the twenty-eight southern Carolinian element tree and herbaceous species found in the Rouge Valley Park have their northeastern range limits here, a clear indication that it is close to where the Carolinian abuts with the mixed forest (Varga et al., 1991). These twenty-eight species include *Platanus occidentalis* and *Bromus pubescens* (Canada brome). To the southwest are small near-pristine remnant stands, such as at Bronte Creek Provincial Park, but much of southern Ontario's broadleaf deciduous forest has been reduced to small remnant stands and woodlots. For a review of the locations of significant remnants, see Allen et al. (1990).

Along the Niagara Escarpment, many homeowners with large lots have fortunately retained a relatively natural forest cover. Where the escarpment cliff edge is steep, species richness–diversity indices increase, and there is often a pronounced transition from dense canopy *Acer saccharum* and *Quercus rubra* dominated forest on the plateau above to a broken *Thuja occidentalis* canopy on the cliff proper (Bartlett et al., 1991; Larson et al., 1989). This old-growth, little-disturbed forest, with stunted *Thuja,* is found on cliffs all along the escarpment and contains some cedar individuals which exceed 1,000 years in age (Larson and Kelly, 1991). Such stunted cedar growth may well be related to moisture deficiencies, because it does not seem to result from lack of mycorrhizal inoculation or deficiencies in macronu-trient supply (Matthes-Sears et al., 1992). Longevity on such rocky, xeric sites may well be related to reduced competition, slow growth rates which maintain structural integrity, and reduced chances for severe fires, all reasons noted above for the pos-sible longevity of 900-year-old cedar on xeric rock outcrops in the boreal of north-western Quebec (Archambault and Bergeron, 1992).

A characteristic of many deciduous forests floor is the presence of numerous shrub and herbaceous species. Common shrubs in Ontario hardwood forests include *Zanthoxylum americanum* (northern prickly ash), *Juniperus virginiana* (red cedar), *Rhus aromatica* (fragrant sumac), *R. vernix*, and Carolinian species such as *Crataegus mollis* (hawthorn), *Vitis aestivalis* (summer-grape), *Cornus drummondii* (roughleaf dogwood), and *Vaccinium stamineum* (deerberry). Many of the forbs and some of the grasses, such as *Aristida dichotoma* and *A. purpurascens*, are also not found elsewhere in Canada (Scoggan, 1978). Within the herbaceous layer of typical *Acer saccharum* forest, aspect influences dominancy (Goodwillie, 1975). *Medeola virginiana, Pyrola* sp. (wintergreen), *Trientalis borealis* (American star-flower), *Aralia nudicaulis,* and *Clintonia borealis* (bluebead-lily) are generally restricted to the north-northwest aspects, while *Trillium undulatum* (painted trillium) and *Lycopodium lucidulum* (shining club-moss) are primarily facing north-northwest and northeast. *Brachyelytrum erectum* (bearded short-husk), *Oryzopsis canadensis, Prenanthus altissima* (rattlesnake-root), *Lycopodium obscurum* (ground-pine), *Thelypteris novaboracensis* (New York fern), and *Aster cordifolius* occur mainly on the south side, where species diversity is also the highest (Goodwillie, 1975).

Concern has been expressed about the impacts of forest clearance and drainage on the long-term conservation of this Carolinian forest element in Canada (Allen et al., 1990). The northern half of Ekfrid County near London had in excess of a 70% forest cover in 1869, but by 1980 less than 5% forest cover remained (Hilts and Cook, 1982). So bare and open has the landscape become that many farmers no longer have their own woodlot and are forced to buy cordwood (Warkentin, 1968), and so small are many of the remaining groves that the long-term survival of remaining narrow-distribution, southern-element species is in question. The issue of scale invokes the implications to species equilibrium outlined in the island biogeog-raphy theory (Eagles, 1990). It could well be that the paucity of fruits on some *Gymnocladus dioica* (Kentucky coffee-tree) today is the result of long distances to trees in other isolated groves (Ambrose and Kevan, 1990). Along the Great Lakes, the swampland tree species *Nyssa sylvatica* (black gum), which was always rare in Canada, is now even rarer as it faces regeneration problems following increased drainage of its habitat (McCaw, 1985).

7.4.2 United States of America

This temperate deciduous forest is the largest single forest of its kind in the world. As shown in Figure 7.3, it can be conveniently subdivided into six major forest regions, as well as minor areas of floodplain forest in the prairies and a number of mixed hardwood-conifer types which are found to the north, southeast, and south. It should also be noted that the great variations in elevation, topography, and geology that exist throughout this zone, particularly in the Appalachians, give rise to a complex mosaic of communities (Whittaker, 1956). In addition, broadleaf forests mix with tall-grass prairie to the west to form parklands. In the warmer parts of this distribution, phanerophytes comprise as much as 17.6% of the total vascular species count, while chamaephytes constitute less than 2% (Scoggan, 1978). Although the discussion which follows primarily concerns itself with regional differentiation based on tree species, the important role of herbaceous species is acknowledged. For a summary of variations in herbaceous dominants throughout the region, see Greller (1988).

7.4.2.1 Temperate Deciduous Forests

Beech-Maple Forests   These forests are described above in section 7.4.1 because this cover type is also prominent in southern Ontario. In the USA it is found along the southern shores of Lake Ontario and Lake Erie and extends southwest through much of Ohio and Indiana. Forests tend to be remarkably uniform throughout the region, bearing in mind that somewhat different combinations of tree species dominate upland forests as compared to floodplain stands. In southwestern Ohio the two dominants, *Fagus grandifolia* and *Acer saccharum*, comprise between 80 and 90% of the canopy on uplands even though as many as fifteen tree species comprise that canopy (Braun, 1967). Along the margins of the fragmented remaining stands of old-growth forest, however, *Carpinus caroliniana* (ironwood), *Fraxinus americana, Ostrya virginiana* (hop-hornbeam), *Quercus rubra,* and *Tilia americana* (basswood) show an increased importance (Whitney and Runkle, 1981). In south-central Indiana, typical dominants are *Acer saccharum, Fagus grandifolia, Tilia americana,* and *Ulmus fulva* (slippery elm). Other components include *Ulmus americana* (American elm), *Aesculus glabra* (Ohio buckeye), *Carya ovata* (shellbark hickory), *Fraxinus americana, Liriodendron tulipifera* (tulip tree), *Juglans nigra* (black walnut), *Prunus serotina* (black cherry), and *Quercus rubra* (Küchler, 1964).

While *Ulmus americana* is not as significant a component of this vegetation cover as in the Elm-Ash forest region, many elms have become infected with the previously noted Dutch elm disease (see section 5.4.2.1). Although the disease spread quite rapidly across the deciduous forest region of North America, it fortunately has had neither the rate of infection among individuals nor the cessation of reproduction effects that the chestnut blight has had on *Castanea dentata*. While many elms have died, particularly in cities, reproduction has continued under natural conditions. In 1976, elm mortality-regeneration was studied in an east-central Indiana Beech-Maple forest in which elms formed an important part in an original 1926 survey. Results of the 1976 survey showed that of the 172 individuals tagged in 1926, 94.5%

had died. The 1976 survey also showed that there were now 633 elm individuals in the same study area, indicating a threefold increase in elm numbers (Parker and Leopold, 1983). It would appear that early reproductive maturity has allowed the elm to persist, although its presence is limited to smaller diameter individuals.

Elm-Ash Forests Bordering the shores of southwestern Lake Erie and southern Lake Huron are mesic lowland deciduous forests dominated by *Ulmus americana* (American elm) and *Fraxinus pennsylvanica* (green ash). Soils are typically poorly drained Peaty Gleysols and groundwater Podzols. Because of the poorly drained nature of this forest type, early settlers gave the name "black swamp" to the large area bordering southwestern Lake Erie. Dense forest covered the black swamp region until the latter part of the nineteenth century, but subsequent clearing has left only 15% in forest today (Boerner and Cho, 1987). While the most poorly drained sites within the black swamp are dominated by *Acer saccharinum* (silver maple), *Fraxinus nigra* (black ash), *F. pennsylvanica, Ulmus americana, Quercus macrocarpa,* and *Carya laciniosa* (big shellbark hickory), intermediately drained sites have a characteristic cover of *Fraxinus americana, Tilia americana* (basswood), and *Acer rubrum* (red maple). Better-drained stands within the black swamp are essentially Beech-Maple forests (Boerner and Cho, 1987).

Appalachian Oak Forest The northern portion of this large forest region extends south across the northern Appalachian Mountains from southern New York State and New England to New Jersey, and southwest through Pennsylvania to eastern Ohio and northern West Virginia. A major southern extension is also found at higher elevations along the spine of the Blue Ridge region of the Appalachians from western Virginia through North Carolina to northern Georgia. In the past this region was known as the Oak-Chestnut forest, with chestnut comprising up to 40% of the basal area of stands within the Appalachians (Braun, 1967). Today, following decimation of *Castanea* by the chestnut blight, use of this name must be qualified. While many dead chestnut trees sprout from their bases to form clones, there are no mature chestnut trees, and the disruption to the forest ecosystem has been so profound that it has not yet stabilized. In northeastern Massachusetts, chestnut sprout densities range from one to more than two hundred, living clones per hectare. Here the original trees died before 1922, and over 95% of the sprouts show no connection to pre-blight trees, but originated as suppressed seedlings before 1922. It would appear that no new chestnut seedlings are becoming established and that many have been through a series of blight cycles. The chestnut story is far from over, however, because under normal precipitation regimes, sprouts have tended to increase in number and size and are often one of the dominant woody components in the early revegetation of clear-cuts in the southern Appalachians (Paillet, 1984). A recent precipitous decline in chestnut clones in southwestern Virginia is attributed to the interaction between the blight and growing season droughts during the eighties (Parker et al., 1993).

Russell (1987) reports occasional fruiting individuals in north-central New York, eastern Pennsylvania, and northern New Jersey, while some isolated, planted mature individuals still stand (Cochran, 1990).

Arboreal dominants within these tall deciduous forests are *Quercus alba* (white oak) and *Q. rubra* (northern red oak). Other major components include *Acer rubrum, A. saccharum, Betula lenta* (sweet birch), *Carya cordiformis, C. glabra, C. tomentosa, Fagus grandifolia, Liriodendron tulipifera,* and several additional species of *Quercus* (Küchler, 1964). Although there is little historical record of pre-colonial intentional forest burning (Russell, 1983), it was likely that these eastern hardwood forests were relatively frequently burned by the Indians (Abrams, 1992). Pre-colonial oak forests of southern New England, together with other forest areas such as the Mid-west, the mid-Atlantic region, and the southern Piedmont, probably had a 50–100-year fire frequency which promoted oak dominancy (Abrams, 1992). To the west, in central Pennsylvania, valley floor forests were dominated by *Q. alba* and *Pinus strobus*, but for a time following settlement, fire and charcoal production favoured oak, while more recently, logging has accelerated the rate that *Acer* spp. reach dominancy (Abrams and Nowacki, 1992). At the time of colonial settlement, *Fagus grandifolia* was the forest dominant in Massachusetts; however, settlement brought about a rapid decline in beech, a situation taken advantage of by *Castanea*. Now that the chestnut has been decimated, species of *Quercus, Betula,* and *Acer* have the advantage in forest succession (Bradshaw and Miller, 1988). Evidence indicates that *Quercus* spp. are not typical dominants in late successional forests unless there is periodic burning or extreme edaphic or climatic limitations (Abrams, 1992).

At several points along the spine of the Blue Ridge Mountains, oak forest gives way to outliers of mixed forest dominated by *Acer, Betula, Fagus,* and *Tsuga*. At the highest elevations, such as on the summit of Mount Rogers in southwestern Virginia, *Picea rubens* (red spruce) and the southern Appalachian endemic *Abies fraseri* (Fraser fir) are also encountered (Stephenson and Adams, 1984). Like the chestnut and American elm, this fir has undergone recent rapid decline due to the insect *Adelges piceae* (balsam woolly adelgid), which was introduced to North America from Europe prior to 1908 (Busing and Clebsch, 1988). It is also possible that air pollutants such as ozone may be increasing mortality (Steiguer et al., 1990). The net effect of this is that the proportion of dead standing trees is higher in these mixed stands than in deciduous stands (Tritton and Siccama, 1990). In the Appalachians of central West Virginia and northern Virginia are the most southerly stands of *Abies balsamea* (balsam fir). This fir is normally associated with the co-dominant *Picea rubens* and with some *Tsuga canadensis* and *Betula lutea* (Stephenson and Adams, 1986). Stretching north from Springer Mountain, Georgia, along the mountain spine to Mount Katahdin in central Maine runs the 3,760 km hiking trail known as the Appalachian Trail (Keyser, 1988). This is the longest marked footpath in the world, and it provides the hiker with magnificent views of these hardwood, mixed, and coniferous forests.

Mixed Mesophytic Forests   As this subregional name suggests, these deciduous forests are found on the relatively well drained soils which characterize the unglaciated uplands of the Cumberland Plateau in eastern Tennessee and western Virginia, and in the Allegheny Mountains in eastern Kentucky, southeastern Ohio, and West Virginia. Broadleaf dominants include *Fagus grandifolia*, *Liriodendron tulipifera*, *Tilia heterophylla* (white basswood), *T. caroliniana*, *T. americana*, *Acer saccharum*, *Aesculus octandra* (sweet buckeye), and *Quercus* spp., while *Betula alleghaniensis*, *Magnolia acuminata* (cucumber-tree), *Prunus serotina* (black cherry), *Acer rubrum* (red maple), and *Fraxinus americana* (white ash) are often locally abundant (Braun, 1967). In southeastern Ohio the most important canopy species in old-growth stands are *Quercus alba*, *Acer saccharum*, *Liriodendron tulipifera*, and *Q. prinus*, while *Acer saccharum* and *Cornus florida* (dogwood) are the most important tree species in the sapling stratum (McCarthy et al., 1987).

Oak-Hickory Forest   Found in the west and west-central portion of the North American temperate deciduous forest, this is the largest of the six subregions identified as typical of broadleaf deciduous forest. It stretches from the border with Canada all the way south to Texas. On its drier western margin, it is bordered by oak-grassland parklands, while on its eastern edge, it is bordered by Mixed Mesophytic Forest. Here, moisture indices are modestly positive, and sufficient moisture is provided for closed forest formation. Braun (1967) calls the wetter eastern portions of this Oak-Hickory forest found east of the Mississippi River the Western Mesophytic Forest, an area elevated to temperate forest subregional status by others (e.g., Barnes, 1991).

On the better-drained soils throughout the great latitudinal range of the Oak-Hickory forest west of the Mississippi are found *Quercus alba* (white oak), *Q. rubra* (northern red oak), *Q. velutina* (black oak), *Carya cordiformis* (bitternut hickory), and *C. ovata* (shellbark hickory). *Quercus ellipsoidalis* (northern pin-oak) is essentially restricted to the northern portion of this western Oak-Hickory forest, where *Quercus macrocarpa* also assumes importance. In the southern portion, the dominant oaks include *Q. marilandica* (blackjack oak), *Q. stellata* (post oak), and *Q. shumardii* (Shumard red oak), while *Carya arkansana* and *C. villosa* are the dominant hickory species (Braun, 1967). In the eastern portion, east of the Mississippi, oaks and hickories are joined by *Tilia heterophylla* and *Aesculus octandra* (sweet buckeye), two species more typical of the mesophytic forest region to the east. In addition, upland sites in western Kentucky, Tennessee, and adjacent states often include *Fagus grandifolia* and *Liriodendron grandifolia*, two species also characteristic of the forests to the northeast, as well as *Pinus echinata* (short-leaf pine) and *Caraya tomentosa* (mockernut), two species more characteristic of the south (Franklin et al., 1993). In Kentucky are also found extensive areas of parkland locally called the "bluegrass region" after *Poa pratensis* (bluegrass). These bluegrass parklands were no doubt originally forest, but were modified by fire. Not shown on Figure 7.3 are the large cedar glades of the Nashville Basin, central Tennessee. Here the base-rich flatlands (overlying limestone) are dominated by *Juniperus virginiana* (red cedar), while *Quercus stellata, Carya ovata* (shellbark hickory), *Cercis canadensis* (redbud), and *Ulmus alata* (winged elm) are associates (Braun, 1967).

The dominance of oaks throughout this forest region, and especially in the oak parklands close by, may be related to past land use and frequent disturbance. *Quercus rubra* (northern red oak) appears to have been greatly encouraged by frequent fires and heavy cutting because of its ability to sprout. Following fire, oaks show much more sprouting success than do shade-tolerant hardwoods, so if fires are relatively frequent, oak can increase its importance. Adaptations by oak to moisture stress may also help account for their dominance, both here and in the oak-prairie parkland to the west. These adaptations include reduced dark respiration under stress and an ability to acclimate photosynthesis to a broad range of leaf temperatures and continue photosynthesis under stress conditions (Hicks and Chabot, 1985). In eastern deciduous forests, the practice of producing even-aged stands through clearcutting has proved disadvantageous to the re-establishment of oak (Crow, 1988). In old-age stands, however, oaks are present, having taken advantage of openings created by the death of chestnuts or by windthrow (Barden, 1981; Bradshaw and Miller, 1988).

Maple-Basswood Forests   These forests are transitional between true temperate deciduous forests to the southeast and the riverine communities of the aspen parkland in the continental interior (Figure 7.3). At the same time, they are bordered to the north and east by mixed forest and boreal forest outliers and to the south and southwest by tall-grass prairie, and are frequently broken up by areas of derived oak-prairie parkland. Occupying primarily the Driftless area of Wisconsin and scattered throughout southern and central Minnesota, these forests are dominated, as the forest name suggests, by *Acer saccharum* and *Tilia americana* (basswood). *Carya cordiformis* (bitternut hickory), *Quercus macrocarpa,* and *Q. rubra*, which are typical of forests to the southeast, are also present. Likewise, the presence of *Acer negundo, Fraxinus pennsylvanica,* and *Ulmus americana*, species also typical of riverine communities in the northeastern prairies, attest to the transitional nature of this forest type. Increasing aridity towards the west is reflected in the decline of *Acer* and the increased importance of *Tilia*. Sugar maple reaches its limit in west-central Minnesota, while basswood extends into the Dakotas and southern Manitoba (Barnes, 1991). *Tilia americana* is therefore an important species in two parts of Canada: in the prairie riverine communities of southern Manitoba (see section 5.4.1.2) and in the beech-maple forests of southern Ontario.

7.4.2.2 Mixed-Forest Types

There are three major mixed-forest types bordering true deciduous forests in addition to the northern hardwood forest (the mixed forests of the boreal-deciduous ecotone discussed in section 4.4.3). In terms of size, the largest zone is the Oak-

Hickory-Pine forest, which covers a great swath across the southern and southeastern USA from eastern Texas to New Jersey. Intersecting this forest along floodplains is a second mixed forest known as the southern floodplain forest (Figure 7.3). The final mixed forest community is the southern mixed forest, but as this latter type includes broadleaf evergreen, broadleaf deciduous, and evergreen coniferous tree species, along with some areas of truly evergreen forest, it is not discussed here.

Oak-Hickory-Pine Forest   This mixed-forest region includes much of the eastern seaboard of the USA, together with the northern halves of Alabama, Mississippi, and Georgia. Here, pines are a frequent part of succession, and although they do not normally constitute climax forest, except on the better-drained, sandy soils, frequent disturbance ensures their importance and prevents the establishment of pure broadleaf deciduous forest. Fire disturbance regimes of the past significantly increased the importance of conifers throughout the region (Barnes, 1991). Today, dominants include *Carya* spp. (hickory), *Pinus taeda* (loblolly pine), *P. echinata* (short-leaf pine), *Quercus alba,* and *Q. stellata* (Küchler, 1964). In addition to the pines, *Oxydendrum arboreum* (sourwood) and *Nyssa sylvatica* (black gum), two tree species not typical of the deciduous Oak-Hickory forests to the northwest, help distinguish this region from other forest types (Braun, 1967). While *Quercus alba* is the most abundant tree throughout the Piedmont region of Virginia, *Carya* spp. are most abundant on the calcium- and magnesium-rich Triassic sediments and are of relatively low importance on more acid soils (Farrell and Ware, 1991). Pine forests were once much more extensive throughout the region.

There are many areas along the eastern seaboard where low-nutrient-retaining outwash sands and gravels of the Wisconsin Glaciation support a xeromorphic, fire-adapted cover, frequently referred to as "pine barrens" (or sometimes "pinelands" or "pine plains"). These barrens are characterized by an open canopy of *Pinus rigida* (pitch pine), with an understorey of scrub oaks such as *Quercus ilicifolia* (bear oak) and *Q. prinoides* (willow oak) and ericaceous shrubs. Some oak-dominated barrens are called "oak barrens" (Greller, 1988). While most of the barrens are encountered in the oak-hickory-pine region, they can also be found in such areas as the Hudson Valley sand plain, surrounded by deciduous forest. *Pinus rigida* – dominated sandy glacial soils are encountered as far north as Cape Cod (Tzedakis, 1992). Barrens frequently experience burning episodes, but if fire is excluded, composition often changes as shade-tolerant, fire-intolerant species of oak, maple, and ash become established (Milne, 1985). At the Brookhaven National Laboratory on Long Island, typical forest dominants include *Quercus alba* (white oak), *Q. coccinea* (scarlet oak), *Q. velutina* (black oak), and *Pinus rigida* (pitch pine). Stands also frequently include some *Carya glabra* (pignut hickory) and *Sassafras albidum* (white sassafras) (Whittaker and Woodwell, 1969).

The 520,000 ha southern New Jersey pine barren region has a typical pitch pine and oak cover and is subject to both wildfires and prescribed burning as a management tool (Little, 1979; Boerner, 1983). Here, community diversity from transitional pine-oak forest on uplands to mixed pine-maple swamp in lowlands is most closely correlated with the soil hydrologic regime (Zampella et al., 1992). In northern Virginia are found stands of sparse-canopy pitch pine–bear oak, where *Pinus rigida* and *P. virginia* (Virginia pine) are joined by occasional stunted *Quercus marilandica* (blackjack oak) and *Q. prinus* (chestnut oak). Here the undergrowth is frequently a dense thicket of *Q. ilicifolia* (bear oak), ericaceous shrubs, and frequently, dense *Kalmia latifolia* (mountain-laurel) (Olson and Hupp, 1986). In the southwestern part of this forest region is a great crescent-shaped area commonly called the Alabama Blackbelt (Figure 7.3). In reality, this region is not distinctive, for it is primarily forest with some prairies and therefore much the same as the surrounding forest areas (Rostlund, 1957).

Southern Floodplain Forest   The largest single area of these mixed hardwood-conifer floodplain forests is that along the southern Mississippi (Figure 7.3). As should be expected in a floodplain environment, minor land-form variations impact greatly on flooding regime, soil moisture conditions, and forest species dominancy. Based on freshwater flooding regimes, three generalized swamp types can be recognized: deep swamps, shallow swamps, and peaty swamps (Barnes, 1991). Floodplain deep-swamp forests such as those along the Savanna River in South Carolina (Figure 7.3) are dominated by *Taxodium distichum* (bald cypress) and *Nyssa aquatica* (water tupelo). Adjacent shallow-swamp bottomland mixed-forest communities include *Nyssa sylvatic* var. *biflora* (swamp tupelo), *Quercus* spp., *Liquidambar styraciflua* (sweetgum), and *Pinus taeda* (loblolly pine). Peaty swamps are generally dominated by evergreen phanerophytes. Normal winter flooding promotes hydrochory (seed dispersal by water) for both bald cypress and water tupelo, while the occasional short-term, high-discharge flood raises water levels one to two metres and does the same for bottomland mixed-forest tree species (Schneider and Sharitz, 1988). For more information on bald cypress swamps, see section 8.6.

7.4.3 Southern Mexico and Central America

Significant areas of temperate deciduous forest are found in San Luis Potosi, Veracruz, Oaxaca, and Chiapas states in Mexico and in parts of Guatemala and Belize. The influence of the subtropical latitude on local climate is such that the leafless season is quite short, and during milder winters, some foliage may be retained even until the spring leaves emerge (Röhrig, 1991b). Although few areas are entirely deciduous, the region is considered deciduous temperate rain forest because so many genera typical of the Carolinian deciduous forests are present here. Species such as *Liquidambar styraciflua, Carpinus caroliniana, Nyssa sylvatica, Prunus serotina,* and *Ostrya virginiana* are the same in both regions, and *Magnolia grandiflora, Fagus mexicana, Acer skutchii,* and the evergreen *Liquidambar macrophylla* are close relatives of species farther north. While there are large numbers of oak species, this region differs in that many are evergreen. As in the Near East and North

America, *Quercus* is also the genus with the most growth forms. Unlike typical deciduous forests to the north, however, deciduous species form the upper canopy layer, while evergreen tropical species form an understorey level. In addition, these mixed forests are often rich in epiphytes and lianas (Röhrig, 1991b).

7.5 Europe

It must be recognized from the outset that most of the European deciduous forest has been cleared for agriculture and that remaining stands have been greatly modified from their original form by humans and domesticated animals (Jahn, 1991; Walter, 1979; Pennington, 1969). As early as 5,000 BP, *Ulmus* started to decline probably because its leaves were used as fodder for livestock. Soon after, a general forest decline reflected the *landnam* clearance (clear-cutting) by Neolithic farmers (Pennington, 1969). The practice of burning associated with this agricultural expansion also promoted the development of extensive heathlands. In addition, timber and charcoal production continued the forest decline. Today, even larger percentages of the native forests of Europe have been cleared than is the case with North American deciduous forests, but cleared areas in both continents have developed into quite similar cultural landscapes. Just as in North America, European deciduous stands are dominated by species of *Quercus, Fagus, Acer, Ulmus,* and *Fraxinus* (Figure 7.4), although in Europe fewer tree taxa survived the Pleistocene. Compared with North America, however, *Carpinus, Alnus, Taxus, Pyrus, Sorbus*, and *Corylus* can reach 100–1,000% greater height in Europe, a phenomenon which may be a type of ecological substitution in the absence of competition from now-extinct genera such as *Liriodendron, Carya, Diospyros, Tsuga*, and *Thuja* (Campbell, 1982). On the basis of tree species dominancy, European temperate deciduous forests can generally be divided into three major forest regions: (1) Atlantic, (2) central European, and (3) eastern European.

7.5.1 Atlantic Deciduous Forest

This region is characterized by oak, beech, and damp woodlands, while areas of wetlands and mixed forests are found on its western periphery. Climate is humid, with a moisture surplus usual in most months except perhaps at the height of the growing season, when the rapidly transpiring cover leaves little excess water for percolation (Figure 7.2A). The oak woodlands are dominated by *Quercus petraea* (sessile oak), *Q. robur* (pedunculate oak), and the shrub *Corylus avellana* (hazel), while in the southwest these dominants are joined by *Q. pyrenaica* (Pyrenean oak) (Polunin and Walters, 1985). *Quercus petraea* favours the well-drained, somewhat acidic silicious soils of the more humid west and north, and typical trees and shrubs in an oak forest in the west of Ireland would include *Q. petraea, Taxus baccata* (yew), *Betula pubescens* (downy birch), *Ilex aquifolium* (holly), and *Sorbus aucuparia* (rowan). This forest floor would comprise ferns such as *Pteridium aquilinum*, a rich variety

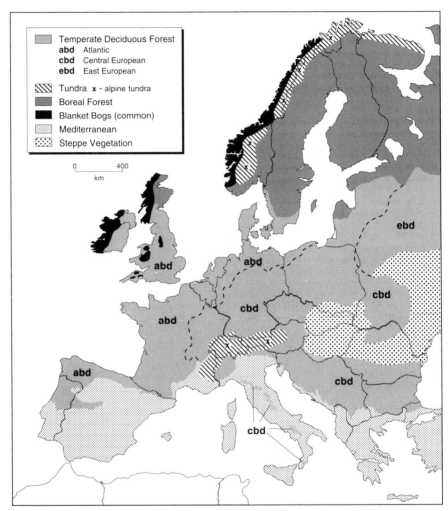

Figure 7.4
Temperate broadleaf deciduous forests in Europe (adapted from Polunin and Walters, 1985)

of mosses, and such sub-shrubs as *Calluna vulgaris* (ling or Scotch heather). *Quercus robur* thrives on the heavy, fairly moist, neutral-to-alkaline soils in the more continental portions of the Atlantic region. A typical pedunculate oak forest in southeastern England would include *Q. robur, Fraxinus excelsior* (European ash), and *Acer campestre* (field maple), with perhaps some *Populus, Betula, Alnus, Ulmus,* and *Carpinus* species. *Salix* spp. and *Corylus avellana* thrive in the shrub layer, while these may be joined by such shrubs as *Cornus sanguinea* (dogwood) and *Rhamnus catharticus* (buckthorn) on strongly calcareous soils (Polunin and Walters,

1985). Favouring acidic soils, *Q. pyrenaica* is found only in southwestern France and in the Iberian Peninsula. Many of these forests have been overgrazed and replaced by heather-gorse heaths. In the Basque country of northern Spain, semideciduous forests dominated by *Q. faginea* and *Q. pubescens* are found on more base-rich soils and essentially represent the transition forest between true deciduous and Mediterranean types (Loidi and Herrera, 1990).

Although *Fagus sylvatica* (common beech) forests are more typical of central Europe, they are also frequently encountered on soils that are less favourable to oak in the Atlantic region. These soils vary in reaction, are generally shallow, and as beech can not tolerate waterlogging, are usually well drained. While beech like richer, deeper soils, competitive exclusion by *Quercus* restricts them to less-fertile soils. Beech frequently follow *Fraxinus excelsior* in natural succession. They develop canopies which allow so little light penetration that ground cover is restricted, especially on the more acidic soils. Atlantic montane beechwoods are also typical of portions of the Cantabrian Mountains in northern Spain. Here *Fagus sylvatica* is frequently almost the only constituent, although in moister soils, *Quercus petraea* takes on importance, and locally or sporadically encountered are *Sorbus aucuparia* (rowan), *S. aria* (whitebeam), *Ilex aquifolium,* and *Acer pseudoplatanus* (sycamore) (Garcia-Gonzalez, 1988). In northern Britain, where climatic conditions are less favourable to beech, these less-fertile soils frequently have a *Fraxinus excelsior* climax (Polunin and Walters, 1985). Where soils are waterlogged or near waterlogged, such as around the margins of lakes or on floodplains, *Alnus glutinosa* (common or black alder) and *Fraxinus excelsior* replace both the oak and the beech (Polunin and Walters, 1985).

Heathlands are found in the wet, mild climate of the western fringe of Europe, where soils are so acidic an ericaceous (*Calluna vulgaris*) humus layer develops on the Podzolic soil profile (Gimingham, 1972). Although their post-glacial origin appears to have coincided with both a significant climatic shift and increased human interference, Gimingham et al. (1979, p. 378) conclude that "the evidence for ascribing the origins of heathlands very largely to human influence, throughout the greater part of the region, is now overwhelming." While the original term "heath" (from the German word *Heide*) referred to an 'uncultivated stretch of land"(Specht, 1979, p. 1), the North American literature generally restricts usage of the term to communities in which members of the family Ericaceae (or the order Ericales) form an important component of the cover (Whittaker, 1979). As noted in section 4.5.1, *Pinus sylvestris* (Scots pine) is sometimes associated with this European heathland landscape. In areas of derived heath, such as in the Lüneburger Heide in northwest Germany, a return to oak-beech forest is sometimes possible if the succession is allowed to progress. In this example, *Calluna-Pinus* heath is first replaced with *Pinus-Betula* forest and finally with *Quercus-Fagus* forest (Rode, 1993). Temperate forest wetlands include large areas of blanket bog which are dominated by the mosses *Sphagnum* spp., *Scirpus cespitosus* (deergrass), *Molinia caerulea* (purple moorgrass), and *Eriophorum vaginatum* (cotton-grass). Fens are found where base-rich groundwater comes to the surface and are frequently dominated by *Juncus acutiflorus* (sharp-flowered rush). Raised bogs are wetlands where the peat becomes strongly acidic as organic accumulations cause it to rise above regional groundwater, due to paludification (for a fuller discussion of European wetlands, see chapter 8).

7.5.2 Central European Deciduous Forest

In pre-Neolithic times, beech (*Fagus*), oak (*Quercus*), and hornbeam (*Carpinus*) dominated these central European lowland forests, while they were joined by lime (*Tilia*) in the east and conifers in upland regions forming mixed forest. In southwestern Germany there is strong evidence for a culturally mediated (Neolithic) transformation from beech to *Corylus* (hazel) around 3,700 BC (Clark et al., 1989). The central European climate favours summer-green deciduous forests with well-protected winter buds, and Köppen designated this region, together with the Atlantic region, as having a "beech climate" (Ellenberg, 1988). Central European deciduous forests can be generally classified into (1) beech, (2) oak-hornbeam, (3) oak, and (4) wet alder-ash woodlands (Polunin and Walters, 1985). As in the Atlantic region, there have been tremendous changes to this forest, and today, agricultural landscapes dominate. While some of the valleys in the Polish Carpathian foothills were 80% or more forested with oak and beech in the fifteenth century, today only 10% or less remains (Dzwonko and Loster, 1988). Surviving upland forests in the western Sudetes Mountains of Poland also exhibit serious damage from air pollutants and in places, approach total destruction (Mazurski, 1990).

Unlike in the Atlantic forests, *Fagus sylvatica* is probably the natural climax species throughout central Europe, even on the better soils with adequate drainage. Beech does not like drier, sandy soils and is outcompeted by *Alnus* spp. (alders) on wet valley soils. Today, most of the natural beech forests are restricted to mountains, although even here their valuable timber encourages extraction. Generally, oak-hornbeam forests are more often found on the more acidic soils than are the beech forests and may also owe much of their composition to both selective timber extraction and the influence of grazing livestock. These stands are dominated by *Carpinus betulus* (common hornbeam) and either *Quercus robur* or *Q. petraea*, and they commonly include tree species such as *Fraxinus excelsior* (European ash), *Acer pseudoplatanus,* and *Tilia cordata* (small-leaved lime). Oak woods dominated by *Quercus petraea* are usually found on acid, siliceous soils in foothills up to elevations of 550 m. In southern Sweden and around the southern Baltic, *Quercus petraea* tends to dominate in place of the sessile oak on similar soils, while to the south, in the Polish Carpathian foothills, *Q. robur* dominates, together with *Tilia cordata, Carpinus betulus,* and *Fagus sylvatica* (Dzwonko and Loster, 1988). In the foothills of the Alps in northern Italy, *Quercus petraea, Q. pubescens,* and *Q. cerris* are important dominants, while in the northern Appennines, *Q. cerris* and *Q. pubescens* are joined as dominants by *Fraxinus ornis* and *Ostrya carpinifolia* (Jahn, 1991). In northern Croatia, montane forests are dominated by *Quercus petraea, Carpinus,* and

Fagus, while better-drained flatlands are dominated by *Q. robur* and *Carpinus* (Vukelić, 1991). Two other tree species, which have since become very widely distributed throughout western Europe, are found native to southeastern Europe. These are *Castanea sativa* (sweet chestnut) and *Aesculus hippocastanum* (horse chestnut).

As in North America, here in north-central Europe in areas of glacial outwash sands, conditions suitable for conifers are created, and mixed forest results. The major conifer in this region is *Pinus sylvestris*, although conifers also include *Larix decidua, Picea abies*, and other pines (see section 4.5.1). As with the Atlantic region to the west, there are sizeable areas of wet lowlands, usually along river floodplains with gleyed or peaty soils, where *Alnus glutinosa* (common alder), *A. incana* (grey alder), and *Fraxinus excelsior* (ash) dominate. Unlike the Atlantic region, however, to the southeast is the somewhat drier climate of the Hungarian Puszta (the Pannonic region of Polunin and Walters, 1985). Here, to the east of the Alps and south of the Carpathians, *Quercus* species dominate the rain-shadow steppe woodlands, much as they dominate the transition from deciduous forest to prairie in North America. Important tree species in this region include *Q. cerris* (Turkey oak), *Q. frainetto* (Hungarian oak), *Q. pubescens* (white oak), *Tilia tomentosa* (silver lime), *Carpinus betulus, C. orientalis* (oriental hornbeam), and *Acer tataricum* (Tatarian maple) (Polunin and Walters, 1985). On the moister soils of valley floors can also be found *Quercus robur* and *Q. petraea*.

7.5.3 East European Deciduous Forest

Temperate deciduous forests are found all the way east to the southern Urals, but as shown on Figure 7.1, this forest decreases in latitudinal extent as it is pincered between the boreal forest to the north and the increasing aridity of the steppes to the south. The western edge of this region is where the climate becomes too continental for *Fagus sylvatica* to dominate and where it is gradually replaced by *Carpinus betulus* (common hornbeam). In the eastern portion of the region, including the southern Urals, *Tilia cordata* (small-leaved lime) dominates. As with the other deciduous forest regions, wetlands are present. In addition, to the south, there is a transitional *Quercus-Carpinus* forest leading to *Quercus-Prunus* steppe-parkland (see section 5.5).

7.6 Asia

Significant areas of temperate deciduous forest are found in the Near East along the southern and eastern shores of the Black Sea, in the eastern Caucasus, from southern Azerbaijan, through the Elburz Mountains, and along some of the larger rivers draining into the Aral Sea in Turkestan (Figure 7.1). In addition, sizeable areas are found in northern and east-central China, western Korea, and central Japan (Figure 7.5).

Due to the fact the Near East served as a refugium for European species during the Pleistocene, it has a more diversified temperate deciduous forest than Europe.

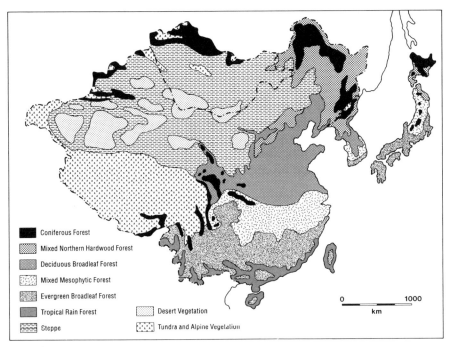

Figure 7.5
Temperate broadleaf deciduous forests in east Asia (generalized from Whyte, 1968)

The region is home for more than twenty species of *Quercus*, and the common beech so dominant in Europe is replaced by *Fagus orientalis* throughout. In addition to numerous endemics, not a single European tree genera is absent from the deciduous forests of the region (Röhrig, 1991c).

While China has the largest area of potential deciduous forest in Asia, it is also the East Asian country that has experienced the greatest degree of forest clearing for agriculture. There are two areas of deciduous forest in China: a smaller region in Liaoning and Kirin provinces in the northwest and the major region stretching from Beijing south to northern Hunan Province (Figure 7.5). In addition, deciduous forest stands are common within the Mixed Mesophytic (broadleaf deciduous with broadleaf evergreen) forests to the south. The climatic data for Nanjing, Jiangsu Province (Figure 7.2B), are more representative of the southern portions of the temperate deciduous region close to its southern boundary with Mixed Mesophytic forest. In terms of moisture indices, China's broadleaf deciduous forest is found in the im 0–60 range (Fang and Yoda, 1990).

Unlike Europe, but similar to the situation in North America and the Near East, there was greater Pleistocene survival of arboreal taxa in China, probably because of the continuous mountainous region from north to south (Campbell, 1982). Today, however, even these mountainous areas have experienced forest removal on a vast scale, with lowland provinces such as Jiangsu having practically no unaltered deciduous forest left. Unlike in North America, therefore, the vegetation has been extensively disturbed and modified over the last four or five millennia (Liu, 1988). The consequences of the last 2,000 years of intense exploitation has simply left few natural stands and only clues as to the composition and structure of former deciduous forest (Ching, 1991; Tuan, 1969). Unfortunately, while destruction of native forests is almost universal within the temperate zone and the Chinese have traditionally been among the most proficient practitioners of agroforestry, their afforestation efforts have not kept pace with forest destruction (Smil, 1984).

As with North America and western Europe, the original deciduous forest cover in eastern China was dominated by oaks. Dominants were probably *Quercus aliena, Q. variabilis* (oriental cork oak), and *Q. dentata* (Daimyo oak), together with some *Fraxinus, Ulmus,* and *Celtis sinensis* (Chinese hackberry). In the Chang Jiang (Yangtze) valley to the south is a mixed mesophytic forest which includes the deciduous species *Quercus acutissima, Carpinus fargesii, Castanea sequinii,* and *Albizia kalkora,* along with a number of Tertiary relics such as *Ginkgo biloba* (maidenhair tree) and *Metasequoia* spp. (water larch), large areas of managed bamboo thicket, and many evergreen broadleaf tree species (Ching, 1991). Small remnant deciduous forests can still be found in parts of upland Shanxi and Shandong provinces, but native cover, except perhaps along water courses, is entirely lacking in the north China plain, although there are numerous small plantations and semi-natural stands under human protection. The numerous river courses, with their moister Gleysolic soils, support species such as *Salix babylonica* (weeping willow) and *S. matsudana* (corkscrew willow), together with *Populus* spp. (Tuan, 1969).

In the badly deforested loess plateau of Shaanxi province in western China, recent small-basin efforts at afforestation have effectively reduced large amounts of sediment from entering the Huang He (Yellow River) and have demonstrated that vast areas of this plateau are suitable for reafforestation (Liu and Wa, 1985). To the north, around Harbin and Tsitsihar (formerly northern Manchuria), deciduous forests tend to be mixed northern hardwoods dominated by *Betula,* while between this region and the central Korean peninsula are mixed northern hardwood forests of *Acer, Tilia,* and *Betula.* In the south-central Korean peninsula, *Quercus*-dominated forests similar to the warmer deciduous forest in China are encountered.

In Japan, Mixed Mesophytic temperate deciduous forests cover southwestern Hokkaido and northern and west-central Honshu north of about 37°N and are limited to the north and south roughly by the 6° and 13°C mean annual isotherms respectively (Satoo, 1973). Here, forest climax is normally dominated by *Fagus crenata,* often with an understorey of dwarf bamboo such as *Sasa kurilensis,* while the multistemmed *F. japonica* joins it to become climax on the Pacific coast (Hukusima and Kershaw, 1988). This multi-stemmed characteristic and the ability to reproduce from sprouts appear to result in *F. japonica* dominance here (Peters and Ohkubo, 1990), a situation not unlike the success of the multi-stemmed *Tilia americana* (basswood) along some river terraces in parts of southern Canada. *Quercus mongolia* var.

grosseserrata, Sasa spp., and *Castanea crenata* (Japanese chestnut) are also common, while *Acer mono* var. *glabrum* and *Tilia japonica* are frequently represented. On low-lying wetter ground, such as in northwestern Honshu, *Salix* spp. forest dominates and includes *Alnus japonica* and *Ulmus davidiana* var. *japonica*. It is estimated that less than 30% of Japan's temperate deciduous forest remains and that some of this is substitutional forest dominated by *Castanea crenata* and *Q. mongolia* var. *grosseserrata* (Geographical Survey Institute, 1977). For a more detailed review of the East Asian deciduous forests, see Ching (1991).

7.7 Southern Hemisphere

Only a few small areas within the southern hemisphere can reasonably be included in the temperate broadleaf deciduous forest zone. These forests are found in central-southern Chile, southwestern Argentina, and in parts of the South Island of New Zealand. Unlike their northern hemisphere counterparts, not only are they dominated by quite different species, but the pronounced oceanic influences encourage a mixing with evergreen broadleaf and coniferous species as well. Although separated by large distances of ocean, the forest floras of Chile and New Zealand have much in common. Dominants in both forests include representatives of the southern hemisphere beech genera *Nothofagus* in mixed deciduous-evergreen broadleaf forest stands, which also often include conifers.

The temperate deciduous forests in Chile are located between 37°35' and 40°40'S, and they are bordered to the north by the scrub and sclerophyllous woodlands of the Matorral and to the south by the evergreen broadleaf Valdivian forest (Innes, 1992). Due to the extreme maritime climate, evergreen and deciduous hardwood species are found together and are often joined by conifers. Seven *Nothofagus* spp. (southern hemisphere beech), two of which are evergreens and five of which are deciduous, constitute about two-thirds of the forest population (Quintanilla, 1977). While mixed deciduous-evergreen hardwood forests dominate, some areas, especially where rainfall is high, are dominated by the evergreen *N. dombeyi* (coigue) or by *Araucaria araucana* (Chile pine, or monkey-puzzle) and *Pilgerondendrum uviferum* (a cypress). In the drier north, *Nothofagus obliqua* and *N. glauca* dominate at elevations up to 600 m, *N. alpina* and *N. dombeyi* combine to form a transitional zone between about 600 and 900 m, and *N. alpina* dominates between 900 and 1,200 m. *Nothofagus antarctica* and *N. pumilio* generally dominate between 1,300 and 2,000 m. In the Andes, *Araucaria araucana* often dominates on the warmer, drier slopes and plateaux (Schmaltz, 1991). While there are eight indigenous conifers in Chile, those in the north are generally found at higher elevations, while in the south they are found at all elevations, particularly where it is humid (Quintanilla, 1977). On the leeward slopes of the Andes in southwestern Argentina, in areas such as along the western shores of Lake Nahuel Huapi near Bariloche, are also mixed forests of conifers and *Nothofagus* spp., often with a thick *Chusquea quila* (bamboo) understorey.

Table 7.1

Preliminary forest productivity estimates for temperate deciduous forests. Values are for dry matter in g m^{-2} yr^{-1}.

Species	Gross Primary Productivity	Net Primary Productivity
Beech[1]	2,360–4,250	1,400–1,780
Oak[1]	2,890–4,660	1,400–1,944
Other[1]	3,360–3,530	1,490–2,100
Deciduous forest world average[2]	–	1,200
Warm temperate mixed forest[3]	–	1,000
Temperate forest[4] (average)	–	1,300

[1] Based on 1972 IBP Woodlands Project Reports by Reichle et al. (1973).
[2] Whittaker and Likens (1975).
[3] Lieth (1975).
[4] Whittaker (1970).

On the South Island of New Zealand are some areas of mixed deciduous-evergreen forest. Here, deciduous tree species include *Plagianthus betulinus*, which grows mainly on fertile alluvial flats, *Fuchsia excorticata* (fuchsia), a pioneering woody species, and *Hoheria glabrata* (ribbonwood), a small tree which dominates moist subalpine talus slopes (Wardle et al., 1983).

7.8 Primary Production and Phytomass

When temperate deciduous forest productivity is compared to that of boreal forest, the temperate shows consistently higher productivity rates and biomass totals per unit area (Olson, 1975). At the elemental level, deciduous forests accumulate more calcium than boreal forests, and this is stored primarily in the living vegetation (Cole and Rapp, 1981). On the other hand, while temperate and boreal forest ecosystems store similar amounts of nitrogen, the boreal stores a greater percentage of its nitrogen accumulation in litter. On the basis of limited comparisons, Cole and Rapp (1981) conclude that there is no clear relationship between occurrence of a deciduous or coniferous species and the ability for those ecosystems to accumulate nitrogen. They do conclude, however, that demand and rate of nitrogen recycling is substantially higher in deciduous forest. This greater demand for nitrogen and minerals such as calcium and magnesium is primarily due to the deciduous habit which necessitates higher rates of recycling. To some extent, this demand is met by the much more rapid forest-floor detrital turnover than under conifers (Onega and Eickmeier, 1991). In young stands, as much as 10–15% of the above-ground nutrient pool may be returned to the litter layer through litterfall each year (Likens et al.,

Table 7.2
Typical biomass values for temperate broadleaf deciduous forest. Values in t ha^{-1}.

Community	Tree, Above Ground	Roots	Litter	Soil o.m.	Living[a] Biomass Only
Liriodendron[1] tulipifera	124.7	36.0	6.0	159.0	168.8
Oakwoods,[2] Minnesota	163.1	14.9	36.7	–	178.7
Fagus,[3] Europe	–	–	–	–	370.0
Fagus sylvatica[4]	315.0	43.4	–	–	358.4
Prunus pensylvanica,[5] USA	–	–	–	–	239.8
Oak-pine,[6] Brookhaven	65.6	36.1	–	–	101.9
Western Europe[7] (average of 8 studies)	324.4	–	–	–	–
World temperate deciduous (mean)[8]	–	–	–	–	300.0

[a] includes understorey vegetation.
[1] rounded from Cole and Rapp (1981).
[2] Ovington (1963).
[3] Rodin and Bazilevich (1967).
[4] Nihlgard and Lindgren (1977).
[5] Art and Marks (1971).
[6] Whittaker and Woodwell (1969).
[7] Röhrig (1991d, Table 8.3).
[8] Whittaker and Likens (1975).

1977), while this percentage decreases as the above-ground trunk biomass increases. As can be seen from a comparison of Tables 7.1 and 4.1, net primary productivity levels for temperate deciduous forest are also higher than in boreal forest. For a comprehensive review of temperature forest "ecochemistry," see Khanna and Ulrich (1991).

Litterfall in typical deciduous forest can be considerable. Khanna and Ulrick (1991) report a survey of previous research which indicates this litter production is normally between 3,500 and 5,000 kg ha^{-1} yr^{-1}. It is noted from the data in Table 7.2 that there are large organic matter reserves in forest-floor litter even though the deciduous foliage component decomposes relatively rapidly. One consequence of forest clear-cutting is that these litter reserves are decomposed even more rapidly than under mature forest and that successional species are not yet in place to reduce losses to the released nutrients (Likens et al., 1978). Bird and Chatarpaul (1988) conclude that the more rapid decomposition following harvesting of mixed conifer-hardwood forest on the Canadian Shield may also result in both nutrient and long-term productivity losses. Johnson et al. (1985) conclude the same for Appalachian Oak forest and that the problem is exacerbated by the now-common practice of whole-tree harvesting. From an ecosystem perspective, however, it is fortunate that temperate deciduous forests in general and those in eastern North America and Europe in particular are increasing their biomass once again. From a global carbon-cycle perspective, this may also be fortunate, since it slows the rate of anthropogenically induced increase in the atmospheric carbon pool (Kauppi et al., 1992).

CHAPTER EIGHT

Wetlands

8.1 Introduction

The term "wetland" can be defined as "land that has the water table at, near, or above the land surface or which is saturated for a long enough period to promote wetland or aquatic processes as indicated by hydric soils, hydromorphic vegetation, and various kinds of biological activity that are adapted to the wet environment" (Tarnocai, 1980, p. 10). This broad definition therefore includes those environments inherent in other, narrower traditional descriptive terms such as "moor," "muskeg," "mire," and "peatland." In effect, wetlands are "neither 'firm' lands in the conventional sense nor bodies of water; hence they occupy a transitional position between land and water" (Zoltai, 1988, p. 3). In addition, wetlands are found scattered from the high Arctic to the prairies, and from the temperate deciduous forest zone to the evergreen tropical rain forest (Figure 8.1). As they are clearly found within any of the major ecoclimatic provinces or vegetation zones, collectively they can be described as intrazonal. Just as Gleysols or Organic soils can be found in practically any climatic region and are therefore described as intrazonal soils, so wetlands deserve distinction. Likewise, Gleysols and Organic soils develop in environments that are waterlogged much of or all of the time, and wetlands occupy a similar position in the landscape.

The Convention on the Conservation of Wetlands of International Importance, or Ramsar Convention, has acknowledged the major value of wetlands worldwide as areas of high biological productivity and international significance. Named after Ramsar, Iran, where the convention was formulated in 1971, this important international agreement came into force on 21 December 1975 (Gillespie et al., 1991). As of March 1994, the Ramsar Convention has been ratified by 81 nations, including Canada, and the strength of the Ramsar designation given to wetland areas lies in its perceived value of international concern. Wetlands are particularly significant within the Canadian landscape since 14% (127,199,000 ha) of the country is wetland, and major portions of the boreal forest and forest-tundra have more than 50%

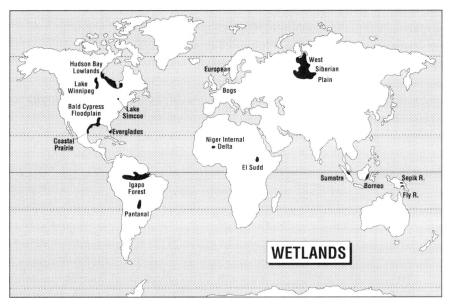

Figure 8.1
World map showing the location of some of the wetlands discussed below

wetland cover (National Wetlands Working Group, 1988). They also comprise size-able percentages of the prairie, mixed-forest, and temperate deciduous forest zones, where they serve as important wildlife habitat or groundwater recharge areas or have been drained for agriculture (Figure 3.5, Table 8.1). In 1981 Canada designated its first Ramsar site at Quebec's Cap Tourmente National Wildlife Area. As of March, 1994, 648 sites covering 43,400,000 ha had been designated Ramsar sites in 81 countries. Thirty-two of these sites are in Canada and include such areas as the 617,000 ha Old Crow Flats in the Yukon, the 23,000 ha Delta Marsh in Manitoba, and the 26,800 ha Southern Bight–Minas Basin in Nova Scotia. There are Ramsar-designated wetlands in every Canadian province, and in total they comprise 13,000,000 ha, or 30% of Ramsar wetlands worldwide (Clay Rubec, 1994, personal communication).

Because wetlands are intrazonal in nature, this chapter differs in format from pre-vious ones. While it retains a general overview section on climate, readers are di-rected to chapters dealing with the zonal vegetation types in which a particular wetland is found for greater detail on local climate. There is also a brief review of wetland soils because of their dependence on similar conditions to those favouring hydrophytic/hydromorphic plants. Subsequent sections deal primarily with fresh-water wetlands (the helobiomes and hydrobiomes of Walter, 1979), with only minor reference to brackish or saline wetlands (halobiomes). Saline wetlands, such as the mangrove forests of low-latitude coastlines, are not treated here. As with previous

Table 8.1
Occurrence of wetlands and peatlands in Canada (as summarized by National Wetlands Working Groups, 1988)

Province or territory	Wetland Area		Peatland Area	
	ha X 10³	% of land area	ha X 10³	% of land area
Alberta	13,704	21	12,673	20
British Columbia	3,120	3	1,289	1
Manitoba	22,470	41	20,664	38
New Brunswick	544	8	120	2
Newfoundland-Labrador	6,792	18	6,429	17
Northwest Territories	27,794	9	25,111	8
Nova Scotia	177	3	158	3
Ontario	29,241	33	22,555	25
Prince Edward Island	9	1	8	1
Quebec	12,151	9	11,713	9
Saskatchewan	9,687	17	9,309	16
Yukon Territory	1,510	3	1,298	3
Canada	127,199	14	111,327	12

chapters, Canadian wetlands will receive particular attention, and use will be made of the Canadian Wetland Classification System (Zoltai et al., 1975; Tarnocai, 1980; National Wetlands Working Group, 1987).

8.2 Climate

Although wetlands can be found in almost any climatic region their various dependence on (1) water supply, (2) evapotranspiration rates, and (3) decomposition rates suggests correctly that certain climatic environments are much more conducive to their formation than are others. An examination of the distribution of wetlands in Canada (Figure 3.5) indicates that the greatest concentration of land where wetlands comprise more than 75% of the landscape is in the region peripheral to southern Hudson Bay. Here moisture indices are strongly positive, potential evapotranspiration rates moderate, and decomposition rates slow, and as a result, wetlands are found in uplands as well as on lowlands or flatlands. The prevalence of wetlands is encouraged here by such non-climatic factors as deranged drainage patterns and low-lying terrain. However, even if these geomorphic variables were quite different, the density of wetlands here would probably not be appreciably altered because wetlands can be found in almost any topographic position when climatic conditions are per-humid. The tropical swamp forests of the Darien in Panama and the aguaje palm (*Mauritia* spp.) swamps of the western Amazon Basin to a large extent also result from year-round per-humid conditions, although their extent is locally encouraged by level terrain.

An examination of Figure 3.5 also shows that wetlands can be found in such ecoclimatic regions as the Subhumid Transitional Grassland (Gt on Figure 1.4). Here, however, wetlands are more the product of geomorphic phenomena than of local climate. As an example, the large Delta Marsh at the southern tip of Lake Manitoba has a negative moisture index; it owes its origins to, among other things, a combination of gentle grade, heavy lacustrine sediments, reception of runoff from a large area to the south, and impoundment by a barrier beach. Likewise, such wetlands as the Holland River fen near Lake Simcoe, Ontario, the shorefen at Port Royal, Ontario, and wetlands along the upper St Lawrence River owe their origins much more directly to topographic considerations than climate. Even the Everglades of Florida and the Sudd swamps of the southern Sudan derive more from geomorphology than from local climate.

8.3 Soils

Three important soil orders are associated with wetlands. These are the Gleysolic, Organic, and Cryosolic (the Organic Cryosol great group). The intrazonal distribution of Gleysols and Organic soils along with wetlands is easily explained because both orders depend on greater or lesser degrees of anaerobic (reducing) soil profile conditions. Gleysols do not form unless the water table is close to the surface and anaerobic conditions are dominant within the soil profile for most of the time. The reducing conditions that result are sufficient to give the characteristic grey colour to the lower, and perhaps upper, mineral layers, but conditions are such that any organic layer developing on the soil surface is thinner than needed to reclassify the profile as an Organic soil. This means that on their better-drained margins, Gleysols border with Gleysolic subgroups of better-drained soil orders, while on their more poorly drained margins, where the water table is close to, or above the surface, organic (peaty) layers develop to a thickness where the profile must be classified as an Organic soil. Clearly, between these two extremes, Gleysolic great groups exhibit considerable variety from those which have an illuvial layer (Luvic Gleysol) to those with little in the way of an Ah horizon (Gleysol) and those which do have an Ah horizon that exceeds a thickness of 10 cm (Humic Gleysol). In addition, peat layers can accumulate to depths of 40–60 cm before the Peaty-phase Gleysol profile must be renamed an Organic soil. See Figure 3.4 for Gleysol and Mesisol profiles and Figure 2.4 for Organic Cryosols and Fibrisols. In Canada as a whole, there are some 127,199,000 ha of wetlands, of which 111,327,000 ha are peatlands (Organic, Organic Cryosols, and Peaty-phase Gleysols) (National Wetlands Working Group, 1988). By definition, therefore, the term "peatlands" is restricted to environments where organic debris is accumulating to sufficient depths to form peat. Sometimes, the term "peatscape" is used to describe landscapes dominated by organic soils, mosses, sedges, and open, scrubby trees. The Algonquin Indians used the term "muskeg" to describe the same landscape (Stanek, 1977).

One of the most significant variables associated with wetland soils is pH. In addition to signifying the degree of base saturation and the availability of nutrients, acidity (pH) is of importance in decomposition processes and to competitive strategies between plant species. Soil acidity is a complex variable, with values that can vary widely both spatially and temporally. Where minerotrophic waters (eutrophic, or base-rich, and mesotrophic, or moderately base-rich) enter wetlands, basophilous plant species dominate, higher pH levels promote modest decomposition rates, organic soils are usually Mesisols or Humic Mesisols, and organic deposits are often described by the term "fen." Where a wetland is either supplied entirely by rainwater or by ombrotrophic (oligotrophic, or base-poor) groundwater originating from non-calcareous parent material, then the soil solution is acidic, acidophilous plant species dominate, acidic conditions slow decomposition processes, soils are usually Fibrisols and Fibric Mesisols, and bog materials accumulate. The main source of nutrient supply in these acid bogs is from wet and dry deposition, with one northern Minnesota study showing that the atmosphere potentially supplied 3% and 20% of the annual bog plant uptake of K and Mg respectively (Grigal, 1991). There are no indications of any surface supply of inorganic plant nutrients from the underlying peat (Malmer, 1988), due primarily to the fact that at depth, peat acts much like clay in terms of hydraulic conductivity, so that water movement is almost entirely surficial (Heinselman, 1975). Below 35 cm the hydraulic conductivity is extremely slow (<0.03 cm h^{-1}) and correlates well with the von Post scale of peat decomposition (Gafni and Brooks, 1990). It is therefore not surprising to find major differences in pH with depth, because of limited vertical interchanges. Surface soil pH values correlate so closely with the habitats of specific plants that species can be ranked in the pH groupings in which they are most frequently encountered. Species adapted to pH values below 3.2 are strictly ombrotrophic, while between a pH of 3.2 and 3.5, species can be described as transgressive ombrotrophes. Species preferring pH values greater than 3.5 are considered as poor, intermediate, or highly minerotrophic (Gerardin and Grondin, 1987). Nutrient status is therefore of underlying importance to the development of trends in vegetation cover (Kenkel, 1987).

It is common to have alkaline (fen) conditions at the edge of a large wetland, while towards the centre of the same wetland, soil reaction is much more acidic (e.g., a domed bog). These bogs have developed through a process called paludification, which involves both upward accretion of peat caused by a water table which rises sufficiently to be above the immediate influence of base-rich groundwater and also outward accretion of peat over fen or previously dry land (Figure 8.2). The first stage in the development of many bogs is probably a brown moss–sedge fen which accretes upwards and levels off the primary topographic relief. As fen peat accumulates, the water table rises, nutrient-rich groundwater movements towards the centre and surface of the bog decrease, conditions became more acid, and eutrophic-mesotrophic conditions gave way to oligotrophic where nutrients and water are entirely rain fed. This acidification and nutrient-reduction process not only gives rise to a second stage which is dominated by *Sphagnum* spp., but decomposition rates are slower at the bog centre than at the outwardly expanding mesotrophic margins (Figure 8.2). *Sphagnum fuscum*, the most acid-loving of the *Sphagnum* genus, is not

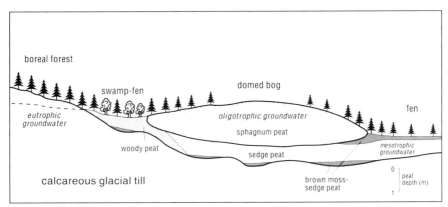

Figure 8.2

A domed bog which has developed through paludification. In this process, eutrophic-mesotrophic fen peat (brown moss and sedge peat) can be capped by oligotrophic bog peat (*Sphagnum* peat) as the peat accretes upwards above the influence of the regional nutrient-rich water table. The resultant domed bog expands laterally as well as vertically and retains base-rich fen along its margins (In this figure tree heights are not drawn to the same scale as peat thickness.)

only the major circumpolar peat-forming species of moss, but also appears to be one of the most decay-resistant species (Johnson, 1987). A convex or copula-like bog relief results where the acidic bog centre becomes domed several metres above the more nutrient-rich (fen) margins, and drainage away from the centre evolves (Ivanov, 1981).

This paludification process is very common throughout the boreal zone. Even though bog development generally requires that the site pass through a fen stage, this portion of the paludification process may be rapid. Many of the young wetlands developing in forest-tundra on recently exposed land inland of Hudson Bay southeast of Churchill exhibit proto-bog development on land that has only emerged from the bay in the last 1,500–3,000 years. Here it is common to see broad, level fen landscapes dotted by small circular-to-oval bogs which are accreting slowly outwards. Kuhry et al. (1993) report that less than 1,500 years may be needed for continental Mid-Boreal wetlands to pass through the initiation of the fen stage to the *Sphagnum* proto-bog stage. They report a somewhat longer time period of >2,000 years for proto-bog development in the Low Boreal. Pollen records for the early Holocene in northwestern Canada confirm that the early importance of white spruce soon gave way to a paludification-induced succession to black spruce and white birch (Ritchie, 1987). In the drier Mixed Forest section of the boreal forest in central Alberta, fen development only began 6,000 BP following the development of a cooler and moister climate and the replacement of prairie by forest (Zoltai and Vitt, 1990). Here the natural progression of the paludification process from brown moss fen to *Sphagnum* bog is even slower.

In addition to temporal considerations at any particular site, paludification can also be refined on the basis of whether or not the peat formation cycle is initiated on site (i.e., endogenous pludification) or is forced on it by peat advancing into the site from bog nearby (i.e., exogenous paludification) (Glebov and Korzukhin, 1992). Paludification therefore represents a classic case of allogenic succession, where the moss and ground cover shrub and tree dominants reflect changing nutrient, soil-water, and soil-atmosphere regimes as the process continues. This is not to say that cyclic autogenic succession may not also occur, say, following fire episodes (Payette, 1988).

The concept of paludification is so appealing, and recently formed lands such as those along the southern shores of Hudson Bay show the process at work so quickly, that we must be wary of concluding that its ultimate goal is for the complete dominance of boreal environments by bryophytes. It must be pointed out, however, that there is a counter to paludification, depaludification, a course initiated by changes in local hydrology, climate, and cover dominants, or simply due to bog "senescence." The Holocene peat-stratigraphy record shows that within one region there have been both paludification and depaludification episodes – the process is reversible with changing conditions (Glebov and Korzukhin, 1992). It is also possible that even without changes in external conditions, there is self-regulation as upward accretion of organic material reaches a balance, bog development loses its aggressiveness, and a type of senescence sets in where the bog surface begins to break up into regressive lake-bog complexes leading back to forest phases again. The overall picture within the boreal is therefore one of an equilibrium having developed over time in which there is balance between forest and bog, but a balance that can shift in either direction as a response to, say, climatic change (Glebov and Korzukhin, 1992).

Wetland organic soils (peats) can be viewed as accreting biomass storage systems into which vast reserves of atmospheric CO_2 are entering. In Canada the long-term accumulation rate for these organic soils varies from 2.8 cm/100 yr in northern Ontario to 10.6 cm/100 yr in southern Manitoba. The average of nineteen boreal wetlands is 6.4 cm/100 yr (Zoltai et al., 1988b), a figure which seems quite rapid until it is compared to average peat growth rates in Indonesian peat swamps of 20–30 cm/100 yr (Anderson, 1983). It should be pointed out, however, that some 20 million ha (about 14%) of Canada's wetlands have been artificially drained (C. Rubec, 1990, personal communication). For more details on wetland biomass accumulation and wetland biomass losses following artificial drainage and peat harvesting, see section 8.11 and Lugo et al. (1990b).

8.4 Canadian Wetland Classification

Because of the significance of wetlands to temperate and continental ecoclimatic regions in general, Canadian wetlands are discussed here in some detail. To better appreciate the great variety contained within Canada's wetlands, it is first necessary to review the Canadian Wetland Classification System (Zoltai and Pollett, 1983; National Wetlands Working Group, 1987). For greater detail on wetland types world-wide, see Lugo et al. (1990a) and Whigham et al. (1993).

8.4.1 Canadian Wetland Classification System

The definition for the term "wetlands" as used in this classification is "land that is saturated with water long enough to promote wetland or aquatic processes as indicated by poorly drained soils, hydrophytic vegetation, and various kinds of biological activity which are adapted to a wet environment" (National Wetlands Working Group, 1987, p. 2). The classification consists of three hierarchical levels: I, class; II, form; and III, type. As background to this classification, it must be remembered that Canadian biomes contain both organic wetlands and mineral wetlands. Organic wetlands are peatlands characterized by peat accumulations exceeding 40 cm (essentially Fibrisols, Mesisols, and Humisols). Mineral wetlands can be classified as Gleysols or Peaty-phase Gleysols, as well as soils under shallow open water less than 2 metres deep. In addition, some Organic and Gleysolic soils having wetland potential have been modified by human activity to non-wetland grazing or agricultural status, but would revert to wetlands upon abandonment.

8.4.2 Wetland Classes

Classes are differentiated on the basis of the overall genetic origin of wetland ecosystems and include five: bog, fen, marsh, swamp, and shallow open water. With the definitions given in Table 8.2 in mind, it is clear that the first two wetland classes, bogs and fens, are associated with large land areas, where the former are acidic and frequently treed, while the latter are mildly acidic to alkaline and usually without a well-developed tree cover. Neither are subject to seasonal inundation. The next two classes, swamps and marshes, are more typically associated with seasonally inundated conditions along rivers or shorelines. Swamps are frequently highly productive forest sites associated with river valleys and bottomlands where gently moving surface waters occur seasonally or persist for long periods, with other periods when the water table drops below the surface. Marshes, on the other hand, are associated with shorelines where inundation can be seasonal, episodic or even tidal, to the extent that trees have difficulty surviving and are frequently absent. Shallow open water basically represents a transitional stage between lake and marsh that is still free of emergent vegetation. Examples of each of these wetland classes are detailed below. For greater detail on wetland classification, see National Wetlands Working Group (1988).

8.4.3 Wetland Forms and Types

Within the five wetland classes, some seventy forms and many hundreds of types can be identified. Wetland forms are differentiated using surface morphology, surface

Table 8.2
Definitions for the five classes in the Canadian Wetland Classification System (summarized from National Wetlands Working Groups, 1987, 1988)

Bog	A peatland generally with the water table at or near the surface. The surface may be raised or level with the surrounding terrain, and it is virtually unaffected by nutrient-rich groundwaters, making it generally acidic and low in nutrients (oligotrophic). The dominant organic material is weakly or moderately decomposed *Sphagnum* and woody peat on Fibrisols, Mesisols, or Organic Cryosols. Can be treed or treeless and usually covered with *Sphagnum* spp. and ericaceous shrubs.
Fen	A peatland with the water table usually at or just above the surface. The waters are mainly derived from mineral soils upslope and are nutrient rich (mesotrophic to eutrophic). Dominant organic residues are well-decomposed sedge and/or brown moss peat (Mesisols, Humisols, or Organic Cryosols). Vegetation cover is dominated primarily by sedges, reeds, grasses, and brown mosses, with perhaps a sparse tree cover.
Swamp	A mineral wetland or peatland with standing water or water gently flowing through pools or channels. The water table is usually at or near the surface, there is pronounced internal water movement from the margins or other mineral sources, water is nutrient rich, and soils are Mesisols, Humisols, or Gleysols. Vegetation is usually a dense cover of deciduous or coniferous trees or shrubs, herbs, and some mosses.
Marsh	A mineral wetland or a peatland that is periodically inundated by standing or slowly moving water. Surface water may fluctuate seasonally, and waters are eutrophic and may be fresh to highly saline. Soils are Gleysolic and occasionally Humisols and Mesisols. Vegetation is usually zoned, with pools or channels interspersed with clumps of emergent sedges, grasses, rushes, and reeds. Submerged and floating aquatics flourish in open water.
Shallow Open Water	Characteristic of intermittently or permanently flooded or seasonally stable water regimes, featuring open expanses of standing or flowing water which are variously called ponds, pools, shallow lakes, oxbows, reaches, channels, or impoundments. Mid-summer depths are less than two metres, and if present, living vegetation consists of submerged and floating aquatic forms. It is distinguished from other wetland types by having water zones occupying 75% or more of the wetland surface area.

pattern, water type, and the morphology of underlying mineral soil (Table 8.3). In turn, each wetland form can be differentiated into types on the basis of vegetation physiognomy (but not plant species composition). Up to sixteen physiognomic categories can be used, and they include terms such as "coniferous treed," "hardwood treed," "tall shrub," "low shrub," and so on (National Wetlands Working Group, 1988). With only slight modification, this Canadian system will be used when describing non-Canadian wetlands as well. As most of the area covered by wetlands in Canada are found in permafrost-related and boreal regions, descriptions of the more common wetland forms found there are given below (after National Wetlands Working Group, 987).

Table 8.3
Canadian wetland forms (from National Wetlands Working Group, 1988)

Class	Forms
Bog	Atlantic plateau bog, basin bog, blanket bog, collapse scar bog, domed bog, flat bog, floating bog, lowland polygon bog, mound bog, northern plateau bog, palsa bog, peat mound bog, peat plateau bog, polygonal peat plateau bog, shore bog, slope bog, string bog, veneer bog.
Fen	Atlantic ribbed fen, basin fen, channel fen, collapse scar fen, feather fen, floating fen, horizontal fen, ladder fen, lowland polygon fen, net fen, northern ribbed fen, palsa fen, shore fen, slope fen, snowpatch fen, spring fen, stream fen.
Marsh	Active delta marsh, channel marsh, coastal high marsh, coastal low marsh, estuarine high marsh, estuarine low marsh, floodplain marsh, inactive delta marsh, kettle marsh, seepage track marsh, shallow basin marsh, shore marsh, stream marsh, terminal basin marsh, tidal freshwater marsh.
Swamp	Basin swamp, flat swamp, floodplain swamp, peat margin swamp, shore swamp, spring swamp, stream swamp.
Shallow Open Water	Channel water, delta water, estuarine water, kettle water, non-tidal water, oxbow water, shallow basin water, shore water, stream water, terminal basin water, thermokarst water, tidal water, tundra pool water.

Permafrost-Related Wetland Forms   Lowland polygon bogs are typical of tundra and subarctic lowlands where permafrost causes ice-wedge trenches in the peat to form polygons with high moss-dominated centres (often called "high-centre polygons"). Peat plateau bogs developed initially in non-permafrost environments but have now developed permafrost. Polygonal peat plateau bogs are the result of ice wedges and permafrost-producing polygons in peat that developed initially as permafrost-free northern plateau bogs. Collapse scar bogs are characterized by oval-shaped wet depressions which have collapsed where permanently frozen peat has melted out. Regions where numerous collapsed scar bogs dominate are often described as thermokarst landscapes. Palsa bogs are large ice-filled peat mounds (palsas) surrounded by unfrozen peatlands, while peat mound bogs consist of small frozen peat mounds usually less than three metres in diameter rising up to a metre above permafrost fen. Lowland polygon fens and palsa fens are similar to the lowland polygon bogs and palsa bogs except that the centres are lower in the fen polygons (low-centred polygons) and nutrient status is greater.

Boreal Zone Wetland Forms   Domed bogs have already undergone paludification (Figure 8.2). Normally the first stage in its development is a brown moss–sedge fen, but as time progresses and peat accumulates, the water table rises, conditions become more acidic and eutrophic-mesotrophic conditions give way to oligotrophic "proto-bog" where nutrients and water are entirely rain fed and *Sphagnum* spp. dominate. A convex or copula-like bog relief results because the acidic bog materials decompose slowly compared to the mesotrophic margins. As a result, the acidic bog centre becomes domed several metres above the more nutrient-rich (fen) margins, and drainage away from the centre takes over. Because of their dependence on rainwater and slow decomposition rates, domed bogs (or raised bogs, as they are known in Europe) are associated with the more humid parts of the boreal. They are also associated with pools which may deepen, may expand laterally, can drain through subsurface piping or by surface erosion, and generally reflect the inherently unstable mature bog system (Foster et al., 1988).

Blanket bogs consist of extensive peat deposits which occur more or less uniformly over gently sloping hills and valleys. They are characteristic of boreal regions where moisture indices are strongly positive, humidities are high year-round, and bedrock/parent material is nutrient poor. Basin bogs occur in basins that have essentially enclosed drainage but derive water from both precipitation and runoff, while basin fens are essentially mesotrophic-eutrophic equivalents of basin bogs. There are two forms of plateau bogs. The Atlantic form is raised above the surrounding terrain and with flat-to-undulating surface, often with large scattered pools, while the northern type generally has a flat surface and small wet depressions. String bogs have low, narrow ridges perhaps 1–2 m across and 30 cm high, separated by depressions, or "flarks" (*flarke* is the Swedish term for such a depression) (Ivanov, 1981). Atlantic ribbed fens and northern ribbed fens are essentially the mesotrophic equivalents of string bogs. Topographic gradient greatly influences the morphology and distribution pattern of these ribbed fens. The steeper the gradient, the closer are the ribs or strings, while the more gentle the grade, the more irregular the strings and broader the flarks. These flarks are normally dominated by sedges and non-*Sphagnum* mosses, while the strings support *Sphagnum* spp., dwarf shrubs, and black spruce. Zoltai et al. (1988a) conclude that in the Low Subarctic it is possible for permafrost to develop in the strings to the point that they totally impede downslope drainage. Permafrost then extends into the flarks, which eventually disappear, leaving behind a peat plateau bog. As with domed bogs, the ribbed fen–to–northern plateau bog sequence stresses both the "organism" characteristic of peatlands and the fact that they are often in but the early stages of development.

8.5 Canadian Wetlands

Canada is second only to Russia in terms of total wetland area, and second only to Finland in terms of per cent of national territory covered by wetlands (Ruuhijarvi, 1983). In a country as large as Canada, with its pronounced temperature and precipitation gradients, surface-drainage characteristics, and soil parent material variations, it is easy to appreciate that wetlands vary in their characteristics from region to region and that locally the percentage area covered by wetlands will also differ (Glooschenko et al., 1993). While Figure 3.5 and Table 8.1 summarize this latter point, Figure 8.3 shows the twenty wetland regions (and a few additional subregions) into which Canadian wetlands have been classified (National Wetlands Working Group, 1986). Wetland regions are "areas within which similar and char-

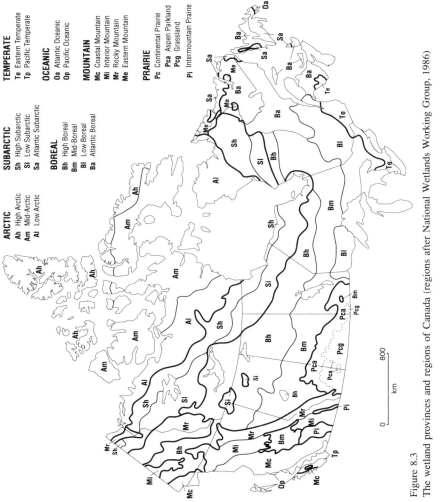

ARCTIC
Ah High Arctic
Am Mid-Arctic
Al Low Arctic

SUBARCTIC
Sh High Subarctic
Sl Low Subarctic
Sa Atlantic Subarctic

BOREAL
Bh High Boreal
Bm Mid-Boreal
Bl Low Boreal
Ba Atlantic Boreal

TEMPERATE
Te Eastern Temperate
Tp Pacific Temperate

OCEANIC
Oa Atlantic Oceanic
Op Pacific Oceanic

MOUNTAIN
Mc Coastal Mountain
Mi Interior Mountain
Mr Rocky Mountain
Me Eastern Mountain

PRAIRIE
Pc Continental Prairie
Pca Aspen Parkland
Pcg Grassland
Pi Intermountain Prairie

0   800
        km

Figure 8.3
The wetland provinces and regions of Canada (regions after National Wetlands Working Group, 1986)

acteristic wetlands develop in locations that have similar topography, hydrology and nutrient regime" (National Wetlands Working Group, 1986). For practical reasons, these twenty regions can be grouped into seven wetland provinces: the Arctic, Subarctic, Boreal, Prairie, Temperate, Oceanic, and Mountain. The designation "province" is not officially used for two or more wetland regions; however, as the discussion below benefits from a term for groups of similar wetland regions (that might not even be coterminous), the term "province" is adopted. Care must be taken, therefore, not to confuse the term "wetland province" with "ecoclimatic province." At first glance, the wetland regions distribution pattern shown in Figure 8.3 does in fact appear to have many similarities with ecoclimatic zonation (Figure 1.4). Closer examination, however, reveals significant differences, particularly where oceanic influences are strong, so the two zonation systems should not be confused. The book *Wetlands of Canada* (National Wetlands Working Group, 1988) details these wetland regions, while the discussion below gives some examples of wetland classes, forms, and types across Canada.

8.5.1 Arctic Wetlands

Appearances can be quite deceptive in the Arctic. While the tundra cover, with its important component of mosses and lichens and its absence of phanerophytes, suggests the presence of large areas of wetlands, the major part of the Arctic has less than a 5% wetland cover. Only in some western parts of Mid-Arctic islands (such as Banks Island), along the western shores of Hudson Bay in southeastern Keewatin District, and in the Mackenzie Delta do values occasionally exceed 50% (Figure 3.5). Of the five wetland classes, bogs and fens are the most frequently encountered in this province.

Wetland vegetation in the Mid-Arctic Wetland Region is dominated by *Carex* and *Eriophorum* species, together with mosses such as *Aulacomnium turgidum* and *Drepanocladus revolvens* (see chapter 2). In the Mid- and Low Arctic portions of the Northwest Territories, permafrost, frost wedging, and cryoturbation all influence bog environments. In the Low Arctic, the elevated central portions of high-centred bog polygons are dominated by *Salix* and *Ledum* spp., lichens, and mosses, while *Sphagnum* spp. are associated with the polygon trenches. Peat mound bogs are also associated with lowland fen (Tarnocai and Zoltai, 1988). In the Low Arctic, low-centred polygon fens are common, and their centres are dominated by *Carex* spp., *Ledum palustre* var. *decumbens,* and *Sphagnum* spp. (see chapter 2), while basin fens occupying topographic depressions are dominated by *Carex* and *Eriophorum* sedges and mosses such as *Drepanocladus aduncus, D. fluitans,* and *Scorpidium scorpioides* (Tarnocai and Zoltai, 1988). Floodplain swamps are also found in the Arctic, where they tend to be dominated by tall willows, sedges grasses, and horsetails. They form distinctive communities on a number of islands in the Mackenzie Delta. Floodplain marshes are found along the channels of the Mackenzie Delta estuary.

8.5.2 Subarctic Wetlands

The High, Low, and Atlantic Subarctic Wetland Regions contain some of the highest concentrations of wetlands in Canada. Most of the High Subarctic Wetland Region has between 6 and 25% of the land area in wetlands, except along the southern shores of Hudson Bay, where the values exceed 75%. In the Low Subarctic, values are proportionately higher, with large areas having more than 51–75% wetlands and some in northern Alberta, northeastern Manitoba, and northern Ontario with more than 75% wetlands (Figure 3.5). Some of these latter bogs and fens are so continuous, being interrupted by only the occasional island of rock or mineral soil, lakes, and rivers, that they have been nicknamed "peat seas" (Sjörs, 1959). Towards the east, in the Oceanic Subarctic of eastern Labrador, Newfoundland, and northern Cape Breton Island, wetlands usually comprise less than 25% of the land cover. Only in the wetter parts of Newfoundland do values rise to between 26 and 50%.

Peat plateau bogs, polygonal peat plateau bogs, and palsa bogs are typical of the Subarctic, as well as parts of the Boreal. Peat plateaux such as those in the Low Subarctic near Gillam, Manitoba, are perennially frozen peatlands raised about one metre above the water table of their associated wetlands and are dominated by an open tree cover of *Picea mariana*, along with the low shrub *Ledum groenlandicum* (the branches of which often act as supports for large growths of *Sphagnum*) and light-coloured lichens. This is an inhospitable environment for higher plant species because of both oligotrophic conditions and the smothering *Sphagnum*, which continuously grows upwards while partially decomposing to peat below. Only dwarf shrubs such as *Ledum* spp. have the ability to "follow up" the *Sphagnum*, thereby avoiding the risk of being buried (Sukhachev, 1973). Polygonal peat plateaux are similar to peat plateaux except that they have polygon margin trenches, are associated more with a cover of lichens (*Cetraria* and *Cladina* spp.), and are usually treeless except for krummholz forms of *Picea mariana* (Zoltai et al., 1988a). Palsa bogs rise 1–4 m above the wetland, have cores of permafrost (ice-filled peat), and generally have a covering of *Sphagnum* peat, lichens, dwarf shrubs, and scattered black spruce.

In the Low Subarctic–High Subarctic transition zone close to the southwestern shores of Hudson Bay, fen dominates large areas, while true bogs are almost totally absent. This reflects the combined influence of carbonate-rich glacial deposits on groundwater and insufficient time since isostatic emergence from the Hudson Bay for paludification to initiate proto-bog development. Here, low-lying fen close to Hudson Bay has had less than 1,000 years to develop. Thirty kilometres inland, at slightly higher elevation, are large areas of somewhat thicker and older mesotrophic fen. Over time, these thickening mesotrophic fens develop areas where nutrient deficiencies and a decreasing surface pH allows for the establishment of acidophilous *Sphagnum* mosses. Once *Sphagnum* "inoculation" occurs, upward peat development is much more rapid, and a proto-bog feature develops. Where this cycle is being initiated, *Sphagnum* mounds may be as little as one metre across, but such mounds rapidly coalesce to form larger, outward-accreting, roughly circular bogs. Bog accretion over fen initiates an exogenous paludification succession in which the ericaceous shrubs, mesotrophic sedges and mosses, and scattering of *Picea glauca* of the original fen give way to oligotrophic species such as *Sphagnum, Ledum, Larix laricina,* and *Picea mariana.* In the proto-bog areas south of the Twin Lakes region of northern Manitoba, the dominant initial *Sphagnum* colonization species is the purplish-red *S. warnstorfii.* Typical pH values in a profile through these Terric Mesic Organic Cryosols would be 3.4 in the living proto-bog surface *Sphagnum* layer, 5.3 and 6.4 at depths of 20 cm and 35 cm respectively in fen peat, and >7.0 at the mineral contact at depths of 50–60 cm just above permafrost. Where paludification has had longer to progress, acidification has intensified, and the initial mosses have given way to darker coloured brown and green species, such as *S. fuscum.*

A number of fen types are encountered in the Subarctic. Northern and Atlantic ribbed fens are associated with gentle gradients where groundwater is moderately nutrient rich. Channel fens formed in topographic channels in mineral soil, bedrock, or frozen organic land forms are also common and frequently associated with the periphery of peat plateaux. Palsa fens are also encountered, with palsa fields often containing several dozen closely spaced, oval mounds 20–35 m across rising 1–2 m above flat-bottomed inter-palsa depressions. It is probable that fen palsas do not grow as tall or large as bog palsas due to the thinner peat layer available to them. Occasionally, hydrolaccoliths are found in fens where artesian water helps form ice-core mounds perhaps 30 m across and 1 m tall. Unlike palsas, these lens-like cores are composed of solid, clean ice and expand upwards a metre or more, stretching the thin insulating fen peat above until it ruptures, forming long cracks. These cracks expose the ice core to melting, and the whole feature rapidly collapses. More typical of the Subarctic are collapsed scar fens, which are often associated with veneer bogs where melting of the permafrost and surface collapse has restored groundwater nutrient supplies. In areas where fen or bog permafrost has experienced major recent melting, thermokarst lakes can occupy as much as 25–30% of the landscape.

Marshes and swamps are not frequently encountered in the Subarctic, except along the unfrozen alluvium of active floodplains such as those in the Mackenzie Delta, Northwest Territories, around the shores of James Bay, or around the larger rivers in northern Manitoba. It is not uncommon to find a marsh zone of *Equisetum fluviatile* (water horsetail) in the more regularly inundated areas which receive large amounts of sediment, while *Carex aquatilis* (aquatic sedge) and *Arctophila fulva* (pendent grass) are also adapted to long periods of inundation but less annual sedimentation (Pearce, 1987). In less frequently inundated areas, *Equisetum-Salix* spp. swamp develops upslope (Gill, 1973). Marshes dominated mainly by *Carex* spp. are also found in Atlantic Subarctic Newfoundland. Here *Scirpus* and *Typha* spp. become more important towards southwestern Newfoundland as sites become more mesotrophic (Wells and Hirvonen, 1988).

Low Subarctic coastal marshes are the norm along the southern shores of Hudson Bay, James Bay, and the Labrador coast, where emergence due to isostasy promotes the lateral expansion of marshes (Price, 1990). Here, marsh zonation seems typical of the classic "succession" from saline to fresh water or brackish to fresh water, depending to a large extent on the salinity of tidal waters. On the west side of James Bay, at Ekwan Bay, vegetation distribution is also closely correlated with the salinity gradient, which decreases from lower salt marshes dominated by such species as *Puccinellia phryganodes*, inland to freshwater meadows and coastal fens dominated by *Juncus balticus, Carex aquatilis, Equisetum arvense,* and *Drepanocladus uncinatus* (Earle and Kershaw, 1989). By contrast, marshes along the southern edge of James Bay are typically dominated by freshwater to brackish species because of the much lower salinity of the water of the bay itself. Here *Eleocharis palustris* (spike-rush) and patchy *Scirpus americanus* dominate on the intertidal flats, while *Carex* and other freshwater species generally dominate inland, except where salinity is present due to release from relict salt deposits found in post-glacial Tyrrell Sea sediments (Price et al., 1988, 1989). East of Churchill, well-drained, but saline coastal spits are colonized above the dead seaweed high-tide line by single vascular species zones of, first, *Honckenya peploides* (sea-purslane) and then, *Elymus arenarius* (sea lime grass). In more sheltered embayments, such as at Bird's Cove, backshore wetlands are dominated by *Puccinellia phryganodes* (alkali-grass or goose grass) and *Honckenya peploides* in more saline areas, by *Senecio congestus* (marsh ragwort), *Hippuris vulgaris* (common mare's-tail), *Elymus arenarius*, and *Rorippa islandica* (marsh yellow cress) in brackish sites, and by willows and sedges in non-saline backshore sites.

8.5.3   Boreal Wetlands

The Boreal Wetland Province is comprised of the High, Mid-, Low, and Atlantic Boreal Regions. Large areas of the High Boreal Wetland Region in northern Alberta, central and northeastern Manitoba, and central Ontario have more than 50% wetland cover. In addition, areas with more than 75% wetlands are found at the western end of Lake Athabaska, the northern end of Lake Winnipeg, and south of Hudson and James bays (Figure 3.5). Much of the remainder of this High Boreal Region has between 26 and 50% wetlands. In the Mid-Boreal, except around Lake Winnipeg, where values exceed 50%, values range between 6 and 50%. The majority of the Low and Atlantic Boreal Wetland Regions fall into the 6–25% wetlands category. Only in the more southerly portions of the boreal are luxuriant coniferous and hardwood swamps encountered (Zoltai et al., 1988b). In the Atlantic Boreal of the Maritimes and Newfoundland (Figure 8.3), wetland cover rarely exceeds 25%, with many wetlands being non-forested, raised-bog heathlands, which exhibit a richer vascular flora than continental raised bogs to the west (Glaser, 1992).

Domed bogs are perhaps the most pronounced wetland form in the Humid Mid-Boreal wetlands of Ontario and western Quebec. Here they take on several distinct profiles, some having linear-convex crests while others have linear crests with swamp margins. The linear-convex crest type is the classic domed form, with *Picea mariana* becoming scattered, stunted, and more shrub-like towards the centre (Figure 8.2). Where tree desity is high, black spruce often share dominancy with *Larix laricina*, while ground cover is normally composed of ericaceous shrubs such as *Ledum groenlandicum*, *Chamaedaphne calyculata* (leather leaf), and *Kalmia polifolia* (bog-laurel), and by *Sphagnum fuscum, S. nemoreum,* and feathermoss (*Pleurozium schreberi*). Where tree cover is absent, cover is dominated either by cottongrass sedges (*Eriophorum vaginatum* ssp. *spissum*) or by a mixture of cottongrass and ericaceous shrubs on a sphagnum carpet. Peat accumulations typically exceed 3 m, and soils are frequently Mesic Fibrisols (National Wetlands Working Group, 1987). In the Atlantic boreal of New Brunswick and Nova Scotia, domed bogs are often as much as 5–8 m thick, and some are excavated for high-quality *Sphagnum* horticultural peat. After Quebec, New Brunswick is Canada's largest producer of horticultural peat (Wells and Hirvonen, 1988).

The linear crest and swamp margin form of the domed bog is often associated with a mineral soil slope against which the level-ground portion of the raised bog is expanding (Figure 8.2). As the original fen gives rise to a domed bog through paludification, a mixed-wood swamp develops along the contact with the mineral soil slope against which the expanding dome is pressing, while more typical fen develops on the sides along which contact with mineral soil slopes does not occur. These domed bogs therefore have a somewhat higher nutrient content than the previous domed bog form because of the input of minerotrophic waters on the upslope margin, and soils are typically Mesic Fibrisols and Fibric Mesisols. The swamp portion is often dominated by *Alnus rugosa* (speckled alder), along with *Picea mariana* (for a detailed description of these forms, see Zoltai et al., 1988b).

Northern plateau bogs have a much flatter surface than domed bogs, although they are still oligotrophic, are rain fed, and have developed by paludification. This plateau form develops best in the more continental Mid-Boreal region as a response to the somewhat drier climate, and it is essentially equivalent to the domed bogs of the more humid portions of the boreal wetland regions. Again *Sphagnum* spp., lichens (e.g., *Cladina rangiferina*), and ericaceous shrubs dominate ground cover, with semi-open to closed stands of stunted *Picea mariana* in the south and sparsely treed black spruce to the north. Lichen-covered *Sphagnum fuscum* mounds create a hummocky terrain on the open bog.

Peat plateau bogs are associated with permafrost and commonly occur as treed islands rising 1–2 metres above surrounding fen. This elevation is attained partly due to the ice and partly due to slower decomposition of acidic materials. While the increased elevation ensures more oligotrophic conditions, it improves drainage over the surrounding fen, and the treed islands are *Picea mariana*–dominated (Zoltai et al., 1988b). Near Flin Flon, Manitoba, all mixes of peat plateaux (with permafrost), palsas, and non-frozen degraded plateaux can be found. This occurrence of both aggrading and degrading plateaux in the same region confirms its transitional position relative to present climate (Zoltai, 1972). After studying peat stratigraphy in the upper Hay River valley of northwestern Alberta, Zoltai (1993) concludes that the cyclical growth and collapse of peat plateau is triggered by fire. Collapse scars form when permafrost under treed bog loses its insulative protection following fire, and then infilling allows for renewed permafrost development and treed-bog restoration. In the Atlantic Boreal of eastern Labrador and northern Newfoundland, permafrost is also important in producing frozen mounds 1–2 m high in Atlantic plateau bogs and palsa bogs 2–3 m tall in northern Labrador (Wells and Hirvonen, 1988).

Basin bogs and basin fens are also commonly encountered within Boreal wetlands. Basin bogs are supplied with ombrotrophic groundwater in essentially closed drainage basins with stunted, open *Picea mariana* cover. Where seepage into such bogs occurs, mesotrophic conditions prevail and swamp margins dominated by better stands of black spruce are found, together with some *Larix laricina*. Where the entire wetland is supplied with minerotrophic waters, basin fens develop. These fens have less than a 10% tree cover of low *Larix* and scattered *Picea mariana*, while the less acidic conditions favour non-ericaceous shrubs species such as *Betula pumila* and sedges, although *Sphagnum* spp. still do well. The high nutrient status conditions peripheral to these fens sometimes allow for the development of narrow *Scirpus*- and *Typha*-dominated marshes, along with willows.

Horizontal fens are common throughout the Boreal, where they often form large areas of flat, featureless, open *Larix* forest, which slope gently in the direction of drainage. In wetter parts of these horizontal fens, sedges (*Carex, Eriophorum,* and *Scirpus*), non-sphagnum mosses, and shrubs such as *Myrica gale* (sweet gale) do well, while on drier ground, better drainage and somewhat lower pH conditions favour *Scirpus* spp., *Equisetum fluviatile*, *Sphagnum* spp., and shrubs such as *Betula pumila* (swamp birch) (Zoltai et al., 1988b). Even though the wetter and drier areas have the same open-canopy *Larix* tree cover, ground-cover differences sometimes give these fens a streaky appearance when viewed from the air. These broad streaks are elongated in the direction of drainage, with the lighter colours representing the drier areas where more lichen is found.

Another important fen form is the northern ribbed fen with its flarks and strings (ribs). Flarks are typically mesotrophic-eutrophic and dominated by sedges and non-sphagnum mosses, while string-ridge cover varies with the degree to which the strings are above the more nutrient-rich water table. On the lower, wetter strings, non-sphagnum and sphagnum mosses are often associated with *Salix* spp., *Betula pumila,* and *Carex* spp., while on the somewhat higher, drier, and more acid strings, *Sphagnum warnstorfii* and *S. fuscum* dominate the moss layer, along with woody species such as *Larix laricina, Betula pumila,* and *Ledum groenlandicum*. Where the strings rise more than 25 cm above the local water table, surface peat is strongly acid and *Picea mariana* is the dominant, often reaching 15 m in height. These spruce-dominated strings often have a ground cover of *Sphagnum fuscum, S. magellanicum,* and light-coloured lichens (Zoltai et al., 1988b). A much smaller fen form, the collapsed scar fen, is found in peat plateau (bog) areas, where it owes its origin to the local melting out of permafrost, perhaps following fire or other surface disturbances which modify the microclimate. This allows for subsidence to the level of the minerotrophic water table, and cover is more typically fen with *Carex* sedges, *Calamagrostis canadensis* (bluejoint), and primarily non-sphagnum mosses.

Marshes are also found in Boreal wetland regions, with perhaps the largest form being delta marshes such as those associated with the Saskatchewan River in west-central Manitoba, which covers approximately 9,200 km^2. As outlined in section 8.4.2, these delta marshes experience periodic inundations which vary in duration and severity. In addition, being associated with river channels, they can differ tremendously in habitat conditions and therefore dominant cover types. These habitats may variously be oxbow lakes, cut-off channels, shallow lakes, level meadows, levees, and even less-flooded areas where treed swamps can be found. For a detailed LANDSAT map showing such wetland variety within the Saskatchewan Delta, see Dixon and Stewart (1984). In deep-water marshes, *Scirpus, Eleocharis,* and *Typha* spp. usually dominate, while in shallower water, *Carex* and *Scolochloa* spp. are common (Zoltai et al., 1988b). Willows, sedges, and horsetails are associated with ground only occasionally inundated, while most other woody species do not fair well in any of the preceding habitats due to the extremes of soil atmosphere switches. Smaller marshes along rivers show distinct gradients in terms of both species and phytomass. Along the Ottawa River, typical marsh community composition varies depending upon duration of flooding, fertility of substrate, litter removal, and exposure. In general, these variables produce four major community types dominated respectively by *Typha latifolia* (common cat-tail), *Scirpus americanus, Eleocharis smallii,* and *Sparganium eurycarpum* along a gradient from higher to lower elevation (Day et al., 1988).

Swamps associated with floodplains are also characteristic of some parts of the Low Boreal Wetland Region under mixed-forest conditions. These swamps develop due to annual inundation by the slowly moving, often nutrient-rich floodwaters of small streams. Peat can develop, but usually soils are Gleysolic. Biological productivity is high, giving rise to a lush forest cover. *Fraxinus nigra* (black ash) up to 30 m tall dominate, while *Betula papyrifera, Ulmus americana, Picea glauca,* and *Abies balsamea* may be present. Shrubs include *Prunus virginiana, Acer spicatum, Alnus rugosa,* and *Cornus stolonifera*, while the humid, shaded forest floor is the home of many ferns and a rich herb flora (Zoltai et al., 1988b).

In Atlantic Canada, wetlands reflect the per-humid climate, with coastal marshes, fens, and bogs being common. Because of the importance of coastal marshes to wildlife, some of the wetlands in Nova Scotia and New Brunswick have been designated as Ramsar sites. Thes include Shepody Bay, Mary's Point, the Chignecto National Wildlife Area, and the Southern Bight–Minas Basin. Inland, bogs are more typically boreal in nature, but are generally thicker as a result of the longer post-glacial time in which peat has been able to accumulate. At Point Escuminac on the north-central coast of New Brunswick, the centre of the >10 km^2 bog has a cover of sedges and mosses (*Eriophorum vaginatum* ssp. *spissum, Sphagnum fuscum,* and *S. flavicomans*) and dwarf shrubs (*Kalmia angustifolia, Gaylussacia baccata, G. dumosa,* and *Empetrum nigrum*) (Warner, et al., 1993).

8.5.4 Prairie Wetlands

The discussion of Canadian prairie vegetation given in chapter 5 concerns itself only with the three Prairie provinces and parts of southern Ontario. The Prairie Wetland Province (Figure 8.3), on the other hand, considers the traditional prairie as the Continental Prairie Region, and those grasslands found in British Columbia as the Intermountain Prairie Region (the Cordilleran Cold Steppe and Savanna Forest region described in section 6.1.7; see Figure 6.1). As should be expected in a region with moderately to strongly negative moisture indices, few areas have more than 25% wetlands, and most have much less. This is particularly true of the Intermountain Prairie Region of the Cordillera and the Mixed-grass Prairie and Fescue Prairie Subregions. The only locations with the potential for higher percentages are in parts of the somewhat more humid aspen Parkland Continental Prairie Subregion (which includes the tall-grass prairie cover type) where drainage is impeded. Such sites include the potholes (kettle holes) of the till plains and moraines and the heavy lacustrine clays of pro-glacial lakes and active deltas. One such area is the lacustrine plain of the Lake Agassiz basin in southern Manitoba, which because of its agricultural potential has seen approximately 100,000 ha drained and converted to agriculture (Elliott, 1978).

Prairie wetlands are generally classified into three groups, palustrine, lacustrine, and riverine (Kantrud et al., 1989). Palustrine wetlands are small in area, are often nicknamed "potholes" or "sloughs," usually have alkaline or saline Gleysolic soils, owe their origins to the ponding of water and not to humid or per-humid climates, and do not possess bogs. Many of these palustrine wetlands are shallow basin and kettle marshes which are less permanent, often going from open water ponds in spring to drying basins in summer. Exceptions to this last generalization would be lacustrine and riverine wetland types such as shoreline marshes and active delta marshes. Active delta marshes usually cover larger areas and, being associated with a river and/or lake system, are subject to more reliable spring submergence and summer drawdown. The Netley and Delta marshes occupy the southern ends of Lakes Winnipeg and Manitoba respectively. Here, as with the broad shore marshes along parts of the lake margins, such as at Grand Marais, *Typha latifolia* and *Phragmites australis* dominate. The hybrid *Typha glauca* (between *T. latifolia* and *T. angustifolia*) is also increasing in importance at Delta Marsh, as it is elsewhere in North America and Europe (Waters and Shay, 1992).

It would appear that there is great variety in both prairie wetland form and species composition owing to variations in precipitation, topography, permeability of the substrate (as it influences water permanency), alkalinity of surface materials, the source of surface waters, and whether the wetlands are recharge, flowthrough, or discharge types (Arndt and Richardson, 1989). In addition, the small size of many prairie wetland marshes and the fact that their water-levels fluctuate widely during the year give rise to distinct wetland subforms which often fashion concentric bands around and within depressions and small lakes. These subforms usually vary within

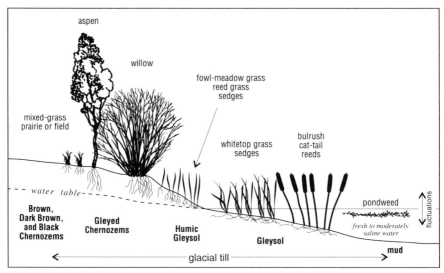

Figure 8.4
Generalized zonation across a prairie pothole lake margin. While annual and seasonal fluctuations
in water-level give rise to lateral movements in the concentric marsh vegetation bands, the aspen-willow
zone remains fairly stationary.

short distances along an increasing moisture gradient: (1) wet meadow, (2) shallow
marsh, (3) deep marsh, (4) intermittent open water, and (5) permanent open water.
Species composition within each of these subforms can be further modified by increasing salinity (Adams, 1988). The following descriptions (together with Figure 8.4) illustrate this subform variety for typical pothole country in the Aspen Parkland Prairie Wetland Subregion, where water is fresh to moderately saline (after
Adams, 1988).

1 Wet Meadows   These meadows are dominated by grasses such as *Poa palustris*
and *Calamagrostis inexpansa*, sedges such as *Carex lanuginosa*, and the rush
Juncus balticus. Willows (*Salix petiolaris*) and *Populus tremuloides* can also occur
in this zone, but would be missing in the drier grassland prairie wetland subregion.

2 Shallow Marsh   This band may form the central subform of very shallow depressions. Here, coarse grasses and sedges such as *Scolochloa festucacea* (sprangle-top)
and *Carex atherodes* dominate, along with some water-tolerant herbs and some
floating forms, such as *Potamogeton gramineus*.

3 Deep Marsh   This marsh subform retains its surface water until at least late summer. It either forms the centre of moderately deep depressions or simply another
band between shallow marsh and intermittent open water. Dominants include

Scirpus lacustris var. *validus, S. lacutris* var. *glaucus, S. maritimus, S. fluviatilis, Typha angustifolia* (narrow-leaved cat-tail), *T. latifolia* (common cat-tail), and *Phragmites australis* (reed). *Potamogeton pusillus* (pondweed) is typical of the floating aquatics.

4 Intermittent Open Water   Usually this subform is stable open water; however, occasionally it may experience drawdown or drying out during a cycle of dry years. This fluctuation causes both emergent and submergent species to fair poorly, and in the open water stage, the community is dominated by floating aquatics such as *Potamogeton gramineus* and *Utricularia vulgaris* (common bladderwort).

5 Permanent Open Water   Water depths are at least 20 cm in September and may exceed 1 m. Aquatic species include *Potamogeton* spp. and *Ruppia occidentalis* (widgeon-grass), while submergent and emergent species are absent.

It is emphasized that these five subforms do not reveal the dynamics of these water-regime dependant wetlands in the sense that at any particular time an area may have been subject to recent drawdown, reflooding, or long periods of relative water-level stability. These wetlands are therefore the subject of short-term and reversible cycles which lead to vegetation decline and rejuvenation. Research on the relationships between these cycles, biological productivity, and feeding by waterfowl has led to techniques in wetland management which can improve wildlife production. It is also emphasized that as salinity increases, so dominants change. In saline wet meadows, *Distichlis stricta, Spartina gracilis* (alkali cord-grass), and *Triglochin maritima* (arrow-grass) are characteristic, while in emergent marsh, *Scirpus maritimus, S. americanus,* and *Puccinellia nuttalliana* normally dominate (Adams, 1988).

8.5.5 Temperate Wetlands

Temperate wetlands are found not only in the Pacific Temperate Region of southwestern British Columbia, but also in the Eastern Temperate Region of the St Lawrence Lowlands of southern Ontario and Quebec and in the Appalachian Highlands of southeastern Quebec and southwestern New Brunswick (Figure 8.3). Neither of these regions contains areas with high densities of wetlands except along the St Lawrence River, where values range between 6 and 25% (Figure 3.5).

8.5.5.1 Pacific Temperate Wetlands

These are found on the drier portions of Vancouver Island and along the mainland coast opposite Vancouver Island. In terms of dry land vegetation, this roughly co-incides with the Coastal Douglas-fir Zone of the Pacific Coastal Mesothermal Forest (section 6.1.4.2). Here, wetland cover is a response to positive moisture indices and the relatively rugged terrain. Small basin bogs up to 10 m thick surrounded by fen are encountered, but slope bogs are absent. These basin bogs probably began as lakes, later becoming sedge fens and finally ending up as *Sphagnum* bogs, some of which have now developed a tree cover dominated by *Pinus contorta* (lodgepole pine) or *Pinus–Tsuga heterophylla* (western hemlock) (Banner et al., 1988).

8.5.5.2 Eastern Temperate Wetlands

These wetlands are best characterized by *Acer saccharum*–dominated swamps, although fens, bogs, and *Typha* marshes are also encountered. In addition, the region contains some of Canada's most important brackish and salt-water marshes. Swamps are associated with peat margins, streams, basins, springs, and shorelines, while fen forms include shore, stream, and channel types, and bogs include flat, basin, domed, and shore bogs. Typical southern Ontario hardwood treed swamps are found on gently sloping ground where small streams overflow and where there is standing or gently flowing base-rich water for long periods of time and soils are peaty Gleysols or Mesisols. In the Beverly Swamp near Hamilton, Ontario, typical forest includes dense maple-ash stands dominated by *Acer saccharum, A. rubrum, Fraxinus nigra,* and *F. pennsylvanica* and *Ulmus americana–Fraxinus* stands which have been disturbed as a result of Dutch elm disease (Glooschenko and Grondin, 1988). Between Cambridge and Paris, Ontario, is found the Sudden Bog with its typical mesic forest, together with minor bogs and swamps. Upland mesic forest cover consists of *Quercus rubra, Acer saccharum,* and *Prunus serotina* (black cherry), grading into drier sites with *Quercus, Carya ovata* (shellbark hickory), and *Pinus strobus.* Swamps include *Cornus stolonifera* (red-osier dogwood), *Sambucus canadensis* (common elder), *Cephalanthus occidentalis* (buttonbush), and *Ilex verticillata* (winterberry), while open bogs are generally dominated by *Chamaedaphne calyculata* (leather leaf), and treed bogs by *Picea mariana, Pinus strobus,* and *Larix laricina* on a cover of *Sphagnum fuscum* (Ministry of Natural Resources, 1990).

Typical conifer swamps dominated by *Thuja occidentalis, Pinus strobus,* and *Larix laricina* are also found in southern Ontario. Shore swamps dominated by *Acer saccharinum* (silver maple) are common along the Great Lakes and the Ottawa, St Lawrence, and Richelieu rivers. Old-growth *Acer rubrum–Tsuga canadensis–Fraxinus nigra–Thuja occidentalis* dominated swamp forest in Jobes' Woods, Presqu'ile Provincial Park, shows the impact of a severe windstorm on these shallow-rooting communities. The tall, one-metre-plus-diameter eastern hemlock were uprooted in large numbers by a 160 km hr^{-1} windstorm in November 1992, and many surviving cedars were snapped by a 45 cm snowstorm the following month. In addition, these toppled shallow-rooting species, especially hemlocks, pulled huge volumes of Gleysolic mineral soil into the air and are compounding the unevenness of the already uneven nature of the swampy forest floor (and no doubt adding Turbic characteristics to the Gleysols). To the east, in southeastern Quebec, peat margin swamps are particularly common where forest develops over thick eutrophic peat.

In the Appalachian Highlands, *Thuja occidentalis* dominates stream swamps, along with hardwoods such as ash, maple, and poplar and the softwood *Abies balsamea* (Glooschenko and Grondin, 1988).

Fens are usually treeless with deep peat layers. Shore fens are dominated by a graminaceous cover that includes *Cladium mariscoides* (twig-rush) and *Carex lasiocarpa*, through which emerge *Salix pedicellaris* (bog willow) and *Myrica gale* (sweet gale) shrubs and the occasional *Larix laricina*. The Holland River fen at the southern end of Lake Simcoe, Ontario, is an example of a stream fen dominated almost completely by *Cladium mariscoides*, together with some *Typha latifolia* and *Myrica gale. Carex lasiocarpa* dominates channel fens, with minor occurrences of *C. limosa, C. chordorrhiza,* and *Phragmites communis* (Glooschenko and Grondin, 1988). Sedges are also found on some of the eastern temperate bogs; however, strongly acid conditions normally promote the development of thick *Sphagnum* peat. Numerous flat and basin bogs formed in kettle holes, often with open *Picea mariana* forest, ericaceous shrubs such as *Chamaedaphne calyculata*, and *Sphagnum magellanicum* and *S. capillifolium*, are typical of moraines in southwestern Ontario. These bogs represent the advanced phase of wetland development for the region since the open-treed bog stage only develops after passing through both the fen and open ericaceous-*Sphagnum* stages (Warner and Kubiw, 1987). In southern Quebec, small shrub-covered domed bogs are dominated by *Chamaedaphne calyculata* on wetter sites and *Ledum groenlandicum* on drier, while *Rhododendron canadense* (rhodora) and *Betula populifolia* (grey birch) shrubs occur more frequently following fire (Glooschenko and Grondin, 1988).

Eastern Temperate marshes display banding or zonation characteristics similar to those of the Prairie Wetland Province, except that many of the species dominants differ. Common shore emergents include the ubiquitous *Phragmites communis, Sparganium eurycarpum* (bur-reed)*, Sagittaria latifolia* (arrowhead), *S. rigida, Typha* spp. (cat-tails), *Scirpus lacustris* var. *glaucus* (bulrush), *Butomus umbellatus* (flowering rush), *Eleocharis smallii* (spiked rush), *E. palustris, Acorus calamus* (sweet flag), and *Zizania aquatica* (wild rice). To these must be added the importance of the exotic *Lythrum salicaria* (purple loosestrife), a hardy perennial which was introduced from Europe to North America in the early 1800s. While noted for its beauty, loosestrife is unfortunately expanding into many marshes and meadows where it forms diversity-limiting monotypic stands. In late summer it is not uncommon to see completely purple hay meadows in the Ottawa region. With few natural predators to keep it in check, loosestrife jeopardizes endangered wetland plants and animals, as well as reducing pasture-carrying capacity (Malecki et al., 1993).

Along the St Lawrence River, in regions such as the Lake Saint-François National Wildlife Area, human disturbances and fire suppression play an important part in marsh and forested swamp dynamics and influence post-disturbance *Typha angustifolia–Lythrum salicaria* community dynamics (Jean and Bouchard, 1993). Tidal conditions can also increase marsh salinity along the lower St Lawrence River. Here *Scirpus americanus* dominates where water is essentially fresh (<0.3% salts), while

Spartina alterniflora (salt-water cord-grass) becomes important in brackish marshes (0.4–1.7% salts) (Glooschenko and Grondin, 1988).

8.5.6 Oceanic Wetlands

Oceanic wetlands can be conveniently classified into two regions, the Pacific Oceanic Region and the Atlantic Oceanic Region (Figure 8.3). In the Atlantic Oceanic Region of southeastern Newfoundland, wetlands cover between 26 and 50% of the land surface. While much of the North Coast Pacific Oceanic Subregion has between 6 and 25% wetlands, the South Coast Pacific Oceanic Subregion has generally less than 6%. The North Coast subregion, consisting of the Queen Charlotte Islands and the mainland along the Queen Charlotte Strait, is the wettest region in Canada, and bogs are the dominant wetland form. The South Coast subregion of northwestern and northern Vancouver Island is more characterized by a mixture of wetland forms which include basin and flat bogs, basin swamps, stream fens, and estuarine or coastal marshes (Banner et al., 1988).

Slope bogs are common in the Northern Pacific Oceanic Subregion, where they often look like blanket bog. Both slope and flat bogs are not sufficiently extensive to be called blanket bogs, even though where two or more coalesce, they appear rather extensive (E. Oswald, personal communication). In the Prince Rupert region, wooded and open slope bogs are common. Here, wooded bogs are generally dominate by stunted *Pinus contorta, Chamaecyparis nookatensis* (Alaska yellow cedar), and *Thuja plicata* (western red cedar) less than 12 m tall, and by *Ledum groenlandicum* and *Gaultheria shallon* shrubs. Ground-level species include *Carex pluriflora, Fauria crista-galli* (deer-cabbage), and *Sphagnum papillosum* and *S. rubellum*. Open shrubby bogs may have scattered shrubby pine and yellow cedar, but are normally dominated by *Trichophorum caespitosum* in more minerotrophic sites and *Sphagnum* spp. on ombrotrophic (Banner et al., 1987).

Flat bogs also cover extensive portions of the Queen Charlotte Lowland, while domed bogs occur there only locally. In the Southern Subregion, particularly on the west side of Vancouver Island, shore and floating bogs consisting of fibrous *Sphagnum* peat, often with *Ledum groenlandicum* and *Myrica gale,* occur along the margins of dystrophic lakes. Occasional slope and small basin bogs can also be found here, as can western skunk cabbage (*Lysichitum americanum*) swamps (Banner et al., 1988). Coastal freshwater and intertidal brackish marshes are also associated with such areas as the Fraser River delta. Here *Scirpus americanus* and *S. maritimus* dominate lower marsh sites, while *Carex lyngbyei* is a better competitor for resources in the high marsh (Pidwirny, 1990; Karagatzides and Hutchinson, 1991).

The Atlantic Oceanic Region consists only of the peninsulas of southeastern Newfoundland. Here, cool summers and cold winters combine with both high humidities and precipitation to promote extensive areas of "climatic" blanket bog which expand outwards through paludification. Extensive slope bogs are found here, together with some slope fens (Wells and Hirvonen, 1988).

8.5.7 Mountain Wetlands

As should be expected of steeply sloping terrain, wetlands do not constitute large percentages of any of the four mountain wetland regions: Coastal, Interior, Rocky, or Eastern (Figure 8.3). Here wetlands are only encountered frequently at high elevation in alpine and subalpine environments and along level valley floors or associated with lake margins. Only the Liard River valley and a few isolated floodplains, such as that along the Bow River in Banff National Park, would be considered to have more than 6% wetland cover (Figure 3.5). Mountain wetlands are also more common on the windward sides of each range than on the leeward rain-shadow sides, and they tend to be small and formed in pockets or depressions. Because of the effects of elevational range on climate, wetlands on higher summits resemble those of the Arctic, at middle ranges they resemble Subarctic forms, and at lower elevations they resemble and interdigitate with adjacent region types. Those of the southern Coast Mountains grade into Pacific Temperate types, while those on the lee sides grade into Intermountain Prairie types (Pi on Figure 8.3). Most Mountain wetland forms are flat or basin bogs and ribbed fens, but frost and permafrost at higher elevations play significant roles in wetland development processes. Snowmelt can be the source of much of the water in high elevation wetlands, particularly those in the northern Interior Mountain Wetland Region where precipitation is low (E. Oswald, personal communication). Wetlands in the Coastal Mountain Wetland Region are outlined in the discussion of Pacific coastal subalpine forest (section 6.1.5). Likewise, *Eriophorum*-dominated alpine tundra wetlands are described in section 2.7.1.1.

8.6 Wetlands in the USA

Wetlands form an important component of the vegetation cover in the coterminous states, but unlike in Canada, where fens and bogs are the dominant wetland types, here marshes and swamps are dominant. Only in Alaska are wetlands comparable to those in Canada in terms of wetland class dominancy. Also, unlike in Canada, a greater percentage of wetlands in the coterminous states have been converted to other uses, with only 46% (or 40 million ha) of the original 87 million ha remaining by the mid-1970s (Wentz, 1988). Bogs and fens are frequently encountered in the Great Lakes–St Lawrence forests, particularly in Minnesota and Wisconsin. Sizeable areas are also found in the New England states and in parts of the southeast from North Carolina through Florida to Mississippi, although these latter states rarely have the classic ombrotrophic bogs typical of the boreal region. Marshes are particularly common throughout the northern Great Plains, where they represent a southerly extension of the Canadian Continental Prairie Wetland Region. Marshes are also quite common along the central eastern seaboard and very common along the Gulf coast and in Florida. Swamps cover large areas in the Great Lakes mixed forests of Minnesota, Wisconsin, and Michigan, along the lower Mississippi River, and throughout all the southeastern seaboard, especially in Florida. Distribution maps for these wetland types can be seen in Hofstetter (1983), and details of the major swamps are given in Thomas (1976).

Alaska can be conveniently divided into the central zone, which includes fens and bogs so typical of the Arctic and Rocky Mountain Wetland Regions of Canada, and the southern-southeastern zone, which is more typical of the Coastal Mountain Wetland Region of Canada. Along the Pacific coast of the coterminous states, wetlands do not constitute a significant part of the landscape except in such areas as the Olympic Peninsula of the Pacific Northwest, and although *Sphagnum* spp. are often present, there are few true bogs. In California the Tule Marshes of the Sacramento–San Joaquin deltas also deserve mention since they originally covered over 200,000 ha. Although the subject of drainage efforts, large areas of the Tule Marshes are dominated by communities of the bulrushes *Scirpus acutis, S. californicus, S. olneyi,* and *S. validus,* while *Phragmites australis, Typha latifolia,* and *Juncus* and *Cyperus* spp. are also encountered (Hofstetter, 1983). In the Cordillera (Western Interior Region), wetlands represent a very small fraction of the landscape. They are normally riparian, associated with springs, or occupy saline depressions. In the prairie region, wetlands again become a significant component in the landscape. As in the Canadian prairies, in North and South Dakota there are vast numbers of small pothole marshes which are so vital to groundwater and aquifer recharge, wildlife, and migratory waterfowl (see section 5.4.2.1).

In the North-Central and Northeastern Regions around the Great Lakes, positive moisture indices combine with the poor drainage characteristics of a recent postglacial landscape to form large areas of both fen and bog, and sizeable swamp forests as well. Freshwater-to-saline marshes are also found in the tall-grass–oak forests along the western edge of this region. Compared to the boreal and mixed-forest zone in Canada, shrub- and forest-dominated swamps are much more common here. In the north are true coniferous swamps, together with some dominated by *Fraxinus nigra* (black ash) and *Thuja occidentalis.* Mixed-forest swamps dominated by *Picea mariana, P. glauca, Pinus strobus, P. resinosa, Thuja occidentalis, Betula papyrifera, Fraxinus nigra, Ulmus americana, Tilia americana,* and *Acer saccharum* are found between the coniferous swamps to the north and the deciduous swamps to the south.

The deciduous swamp forests of southern Michigan, Iowa, and Indiana are generally dominated by *Fraxinus nigra, F. americana, Acer rubrum,* and *Ulmus americana* (see section 7.4.2.1 for details on the Elm–Ash swamp forests, such as the black swamp of Ohio). Wetlands throughout much of this region have been drained for agriculture, and it is estimated that of 125 Ohio peatlands (32,198 ha) first surveyed in 1900, only 2% continued to support typical peatland flora in 1991 (Andreas and Knoop, 1992). Shrub swamps dominated by *Salix* spp., *Alnus* spp., *Cornus stolonifera,* and *Cephalanthus occidentalis* are also common throughout Michigan, Wisconsin, and Minnesota, while *Populus deltoides, Acer saccharum, Fraxinus americana* (white ash), and *Ulmus americana* dominate floodplain swamps to the west and southwest (Hofstetter, 1983). Even though deciduous forest swamps dominate to the south of this region, small basin bogs occupying kettle holes and dominated by *Larix laricina* and *Sphagnum recurvum* are found as far south as Ohio (Andreas and Bryan, 1990).

Along the Atlantic coast and Gulf coastal plains are two major types of wetlands: the coastal marshes along the ecotone between salt and fresh water and the inland freshwater swamps, or "bottomland hardwood forests" (Brinson, 1990). The coastal marshes are particularly well developed along these low-lying emergent coastlines. In addition, peatlands are associated with such large swamps as the Great Dismal Swamp of Virginia and North Carolina, the Okefinokee Swamp of Georgia, and the Florida Everglades (Figure 8.5). Treed swamps are generally classified into two types: deep water and shallow water. In the deeper bald-cypress floodplain swamps along the Mississippi (Figure 8.1), *Taxodium distichum* (bald cypress) and *Nyssa aquatica* (water tulepo) dominate, but where the bald cypress has been harvested or where water becomes really deep, pure stands of water tulepo can be found. At one time almost half the state of Louisiana was wetland and water bottom areas, while today less than half of this remains. It has been estimated that of the 4.8 million ha of alluvial floodplain bottomland forest found along the Mississippi River in 1937, some 55% (2.7 million ha) had been cleared by 1978 (MacDonald et al., 1979). *Nyssa sylvatica* var. *biflora* and *Taxodium ascendens* (pond cypress) swamps are common in upland Gulf coastal plain, while *Magnolia virginiana, Persea borbonia, Pinus serotinia,* and *Chamaecyparis thyoides* (Atlantic white cedar) swamps are common on the uplands of the east coast. Shallower treed swamps are associated with stream and lake margins where *Salix* spp. do well (Brinson, 1990; Hofstetter, 1983).

Relict peatlands are found in the southern and mid-Appalachian Mountains, and they bear close affinities to those in Canada. Cover is invariably dwarf shrub dominated, with *Rubus hispidus* and *Vaccinium oxycoccus* (cranberry) being important (Stewart and Nilson, 1993). The southern Appalachians also contain hundreds of scattered islands of heathland. Many of the rocky points and ridges in the Great Smokey Mountains National Park are best described as "heath balds," bordered below by *Picea rubens* and *Abies fraseri* dominated forest (Whittaker, 1979).

The Great Dismal Swamp is located on acid peats up to 4 m deep and commonly has a *Chamaecyparis thyoides, Pinus serotinia,* and swamp hardwood cover. Fires have given rise to a forest mosaic where surface fires promote *Chamaecyparis thyoides,* shallow peat fires promote pines, and deep peat fires encourage bald cypress and water tulepo. The Okefinokee Swamp (1,600 km^2) is unlike most other southern swamps in that it is at a higher elevation than the surrounding countryside. It actually consists of a mosaic of treed islands with swamp, glades with island fringes, and open marshes. Here, cypress swamps are dominated by *Taxodium distichum* and usually pines and either bald cypress or red maple, while the open marshes are usually dominated by *Nymphaea odorata* (water lily), along with scattered emergents (Hofstetter, 1983). Productivity research by Mitsch (1988) shows that the hydrologic regime of these forested swamps, especially *Taxodium* swamps,

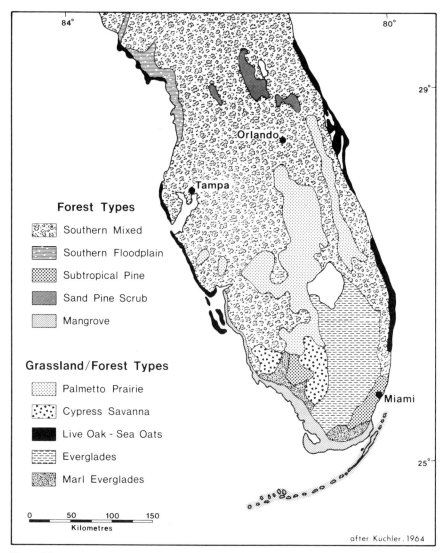

Figure 8.5
The vegetation cover of peninsular Florida (modified from Küchler, 1964)

is very important. He concludes that biomass and net primary productivity values increase dramatically along the hydrologic gradient, stagnant water, flowing, and pulsing (one or two floods per year).

A large percentage of peninsular Florida has been or still is wetland. Typical southern floodplain swamps are found north of Tampa (Figure 8.5), while live oak–sea oats freshwater-to-saline marshes are common along both east and west coasts.

Mangrove forests are also extensive along the southwestern shores and along the Keys, while huge areas of cypress swamps and everglades are found in the vast interior lowlands south of Lake Okeechobee and west of Miami. Historically, the Everglades region covered some 10,000 km^2 and had a graminoid cover dominated almost entirely by *Cladium jamaicense* (saw grass) and with some scattered tree islands. In the shallower water of the marl Everglades (Figure 8.5), *Eleocharis* spp., *Panicum hemitomum,* and *Cladium* are associated with algal mats which blanket the sediments, remove CO_2, raise the pH, and give rise to rapid calcium carbonate (marl) precipitation. Large tree islands are found on low limestone platforms in the southern Everglades. These islands (or "heads," as they are known locally) are dominated primarily by tropical evergreen tree species. West of the Everglades is the large region known as the Big Cypress Swamp, where cover is best described as a cypress (*Taxodium* spp.) savanna. The dwarf cypress trees here are often covered in large epiphytic bromeliads known locally as "wild pine." The twentieth century has witnessed major changes to all of these wetlands as a consequence of highway, canal, and levee construction, artificial drainage, and increased incidences of fire (Hofstetter, 1983).

8.7 Eurasian Wetlands

8.7.1 European Wetlands

Wetlands are particularly characteristic of both northwestern and northern Europe, where they tend to be associated with the subarctic, boreal, and mixed-forest ecosystems. In Europe the term *moor* (German) or *myr* (Swedish) is often used to describe peatlands in general, while the term *heath* is restricted to a formation dominated by ericoid dwarf shrubs (Gimingham, 1972). Though many European wetlands show marked similarities to those found in North America, some differences deserve mention. One striking feature is that Canadian Atlantic coast bogs, such as those in Nova Scotia, are much more similar to continental European bogs than to western European maritime types. The greatest similarities on either side of the Atlantic occur at the ground surface, where both are dominated by *Sphagnum fuscum, S. magellanicum, S. nemoreum,* and *S. rubellum,* while *Eriophorum* has close relatives on both sides. Differences become noticeable at the dwarf shrub level, as European bogs are dominated by *Calluna, Erica,* and less frequently, *Empetrum* spp., while in eastern North America, *Chamaedaphne, Kalmia angustifolia* (lambkill), and *Ledum groenlandicum* dominate (Osvald, 1970). In addition, *L. groenlandicum* is the predominant dwarf shrub from the Queen Charlotte Islands to Nova Scotia, while in Europe, *L. palustre* only forms communities in bogs far away from the coast. Even more pronounced is the fact that coastal bogs in Canada are frequently treed, while west coast bogs in Europe are treeless (Osvald, 1970).

As in Canada, there is a general relationship between wetland type and geographic region. More frequently, fens are associated with mixed-forest regions such as The

Fens in eastern England, while bogs are more characteristic of the northwest and north of Europe. Exceptions to the rule are often found, however, and true patterned fen (ribbed fen) can be found even in northern Scotland, a region more associated with advanced paludification (Charman, 1993). Bog types vary throughout Europe, with blanket bogs more common in the extreme oceanic climate of the British Isles and the west coast of Scandinavia, where, as the name suggests, they cover entire landscapes (Figure 7.4).

The *aapa* mires are string bogs which are more characteristic of northern and central Fennoscandia. Raised bogs (the Canadian domed or plateau bogs) stretch in a band from central Ireland and the Low Countries, through southern Fennoscandia and the boreal zone of Russia, to the Urals. They are generally formed where there were water-filled depressions or in lake basins, so can be found under drier climates than typical blanket bog (Bellamy, 1987). Just south of this raised bog zone in the southern boreal and mixed-forest region of Russia is a band of forest-raised bogs where the drier, more continental climate has allowed invasion by pine. In addition, palsa bogs (peat hummock bogs) are typical of the tundra and forest-tundra zones of northern Scandinavia and Russia, close to the Arctic Ocean (Walter, 1979). As previously mentioned (see section 4.5.1), some of these peatlands resulted from burning and Neolithic forest-clearing episodes, while today some of these same wetlands are now being drained for agriculture, improved forestry, or extracting peat fuel (Gimingham et al., 1979). As of 1979, some 5.8% of Great Britain and 16.4% of Ireland were classified as peatland. Within the United Kingdom, Northern Ireland has the highest value (12.4%), followed by Scotland (10.4%), Wales (7.7%), and finally England (2.8%) (Taylor, 1983).

Classic blanket bog sites are found on mineral soils where peat grows primarily because the climate is per-humid. They are therefore common along the Atlantic coast of Ireland, in upland Wales and England, through Scotland, and along coastal Norway. In Scotland the northern districts of Caithness and Sutherland alone have 4,000 km^2 of blanket bog peat (Charman, 1992). In Ireland, blanket bogs are found in two type locations: near sea level on the west coast (low-level Atlantic blanket bog) and as upland blanket bog (above 200 m). Blanket bog covers some 900,000 ha of Ireland, 37% of which is classified as low-level Atlantic (O'Connell, 1990). It is the main landscape feature of northwestern Ireland, and the Glenamoy IBP peatland site, County Mayo, is typical of the bog type. On the undisturbed flatter or gently rolling bog surfaces, dominants include *Molinia caerulea* (purple moor grass) and *Schoenus nigricans* (black bog-rush), a ground-cover mosaic of mainly four species of *Sphagnum,* the mosses *Campylopus flexuosis* and *C. atrovirens*, mucilaginous algae, and the liverwort *Pleurozia purpurea* (Moore et al., 1975). An inconspicuous dwarf shrub component of *Calluna vulgaris* (ling), *Erica tetralix* (bell heather), and *Myrica gale* is also present. While some of the Glenamoy site has been undisturbed since the late Bronze Age approximately 4,000 years ago, drainage has been improved in recent times by ditching and turf (peat) cutting for fuel, and ericaceous heath has developed. On drainage, *Calluna vulgaris* becomes vigorous, giving a distinctive dark-brown colour to these heaths throughout most of the year and an even more characteristic purplish-red colour during summer flowering. Nearby, where drainage and fertilization are practised for forestry, *Calluna* or *Molinia caerulea* dominate the ground cover under the introduced *Picea sitchensis* (Sitka spruce) and *Pinus contorta* (lodgepole pine) (Moore et al., 1975). Together with *Ulex europaeus* (gorse), the presence of *Pteridium aquilinum* (bracken) in wetlands in the west of Ireland is often a sign of locally improved drainage associated with peat cutting (Scott, 1967). It is also noted that on the naturally better drained upland wetlands along the coast, *Erica tetralix* is frequently the dominant ericaceous dwarf shrub, while inland it is *Calluna vulgaris.*

A somewhat similar blanket bog to that at Glenamoy is found at the IBP blanket bog site of Moor House, which is located above an elevation of 550 m in the Pennines of northern England. This *Calluna-Eriophorum* bog is typical of wetter heaths common in upland England and Scotland. Here *Calluna vulgaris, Eriophorum vaginatum,* and *Sphagnum* spp. generally dominate; however, following fire, *Eriophorum* is quick to recover dominancy, while *Calluna* regains co-dominancy only after twelve to twenty years (Heal et al., 1975). Where burning to promote grazing is frequent, *Eriophorum* spp. increase in dominancy if the fire rotation is ten years, while if the rotation is twenty years, a greater abundance of *Calluna vulgaris* results (Hobbs, 1984). *Calluna* is not a climax dominant in lowland heath bog and would be replaced by phanerophytes through natural succession if it were not for frequent disturbance (Gimingham, 1988). One species which does well on the drier heaths or near existing or felled woodlands is the almost ubiquitous bracken fern (*Pteridium aquilinum*) (Marcuzzi, 1979).

In the southern Peak District, *Calluna vulgaris* and *Vaccinium myrtillus* (dwarf bilberry) are the dominant heath dwarf shrubs, while *Erica tetralix* is uncommon. *Erica tetralix* has, however, invaded the more acid bog portions of the large fens found near The Wash in eastern England. Here a long history of drainage for agriculture has given rise to major fen subsidence. The Holme Post, a metal post inserted into the Holme Fen in 1852, showed a subsidence of 2.1 m in the first eighteen years (Sheail and Wells, 1983). A long history of disturbance is also associated with blanket bogs on the coast of western Norway. On the basis of pollen analyses and radiocarbon dating, Solem (1989) concludes that blanket bog on Hamramsoy Island dates back only about 3,000 years and was primarily caused by human activity in addition to deteriorating climatic conditions.

In contrast to the blanket bogs of uplands and coastal areas, the Central Plain of Ireland has considerable areas of raised bog, with minor areas of fen (Taylor, 1983). These raised bogs (sometimes locally called "red bogs") have recently attracted great attention for peat extraction to generate electricity. The more nutrient-rich fens cover only about 9,000 ha of Ireland today, a reflection of their ease of drainage and their attractiveness to farmers during a long history of agricultural conversion (Bellamy, 1987). In the raised bogs the main peat-forming mosses on mounds are *Sphagnum papillosum, S. capillifolium, S. magellanicum, S. fuscum,* and *S.*

imbricatum, while *S. cuspidatum, S. pulchrum, S. tenellum,* and *S. papillosum* tend to dominate lower-lying surfaces, hollows, and pools (Bellamy, 1987). As in Canada and elsewhere, well-developed raised bogs usually develop a surficial drainage system away from the copula summit. These surface ditches, called soaks in Ireland, reflect the maturing of the copula as the raised bog approaches an equilibrium between peat accretion and peat decomposition. One characteristic of some raised bogs in Ireland, and indeed in other parts of Europe as well, is the bog burst, a sudden and sometimes disastrous out-surging of water trapped in a reservoir under the copula. The Fairy Water Bog in Northern Ireland is a classic example of a *Sphagnum imbricatum*–dominated raised bog which could well cause such a burst, because drilling has revealed a water lens some 2.5 m deep wedged between the bottom fen peat and the lighter bog peat above (Bellamy, 1987).

Raised bogs are also quite typical of the more continental boreal zone from southern Fennoscandia to the Urals. In Scandinavia, large raised bogs are only slightly convex, the relatively steep margins are better drained, and the surrounding minerotrophic fen wetland is locally called a *lagg.* High water contents, suffocating growths of *Sphagnum fuscum* and *S. angustifolium,* and low nutrient status conditions on these raised bogs discourage trees and favour only a few vascular ground-cover species and fruticose lichens. These ground-cover species include the sedges *Trichophorum caespitosum* and *Eriophorum vaginatum* (cotton-grass), together with such dwarf shrubs as *Vaccinium oxycoccus, V. vitis-idaea* (rock-cranberry), *V. uliginosum* (alpine bilberry), *Andromeda polifolia, Calluna vulgaris,* and *Empetrum* spp. (Walter, 1979; Botch and Masing, 1983). Only where artificial drainage is practised or to the south of the boreal zone in Russia are conditions dry enough for forest raised bogs to develop. On these forest raised bogs, *Pinus, Picea,* and *Betula* spp. dominate, together with an ericaceous heath ground cover dominated by *Ledum* on *Sphagnum magellanicum.* Subarctic palsa bog areas are also found to the north of the raised bog zone in Russia.

In southern Sweden, most wetlands were originally fens (eutrophic-mesotrophic), but subsequently *Sphagnum*-dominated raised (or domed oligotrophic) bogs developed on top of many of them. These raised areas are quite oligotrophic, being supplied only by rainwater (Svensson, 1988). Some 30% of Finland consists of peatlands, which can be divided between pine bogs (raised bogs), spruce-hardwood swamps, and treeless fens (Mikola, 1982; Ruuhijarvi, 1983). Pine bogs (*rame* in Finnish) are sparsely wooded raised bogs, usually with a relatively thick peat layer. While the sparse tree layer usually consists of *Pinus sylvestris* (Scots Pine), ground cover is dominated by dwarf shrubs such as *Vaccinium uliginosum, Betula nana,* and *Ledum palustre* or by *Carex* or *Eriophorum* spp. with *Sphagnum* and *Polytrichum* mosses underneath. Spruce-hardwood swamps (*korpi* in Finnish) are similar to Canadian mixed-forest swamps, containing flowing, base-rich groundwater. *Alnus glutinosa* and *Betula pubescens* are characteristic hardwoods in these swamps, but the dominant tree is usually *Picea abies* (Norway spruce).

Anthropogenic disturbances are frequently detected in European wetlands and have given rise to modified plant communities. Major differences between Finland

and Canada are noted in respect to forests wetlands and forest management. In addition to their longer tradition of wetland study (Cajander, 1913) and wetland-forest management, it is noted that some 65% of forested lands in Finland are now owned by individuals and that at least 4 million ha have been drained to improve forestry. After twenty-five to thirty years of drainage, the raised bog ground cover has stabilized, with *Calluna* and lichens reflecting the now drier peat conditions, and conditions for *Pinus sylvestris* have greatly improved. Such drained sites are called *turvekangas* in Finnish (from *turve* for peat and *kangas* for upland) (Mikola, 1982). Finland, like Ireland and Russia, but unlike Canada (as yet), is a major user of peat for fuel. In Ireland, Bord na Móna (the government-sponsored Irish Peat Board) removes large tonnages of bog peat for power generation and to make peat briquettes for domestic use. Left behind are the underlying fen deposits which have limited potential for agriculture or grazing lands (Bellamy, 1987). While Bord na Móna has plans for only 10% of Irish wetlands, much of the rest has been exploited for centuries for turf extraction for domestic use, while some is being afforested. Other anthropogenic disturbances of note include large-scale modification of marshes and acid rain impacts. In the Camargue, an area of some 300 km^2 of marshes and pools in the Rhône delta, southern France, artificial flooding must now be practised to compensate for flood-control dyking along the Rhône and coast (Grillas, 1990). In other parts of Europe the impacts of acid rain are also being felt, and species dominancy changes in Dutch heathlands from a dwarf shrub community dominated by *Erica tetralix* and *Calluna vulgaris* to one dominated by *Molinia caerulea* is in part the result of increased atmospheric inputs of ammonium (Aerts et al., 1990).

8.7.2 Asian Wetlands

One of the largest wetland regions in the world stretches across Asia from the Urals to Kamchatka. In many respects, the western portions of this region has parallels in terms of wetland types and distribution to Canadian wetlands. These western wetlands, however, differ somewhat from European wetlands in that *Calluna* is missing and *Pinus sylvestris* is joined by *P. sibirica*. Farther east, wetland development is less pronounced due to mountainous terrain and reduced precipitation. Siberia contains the world's largest single wetland area, the West Siberian Plain (Figure 8.1). The central portion of this plain consists of raised string bogs, while to the south are extensive areas of reed and sedge fen. To the north, in the Asian tundra are extensive areas of high-centred polygonal bogs and low-centred polygonal fens, while the subarctic of western Asia has large areas of palsa bog. In the forest-steppe and moister steppe zones to the south are eutrophic fens and marshes, while in the drier steppe and desert regions, fresh- and salt-water marshes are found on deltas (Botch and Masing, 1983).

The West Siberian Plain bog stretches some 800 km from the forest-tundra in the north to the steppe in the south, and from the Urals as much as 1,800 km to the Yenisey River in the east (Walter, 1979). One single wetland system, the Vasyugan, covers 5.4 million ha, and in places, bog domes rise as much as 10 m higher than their margins (Botch and Masing, 1983). Here, most environmental conditions seem conducive to wetland formation and the growth of *Sphagnum fuscum* and other *Sphagnum* species. Despite their size, the Irtysh and Ob rivers are ineffective at draining these lowlands primarily due to the fact they flow north. This results in ice jams developing in spring as meltwater comes from the mountains to the south, and floods cover vast areas for most of the summer. Flooding, combined with Podzolic soils and strongly positive moisture indices (the mean annual precipitation is approximately 500 mm, and evapotranspiration rates are only 240–300 mm), exacerbates conditions in favour of string bog development. On the strings are found *Sphagnum fuscum, Pinus sylvestris, Ledum palustre,* and dwarf shrubs such as *Andromeda polifolia, Chamaedaphne calyculata,* and *Oxycoccus microcarpus,* while in the flarks, *Eriophorum vaginatum, Sphagnum balticum,* and *S. majus* dominate (Walter, 1979). One interesting feature of these wetlands is the formation of bog lakes. These often result when escaping methane kills off the local vegetation, giving rise to a pool. As these pools gradually unite and wave action causes erosion, the lakes grow in size (Walter, 1979). Despite the fact that this is one of the most ideal regions for wetland formation anywhere in the world, Glebov and Korzukhin (1992) conclude that a balance between forest and bog now exists.

In eastern Siberia, wetlands are not as numerous, while in northeastern Siberia, valleys tend to be paludified, but peat layers developed from *Sphagnum girgensohnii* and *S. lenense* are not deep. Farther south, *Larix gmelinii* (larch) dominates as the only tree species on the sedge-moss fens. Farther south, towards Mongolia, conditions become quite dry and fens are only found in river valleys. An exception to this is in Kamchatka, where oceanic conditions return. Here in western Kamchatka, up to 80% of the area is under blanket bog, while in eastern Kamchatka, fens are common. On Sakhalin Island, heavy precipitation and cool summers favour *Sphagnum* growth and slow decomposition (Botch and Masing, 1983).

Wetlands are found in warmer parts of Asia as well, but as in North America, they generally cover a smaller area and are less likely to be of the bog wetland type as climate gets warmer. Some 11 million ha of wetlands are found in China, but they are small in individual size and common only in the Ruoergai Plateau of northwestern Sichuan above 3,400 m and on the low-lying Sanjiang plain in northeastern China. *Carex meyeriana* tussocks are characteristic of moist, peaty wetlands throughout cool and warm temperate zones in China, Japan, and Siberia (Tsuyzaki and Tsujii, 1992). The great majority of the fifty-five wetland types identified for China are eutrophic fens, while there are a few marsh and swamp types, but only one type of bog (again dominated by *Sphagnum* spp.) (Liu and Zhao, 1987).

8.8 Central and South American Wetlands

Central and South America have numerous freshwater wetlands, savanna wetlands, and coastal mangrove swamps. Examples of this variety include the wetlands of the so-called Magellanic Tundra along the southern spine of the Andes, the Pantanal swamps of Brazil, and the huge seasonally flooded Llanos de Mojos savanna in Bolivia. The Amazon River, which accounts for approximately 20% of the earth's runoff, also has large areas of freshwater swamps along its lower course and large mangrove swamps in its delta (Junk, 1993).

The low-lying nature of the central Amazon Basin, the broad nature of the Amazon floodplain (*várzea*), and the influences of season (which can give rise to annual river-level changes of as much as 20 m in Peru and 14 m in Brazil) increase the likelihood of swamp development and the variety of swamp types (Junk, 1983). Immediately below the unflooded forests (*terra firme*) of the floodplain margins is the *igapo* (Brazil) or *restingas* (Peru) inundation forest type. This forest type is particularly typical of the low-nutrient, highly organic-stained waters (black waters) that enter the Amazon from the north and the clearwaters (white waters) entering from the northeast and south (Figure 8.1). On the broad (up to 100 km) *várzea* between the *igapo* forests and the main channel are found a series of lakes, smaller river channels, gallery forests along levees, strips of floating grasses, and floodable *várzea-campos* (treeless herbaceous swamps) fed by muddy white waters from the Amazon, which derived this sediment from the Andes to the west. Usually the highest point on the *várzea* is the levee immediately along the sediment-rich main channel covered by gallery forest (Sioli, 1975).

So pronounced is seasonal water-level change in the *várzea* and so lush is the growth potential that the floodable *campo* (areas which dry out in the low-water season but have as much as 6 m of water during flood stage) supports annuals during low stages and perennial aquatics and semi-aquatics year-round. Dry-season annuals include many grasses (*Eleusine, Setaria, Paspalum*) and sedges (*Cyperus, Fimbristylis*) (Junk, 1983). During the high-water season, floating mats of vegetation, which include such common species as *Salvinia auriculata* (water hyacinth), expand rapidly due to their great capacity for vegetative reproduction. Even some perennial grasses such as *Echinochloa polystachya* tolerate the great seasonal changes associated with the nutrient-rich white waters, flowering at high water, having their long (up to 10 m) stems dry our during the dry period, and then resprouting from their bases. It might seem peculiar that *Echinochloa*, a c_4 species which one might suspect should be more associated with tropical and warm-temperate semi-arid zones (because of the high water-use efficiency associated with c_4 species) is growing here at all; however, this high c_4 productivity potential is vital if it is to keep pace with rapidly rising waters and remain emergent (Piedade et al., 1991). Another grass found here is *Paspalum fasciculatum*, which flowers during the dry season but can tolerate up to two years of inundation (Junk, 1983).

Inundation, or *igapo*, forests are seasonally flooded sufficiently to inhibit many of the *terra firme* tree species. These forests are found at the edge of the floodplain and along clearwater rivers which do nor carry sufficient sediment to build high levees, and soils are often sandy (Pires and Prance, 1985). They are normally only flooded for several months, but can survive inundation for as much as three years. Despite these obvious limitations, biological diversity among tree species is high, particularly among the palms, and the forests are lush. Genera commonly represented in these forests include *Salix, Nectandra* (sweetwood), *Cecropia* (umbrella tree), *Ficus, Bombax, Inga,* and the palm genera *Mauritia, Bactris,* and *Leopoldinia* (Junk, 1983). While palms are numerous in seasonal swamp forests in eastern Peru, the number of species involved is generally low (Kahn and de Granville, 1992). Large aguaje palm (*Mauritia* sp.) swamp forests thrive in eastern Peru, where they are known as *aguajales.* Often as many as 450–500 palms ha^{-1} are found here in the Upper Ucayali River swamps (Myers, 1990), with the oil-bearing fruit being harvested from canoes during high-water stage. Many of the tree species found in white-water *várzea* forests around Manaus, Brazil, are widespread in South America. In addition, local site studies reveal that they are not specifically adapted to such seasonal flooding, but appear adapted to seasonality in general (Worbes et al., 1992).

Large masses of aquatic vegetation somewhat reminiscent of the *várzea* swamps along the Amazon are found during the wet season in the Gran Pantanal swamp, which covers some 80,000 km^2 along the northern Paraguay river in southwestern Brazil (Figure 8.1). During January and February, when the swamp is at its fullest, many lakes develop and become dominated by true floating islands of vegetation which includes *Salvia herzogii* (water hyacinth). Under certain conditions, these "islands" are discharged into the main channel and are carried down to Paraguay and Argentina. In Paraguay they are known as *batumes,* while in Argentina they are called *embalsados* (Bonetto, 1975).

8.9 African Wetlands

Africa has numerous large wetlands associated with its humid tropical regions and tropical coasts, but unlike in South America, wetland formation is restricted in the north due to desert conditions and in the south due to seasonal drought (Denny, 1993). The four general types of wetlands most readily identified in Africa are (1) swamp forests, (2) mangrove forests, (3) grass-sedge and reed swamps, and (4) cushion-bog and tussock-sedge wetlands. Most of the major swamp forests are associated with large river systems between the two tropics, such as the Congo Basin swamps in Zaire and the Nile headwaters west of Lake Victoria. Large areas of mangrove forests are found along both humid and subhumid tropical coasts, especially in Mozambique and western Madagascar. Grass-sedge and reed swamps without trees are more characteristic of the region known as the Sudd in southern Sudan, the shores of Lake Chad, the marshes of the lower Nile, the Okavanga Swamp in Botswana, and the Niger Internal Delta Wetland in Mali. These latter swamps are found in drier or even desert climates, where they depend on exotic waters. Cushion-bog and tussock-sedge wetlands best describe the bogs and fens associated with cooler highland areas, such as in Kenya and Ethiopia.

Humid tropical African forest swamps are typically associated with palms, particularly the *Raphia, Phoenix,* and *Hyphaene* genera. In West Africa and along the

Zaire River, *Raphia farinifera* is associated with such broadleaf evergreen tree species as *Ficus congensis, Syzygium guineense,* and *Cleistopholis patens* (Thompson and Hamilton, 1983). In East Africa, around Lake Victoria, *Phoenix reclinata* is the dominant palm along forest swamp edges, while *Raphia farinifera* is important in the lower canopy and dominant broadleaf families include the Leguminosae and Euphorbiaceae. Important broadleaf tree species include *Mitragyna stipulosa, Podocarpus usambarensis,* and the huge umbrella-leaved *Musanga cecropioides* (Lind and Morrison, 1974). Except for the absence of fire ants in their hollow trunks, *Musanga* is almost identical to the genus *Cecropia* in South America. Both genera dominate early succession growth, are short-lived, and play a vital role in trapping nutrients in the almost closed nutrient cycle associated with tropical forest ecosystems (Scott, 1987b). In somewhat drier south-central Africa, *Hyphaene* is more frequently encountered than *Phoenix* spp. and is associated with *Ficus verruculosa* (Thompson and Hamilton, 1983).

Papyrus (*Cyperus papyrus*), the tallest member of the Cyperaceae, frequently dominates the large, treeless floating and bottom-rooted sedge swamps that are so characteristic of low-salinity lakes throughout all of central and east Africa. For some unexplained reason, papyrus is quite scarce west of Lake Chad and is entirely absent from the large Niger Internal Delta Wetland in Mali (Beadle, 1974). The vigorous growth of papyrus is legendry, and it can reach its maximum height of 5 m in ten weeks (White, 1983). Its success throughout much of the continent, particularly in oligotrophic waters, may be partly because of its higher nitrogen-use efficiency, which is due to its c_4 pathway (Jones, 1988). Jones also found that papyrus was more than twice as efficient at using nitrogen as the c_3 pathway species *Typha domingensis,* with which it is frequently associated. *Typha domingensis,* however, also does well in the nutrient-rich inland depression swamps of the Okavanga in Botswana and at higher elevations, particulary under eutrophic conditions. Papyrus is also well adapted to permanent wetlands, having less than 5% of its phytomass in the form of roots. Where water is less permanent, papyrus gives way to the reed *Phragmites australis,* which can tolerate drier periods, having more than 25% of its phytomass below ground (Thompson and Hamilton, 1983).

Papyrus swamps used to be common along the lower Nile, but today *Typha* and *Phragmites* are more important, with papyrus only found under cultivation. The demise of natural stands of papyrus in Egypt may well be related to overuse, agricultural expansion onto the floodplain, flood control, and river-course changes (Lind and Morrison, 1974). The Sudd wetland of the upper Nile headwaters in southern Sudan is a remarkable papyrus swamp that covers some 90,000 km^2, 8,000 km^2 of which are permanent, with the remainder becoming floating during the flood season. This region gets its name from the floating islands of vegetation known locally as *sudd* (Ellenbroek, 1987). *Oryza longistaminata* (wild rice) dominates 13,100 km^2 of Sudd grasslands, where annual flooding depth is between 0.65 and 1.21 m for 135–287 days. Where flooding is less pronounced, *Echinochloa pyromidalis* dominates some 15,800 km^2, while *Hyparrhenia rufa* dominates rain-fed wetlands

(Finlayson and Moser, 1991). The Jonglei Canal was recently built across the Sudd from south to north in an attempt to reduce seasonal flooding and speed wet-season water on its way to the Nile. Lake Victoria also has large areas of floating swamp dominated by papyrus, although other species such as the tall grasses *Miscanthidium violaceum* and *Loudetia phragmitoides*, *Phragmites australis*, and occasional stands of *Typha domingensis* are encountered (Lind and Morrison, 1974). In highland areas, cushion-bogs and tussock-sedge fens in many ways resemble wetlands in higher latitudes, except that *Sphagnum*, ericaceous shrubs, and *Salix* spp. are much less important (Thompson and Hamilton, 1983).

8.10 Austromalesian and Pacific Wetlands

For convenience, this region includes the Malay Archipelago of Malaysia and Indonesia, the Philippines, New Guinea, Australia, New Zealand, and some of the Pacific islands, such as the Hawaiian. Surprisingly, Indonesia ranks fifth (13.7%) among countries with large percentages of its territory covered with wetlands (Taylor, 1983). Vast peat swamps cover some 9.7 million ha of land in coastal east Sumatra, where they are primarily formed on poorly drained alluvial deposits of rivers and deltas which had trouble adjusting to post-glacial sea-level rise (Anderson, 1983). Likewise, peat swamps covering 6.3 million ha are found in Kalimantan, with additional large areas in Sarawak and small patches in the southern Philippines. These lowland peat swamps have striking morphological parallels to domed or raised-bog features in Canada and northern Europe, and they are also oligotrophic and extremely acid (pH 3–4.5). They differ, however, in being made primarily from woody material such as trunks, branches, and stumps, between which is a dark brown to black amorphous soapy material. They also differ in being densely treed, a factor which may account for their thickness (sometimes exceeding 15 m) and rapid peat growth (between 2 and 3 mm yr^{-1}) (Bruenig, 1990; Anderson, 1983); however, similarities between these peats and those in temperate climates outweigh differences (Cameron, 1987). In Sarawak this dense tree cover shows a marked zonation sequence between the edges of treed swamps and their more open centres. Often known as *kerapah* forest in Borneo, the swamps are the home of the massive *Shorea albida* (Bruenig, 1990). A mixed swamp–forest association dominated by *Dactylocladus, Gonystylus,* and *Neoscortechinia* spp. lines the edge of the swamp, while 45–60 m tall *Shorea albida* dominate closed forest on moister ground, and the wettest central areas have an open forest of stunted trees (savanna-like) dominated by *Combretocarpus* and *Dactylocladus* (Anderson, 1983).

Many kinds of lowland swamp are found in Papua New Guinea. Sago palm (*Metroxylon sagu*) swamps are common in the lowlands, where the ground is more or less permanently swampy with an influx of fresh water. While they have fronds up to 14 m long, it is the starch-producing, tall, trunk-like stem (which can reach 20 m) which has made it of value to local populations. Other treed swamp types include pandan (*Pandanus* spp.) swamp woodland, mixed swamp woodland, *Melaleuca* (paperbark eucalyptus) swamp forest, and *Melaleuca* swamp savanna

(Osborne, 1993). *Melaleuca* spp. do well in locations where there is both a fluctuating water-level in back-swamps and occasional fire. Because of fire, it is not clear to what extent *Melaleuca* savannas and swamp forests are edaphic climax communities, fire disclimax communities, or both (Paijmans, 1990). Herbaceous swamp vegetation is also common in Papua New Guinea, particularly along such rivers as the Sepik, Strickland, and Aramia. On levees along the Strickland River and in the Port Moresby hinterland are grass swamps dominated by *Saccharum robustum* and *Phragmites karka*. What makes these grassland swamps different from those of the Canadian Prairies is that *Saccharum* often reaches heights of 12 m!

In Australia the climate is generally arid to subhumid, so therefore does not encourage the development of wetlands. Exceptions are found, however, along the moister Eastern Highlands and southeastern coast, as monsoon swamps along the north coast, near Perth on the southwestern coast, and in Tasmania. Many of the wetlands are marshes or swamps found in regions with negative moisture indices. They owe their origins to floodplain locations or coastal flatlands, and except for the northern coast, do not result from per-humid climatic conditions. Only three genera, the *Casuarina, Eucalyptus,* and *Melaleuca*, are significant within Australia's forested wetlands. Along the northern coast, *Melaleuca* spp. (paperbark eucalyptus) forest swamp covers extensive areas (Specht, 1990; Campbell, 1983), while sedge and grass marshes and swamps also occur. Along the southeastern coast are occasional patches of heath-like vegetation, while swamps are common. One of the more important swamp types is the tea-tree swamp with its dense forest of *Melaleuca quinquenervia*, sedges, and *Phragmites australis*. Low-lying swamps, often with some woodland dominated by *Eucalyptus rudis* and *Melaleuca* spp., are also common in the Swan Coastal Plain near Perth. The temperate maritime climate in Tasmania encourages large areas of tussock-sedge swamps in lowland areas (Campbell, 1983). More typically subalpine moss bogs, cushion bogs, and tussock-grass wetlands are also found at high elevations in both the Eastern Highlands and Tasmania.

Unlike Australia, the mountainous, humid environments of New Zealand are quite conducive to wetland formation. Forest swamps dominated by *Dacrycarpus dacrydioides* (a podocarp tree) are frequently found along river floodplains. Shrub and herb swamps are also encountered and are dominated by *Typha orientalis* and *Carex secta*. Both bog forests and treeless bogs dominated by *Sphagnum* spp. and liverworts are found in areas of poor drainage and high rainfall, and heathlands and the occasional raised bog are encountered at high elevations throughout both major islands. Bogs in the Waikato Basin of the North Island volcanic plateau have experienced major alteration for agriculture, while maximum peat thickness in the still little modified Moanatuatua bog is 14 m (Burrows et al., 1979). Open bogs are also often associated with low *Gleichenia* spp. ferns. In areas with base-rich groundwater, sedge fens are encountered (Campbell, 1983).

Hawaii's best known wetland is the *Metrosideris polymorpha* (ohia)–dominated Alakai Swamp (*alakai* in Hawaiian means "one-file track"), which covers an area of some 75 km^2 at 1,200 m above sea level close to Mount Waialeale on the island of Kauai. Such proximity to Mount Waialeale, the peak with the wettest mean annual average on record, means that the swamp receives approximately 500–650 cm of primarily orographic rainfall each year (Thomas, 1976)!

8.11 Phytomass and Primary Production

To the discussion of primary production and living plant biomass in wetlands can be added the dimension of measuring the accretion of dead peat phytomass. Like most other vegetated ecosystems, wetlands are capable of significant annual amounts of net primary productivity, but unlike most other ecosystems, wetlands possess the potential to store huge amounts of this net production in the form of peat. Tarnocai (1984) has estimated that in Canada alone there are three trillion m^3 of peat, which at a moisture content lowered to 50% would be the equivalent of 507 billion tonnes (Gt)! This is equivalent to 338 Gt of oven-dry organic matter or 189.3 Gt of carbon (using the 56% organic carbon content of peat as determined by Mills et al., 1977). As peatlands worldwide are considered to contain at least 450 Gt of carbon (Warner et al., 1993), Canadian wetlands must serve as an important carbon sink. In comparative terms, this 189.3 Gt is almost identical to the total worldwide carbon release into the atmosphere as a result of burning fossil fuels during the last 125 years! Phillips et al. (1990) report that between 1860 and 1984 the burning of fossil fuels contributed 183 Gt of carbon to the atmosphere, while an additional amount of perhaps 150 Gt resulted from land clearance.

At Rocky Mountain House, Alberta, Zoltai and Johnson (1985) calculated organic matter accumulation rates for fen peat to be 39.1 g $m^{-2}yr^{-1}$ and for treed bog to be 44.8 g $m^{-2}yr^{-1}$. Calculations by Zoltai et al. (1988b) show storage of organic carbon to average 18.9 g $m^{-2}yr^{-1}$, which, if applied over the 52 million ha of wetlands within the boreal wetland province, give rise to an annual storage capacity of 9.8 million tonnes. From this must be deducted losses due to volatilization following drainage for improved forestry (Haavisto and Wearn, 1987) and clearing, accompanied by drainage, for agriculture. Studies in Karelia, northwestern Russia, show that the additional increments in tree-stand carbon following drainage do not compensate for peat carbon losses (Makarevskii, 1992). Some 20 million ha of Canadian wetlands (one-seventh the pre–European settlement total) have already been drained, mostly to provide agricultural land (Clarke et al., 1989). It is noted that in southern Ontario alone, 61% of the wetlands have been drained since 1800 (from 2.38 down to 0.93 million ha) (Rump, 1987). It may also be that atmospheric pollutants are affecting wetland organic carbon accretion. Aerts et al. (1992) suggest that in southern Sweden, high inputs of atmospheric nitrogen do not improve growth in ombrotrophic bogs (because other nutrients are limiting), but do encourage decomposition. It is clear that the lowering of water tables even by 10 cm can inadvertently turn carbon-accreting systems, or sinks, into carbon-emitting systems, or sources (Hogg, 1993).

Net primary productivity within the living vegetation itself has also been measured. On a basis of a literature search, Bradbury and Grace (1983) report shoot productivity values of from several hundred grams to 1,118 g $m^{-2}yr^{-1}$ in North American peat bogs, while they give values of up to 1,660 g $m^{-2}yr^{-1}$ for bryophytes in European and North American *Sphagnum* bogs. They report values of up to 1,900 g $m^{-2}yr^{-1}$ in North American *Typha* marshes, while Thompson and Hamilton (1983) record productivity in a vigorous papyrus swamp (*Cyperus papyrus*) in central Africa can be as much as 30 t/ha/yr (3,000 g $m^{-2}yr^{-1}$). Piedade et al. (1991) report an amazing net primary productivity for the grass *Echinochloa polystachya* in Amazonian white-water *várzea* of 99 t $ha^{-1}yr^{-1}$, a productivity rate which must be one of the highest values recorded for natural plant communities anywhere. Lieth (1975) give an approximate worldwide mean net primary production of 20 t $ha^{-1}yr^{-1}$ for swamp and marsh (including forested types), while Whittaker and Likens (1975) report the value of 30 t $ha^{-1}yr^{-1}$ for the equivalent vegetation types.

Individual studies reveal that the determination of site-specific productivity values for wetlands is, in fact, much more complex than these worldwide averages might suggest. In a *Carex rostrata* wetland in New York State, total living plant productivity is 1,340 g $m^{-2}yr^{-1}$, but of this, 260 g is in roots and rhizomes, 820 g in living shoots, and 260 g in seasonal shoot mortality (Bernard and Hankinson, 1979). At Glenamoy, Ireland, net primary productivity for acid bog is 523 g $m^{-2}yr^{-1}$. Of this, above-ground productivity by vascular species is 125 g $m^{-2}yr^{-1}$, while below-ground production is 162 g $m^{-2}yr^{-1}$. Cryptogams are also important producers here, with algae producing 186 g $m^{-2}yr^{-1}$ and bryophytes producing 50 g $m^{-2}yr^{-1}$ (Moore et al., 1975). At Moor House, England, equivalent average production values are 659 ± 53 g $m^{-2}yr^{-1}$, although locally, production rates vary from 868 g $m^{-2}yr^{-1}$ on a recently burned *Eriophorum*-dominated site to 491 g $m^{-2}yr^{-1}$ on a *Sphagnum*-dominated site without recent fire (Heal et al., 1975).

In terms of plant phytomass, the "intrazonal" characteristic of wetlands demonstrates great extremes. The 1,182 t ha^{-1} value for swamp forests in the Darien, Panama (Golley et al., 1971), is in marked contrast to the 0.25–7.7 t ha^{-1} for Devon Island tundra in the Canadian Arctic (Table 2.2). At the site-specific level, the total peak plant phytomass on virgin bog in the Glenamoy IBP peatland on the coast of northwestern Ireland was 1,600 g m^{-2} (including recognizable dead structures). Of this, above-ground vascular plants make up 83 g, algae 360 g, and mosses 209 g, while below-ground vascular living parts exceed 500 g (Moore et al., 1975). For the Moor House *Calluna*-dominated blanket bog in England, the equivalent summer biomass value is 2,250 ±250 g m^{-2} (Heal et al., 1975). Approximately 72.5% of this is above- and below-ground parts of *Calluna* (Forrest, 1971). One study on cyclical regeneration following fire in a Scottish *Calluna* bog (heath) gave the following values for all vascular plants, including *Calluna*, and bryophytes: pioneer stage 889.2 g m^{-2}, building stage 1,702 g m^{-2}, mature stage 2,305.2 g m^{-2}, degenerate stage 1,560.8 g m^{-2} (Barclay-Estrup, 1970). This regeneration took twenty-four years to reach the degenerate stage.

At first glance, it might seem unreasonable to compare phytomass values for Canadian wetlands with the 1,182 t ha^{-1} value for swamp forests in the Darien, until it is appreciated that Canadian wetland phytomasses greatly exceed this value if one includes all ecosystem organic matter, including peat. The figures quoted above give 338 Gt of oven-dry organic matter for the 127 million ha of Canadian wetlands. This translates into an average of 2,661 t ha^{-1} organic matter and makes Canadian wetlands one of the largest biomass communities worldwide. In case, this figure appears grossly out of proportion to our traditional thinking about ecosystem biomass; 2,660 t ha^{-1} is the equivalent of 266 kg m^{-2}, or an average peat depth of only 190 cm (using a bulk density of 0.14 Mg m^{-3}). Communities where biomass exceed these figures would include coastal redwoods in California (see section 6.3), some *Nothofagus* stands in southern Chile and Argentina, and the deep peat wetlands of Borneo and Sumatra.

With all of this potential energy lying around in the form of peat, it is not surprising that some use is made of it. To the afore-mentioned use of peat for electrical energy production and for household heating in western Europe (see sections 4.5.1 and 8.7.1) can be added such things as *Sphagnum* moss bales for horticultural and garden use and the peat trays and pots which can even be planted with seedlings. In industry, peat is used to aid in the production of bitumens, activated carbon, metallurgical coke, acids, ethyl alcohol, and certain pharmaceutical chemicals (Warner, 1992). Peat acid extracts are even used in Chinese wine-making (Warner, 1992), and one can only speculate as to what true Irish or Scotch whisky might be like if it was not distilled using bog drainage water rich in organic acids!

CHAPTER NINE

Conclusion

The word "conclusion" has a very final ring to it. It is almost like suggesting that the analysis is done and that the truth can now be revealed. When it comes to the discussion of Canada's vegetation cover, however, this is far from being realistic.

Static or Dynamic?

One misconception the preceding chapters might leave is that our vegetation mosaic can be viewed as something static. This interpretation is incorrect for a number of reasons, the first being the constant cycle of natural disturbance and successional renewal brought about by such agencies as fire, flood, insect attack, and windthrow. Biogeographical and ecological studies also tell us of the dynamics and internal struggles going on within ecosystems which constantly promote change. In addition, palynological studies have revealed the many major, as well as subtle, changes to which Canada's cover has been subjected due to natural climatic changes during the Holocene (Ritchie, 1987; Pielou, 1991) and have helped make possible projections as to what changes species distributions and vegetation cover may go through in the future (McAllister and Dalton, 1992; Solomon, 1992; Delcourt and Delcourt, 1991).

Even if we could support a more static view for vegetation formation boundaries, abundantly visible are the alterations to soil-vegetation systems caused by human manipulations for, among other things, food, fuel, and fibre. Whether we like it or not, we are living at a time when the expression "potential natural vegetation" is being rapidly replaced by the more realistic expression "land use–land cover." This realization is increasingly reflected in a new generation of *National Atlas of Canada* mapping projects, many of which are based on contempory data gathered by satellite. While it is natural that the 1:7,500,000 scale *Canada: Land Cover Association* map (Energy, Mines and Resources Canada, 1989) should take such an approach, it is important to note a departure from tradition, with the 1:7,500,000 scale *Canada: Vegetation Cover* map (Energy, Mines and Resources Canada, 1993) actually including categories for built-up areas, cropland, and rangeland and pasture.

Interglacial or Post-glacial?

If we are troubled by the impacts of wetland drainage, expanding cities, the stripping bare of mesothermal rain-forest watersheds, or aspen parkland encroachment on prairie, then describing current climatic conditions as "interglacial" should really get our attention. Is our current vegetation mosaic just a momentary positioning on the accordion of ecoclimatic change, or is it something tangible in the sense that the pattern can last? The answer is of course, yes, it can last. What we have today is really here, and likely to be here into the foreseeable future if we do not wilfully destroy it. What we have is worth preserving, vital to study, and part of Canada's uniqueness. Fortunately, we are as yet a long way from having to travel to Drumheller to study "wooden dinosaurs" (an expression adapted from Lamb, 1979)!

There is legitimate comfort in returning to the idea that some degree of permanency is associated with our ecoclimatic regions, at least as measured in human generations. Berger (1992), basing his opinions on Milankovitch astronomical cycles but avoiding possible anthropogenic disturbances, considers that the long-term cooling trend, which began some 6,000 years ago, will continue for the next 5,000 years. This will be followed by a warmer period around 15,000 AP and a colder interval at 23,000 AP, but a major glaciation may not develop until around 60,000 AP. Even without the immediate intercedence of another glacial, we are still left with the natural perturbations in climate caused by other cycles. Even more pressing is the possibility that technological (anthropogenic) alterations to our atmosphere may promote radiative forcing (Kemp, 1991). Our atmospheric manipulations may be a giant experiment in proving that, yes, we are right, we can change climate, but they may also prove that we are sufficiently lethargic that we are unwilling, or unable, to halt the experiment. One thing is clear: climates change whether we wish them to or not, and whether our activities promote radiative forcing or not. Both sides of the induced climatic change argument agree on this, both see the great advantages of studying the consequences of such changes on ecosystems, and both share the common interest of promoting research which leads to a better understand of our biosphere.

What We See Is What We Get – We Hope

While ecosystems have an inherent tendency to change and adjust, they still form real communities, and we can stroll through them, protect them, modify them, rehabilitate them, destroy them. If we learn nothing more from the biodiversity debate than respect for nature, then we know major portions of our natural ecosystems must be protected. Self-interest alone should bring us to the same conclusion. Reality at any scale tells us that changes are happening whether climates change or not and that if climates do change, then we at least should have some notion as to the consequences of this change on our terrestrial biota. Developing techniques to study the past as well as the present has helped provide us with powerful theoretical tools to look into the future. In human terms, glacials and interglacials are events that have happened, and one day may happen again, but on a scale so grand in terms of decades or generations that essentially what we have now is what all readers of this book are ever going to have, assuming that we do not destroy it by ourselves in the meantime!

The Ecoclimatic Approach

It is important to appreciate the value of both ecoclimatic description and the underlying premise of the ecoclimatic approach. While a simple inventorying of nature has merit, far better is it to study those factors, often climatic, which in combination promote stability or force change within ecosystems. Soils, vegetation, fauna, and climate are so intricately intertwined that an ecoclimatic approach indeed seems appropriate. When we consider that soils are to terrestrial ecosystems as cell membranes are to the living cell, we can better appreciate that they function essentially as the interface between the inorganic and the organic, the dead and the living. Soil properties influence plants, just as plants influence soil profile development and reduce erosion. Both are influenced by climate, but in turn modify microclimate above and below the soil surface. By better appreciating climatic constraints on our terrestrial ecosystems, we are so much better positioned possibly even to forecast ecoclimatic pattern changes in the future (Kemp, 1991).

Research on Ecosystem Interactions

There is an ever-increasing need to research the details of ecosystem interactions so that we can better understand responses to altering our natural environment and to natural and induced changes in climate. The value of such projects as the Model Forests and the BOREAS (BOReal Ecosystem Atmospheric Study) will be appreciated by most only after their results start coming in. It is hoped that one outcome of the nation-wide Model Forests project will be a more sustainable harvesting-reforestation system which reduces the need to harvest yet additional pristine stands and may even allow previously harvested areas to be released from a logging-rotation strategy. This outcome is all the more likely in light of the 1992 National Forestry Strategy, which encourages the participation of all forest stakeholders in decision making (Murphy et al., 1993).

With BOREAS, the direction is more towards understanding the interactions between the atmosphere and Canada's west-central boreal forest, a region which would be particularly sensitive to impacts from possible global warming. In addition, the boreal ecosystem has direct interplay with incoming and outgoing radiation and with the atmospheric contents of two important radiative forcing gasses, CO_2 and CH_4. Warming-induced increased evaporation or reduced precipitation and intentional wetland drainage all promote a lowering of water tables. This would quite probably result in decreased rates of CH_4 release (Bubier et al., 1993), but increased rates of

CO_2 release (Martikainen et al., 1993). In addition, future redistributions of boreal forest, due to either extensive logging or global warming, could initiate important climatic feedbacks (Bonan et al., 1992), and warming also raises legitimate concerns about a potential increase in forest fires (Stocks, 1993).

If climatic changes do occur in the near future, the Arctic will be a region where the consequences will be particularly noticeable (Chapin et al., 1992), as may be advances of the tree line into tundra (MacDonald et al., 1993). Currently used global climatic models all point to a significantly larger increase in mean surface temperatures here than in more southerly latitudes. Consequently, this situation is being monitored by the Global Change Working Group of the International Arctic Science Committee. The recently formed Climate Research Network will open in Victoria in early 1994 to help tackle key components of climatic research on such topics as natural climatic variability, sea-ice interactions, land-air interactions, regional climatic modelling, and palaeoclimatic modelling. Results from these endeavours will help point us in new research directions, so we can better refine our understanding of those ecosystems we often optimistically think we already understand.

Anthropogenic Factors

When it comes to thinking about our "natural" environment, the "split personality" possessed by Canadian society should not strike us as peculiar. On the one hand, we have promoted, and still do promote, the alteration of natural ecosystems on a large scale, while on the other, efforts–often extreme efforts–are directed at protecting those same ecosystems. Most societies are frequently faced with making major environmental decisions that involve direct competition between undisturbed ecosystems and increasing food/fibre productivity. The parkland landscapes of western Europe, the St Lawrence Lowlands, and southern Ontario attest to considerable success in modifying landscapes while maintaining productivity. Much of this success, however, has been achieved at the expense of the near-total elimination of local natural ecosystems. We must recognize that wilderness is on the decline and that, as a non-substitutable, non-renewable common resource, it requires pro-actice, rather than reactive, management if we intend it to survive (Dearden, 1989). The very difficult challenge is therefore to take that extra step beyond interference with productivity maintenance and actually to set aside areas of natural vegetation in as pristine a condition as possible.

Canadian society in general has reached this latter stage and is genuinely interested in the conservation of wilderness (Nelson, 1989). Environment Canada is promoting the national Green Plan, not only through the establishment of model forests and the encouragment of research, but also through direct ecosystem protection. In an effort to meet our international commitment to set aside 12% of our national territory for natural ecosystems, Parks Canada is pursuing the goals of six new parks by 1996 and thirteen by the year 2000 (Dearden and Rollins, 1993). In addition, attempts are being made to ensure national parks are established in each of the thirty-nine terrestrial natural regions identified as needing representation within the parks system. The Canadian Environmental Advisory Council (CEAC) is actively seeking legal protection for ecological reserves; national, territorial and provincial parks; wildlife refuges; and other designated areas, including private land holdings. CEAC's vision is that by the year 2000, a comprehensive network of protected wilderness areas and natural landscapes will have been established and that by the year 2010, this will have been extended to natural seascapes and aquatic ecosystems as well (CEAC, 1991). The Canadian Council on Ecological Areas is also striving towards stewardship and protection of a comprehensive system of unaltered ecological areas. These reserves are essentially for the protection of genetic conservation, research, and educational purposes, and to serve as standards against which disturbed ecosystems can be compared. Provincial governments are also participating by increasing the number of provincial parks and wilderness areas.

Biodiversity

Biodiversity within our ecoclimatic regions is much more than a simple quantification of species numbers or the genetic variety within individual species present. Among other things, it includes structural diversity, function, and interconnectedness (Wilson, 1992). If we see Canada's biotic inheritance only as a string of genes which could, in an emergency, be preserved cryogenically, we are missing the point, and if we see second-growth wildlife diversity only as a listing of summer residents, then we are simply misusing the term (Reimchen, 1991). Our diversity, and the diversity in other parts of the world, must be protected *in situ* and in such a way that ecosystem interaction and connections can be maintained. Preserving any individual Canadian wetland has great merit, but migratory species would prefer the effort be international in scope.

In parts of the open lichen woodland, one can reach out and at the same time touch both the tree species present. "Both" has hardly a diverse ring to it, yet open lichen woodland is indeed a very diverse ecosystem when algae, microarthropods, bryophytes, epiphytic and terrestrial lichens, graminoids, forbs, and shrubs are included in the picture. In addition, its interaction with the pedosphere and atmosphere and with migratory ungulates and waterfowl adds to its vitality and challenges our ability to comprehend it. Understandably, that open lichen woodland can be considered a simple ecosystem is a view maintained only by the uninitiated. Likewise, we should be troubled when someone equates the biodiversity of a 20-year-old second-growth hemlock stand with the 700-year-old rain forest which preceded it.

Canada has a total of 134 native tree species (McAllister, 1991), with no more than 64 in any particular region of the country (Schueler, 1991). These numbers do not seem very impressive when as many as 67 rain-forest tree species can be encountered in a single 900 m^2 quadrat in Amazonian Peru (Scott, 1978). But the point is that ecosystems are more than just tree species – more than just numbers – and the view that open lichen woodland and Amazonian rain forest share much in common and that both are deserving when it comes to conservation is gaining wider acceptance. The protection of both endangered ecosystems and endangered species in Canada is vital to our biodiversity heritage. In Quebec alone, some 20% (374) of the vascular plant species are either threatened (endangered) or vulnerable (Lavoie, 1992). In 1989 the province enacted the *Act Protecting Threatened and Vulnerable Species* to help safeguard provincial biodiversity. Importantly, this act not only relates to all living species, but to their habitat as well.

Future Prospects

Canada's vegetation cover and remaining wilderness areas are currently subject to many different forces that can promote change. Clearing for urban expansion, agriculture, forestry, and outdoor recreational facilities is quite visible. Induced climatic changes are possible (Wheaton, 1992), impacts of acid rain may become more pronounced, and blights or insects may decimate important species and alter ecosystem dominants (Bolgiano, 1993).

Ecosystem alterations for food, fibre, recreation, and human infrastructures have often been necessary, but now must be contained and managed. While "no one group possesses the sole truth about how society should evolve" (Macy, 1991, p. 44), few would argue the need to develop our biological resources very carefully and maintain a balance between conservation and development (Sharma, 1992). "Sustainable development" is just another misused expression if these activities ignore the finite nature of our soil and biotic resources. Sustainability allows for agriculture, forestry, highways, and urban growth; it needs recycling, agroforestry, plantations, heavy-use provincial parks, and managed wetlands; and it requires wildlife preserves, national parks, biosphere reserves, protected natural habitats, and pristine ecosystems. Rewriting history has become such standard practice that we can assume that future generations will judge us more severely than we judge ourselves. But if we can make the effort to preserve for them an environment which will be sustaining, preserve biodiversity, maintain large areas of natural ecosystems, and give them as much pleasure as it gives us, we will, at least in our own eyes, have done well.

Appendix

This appendix identifies genera/species using three systems. If the local name(s) for a tree such as *Pinus strobus* is known, it is given (e.g., white pine). If the tree is known to be a *Pinus*, but the specific local name is unknown, it is identified as a 'pine' (enclosed in single quotation marks). If it is not even known to be a pine, but its physiognomy is known, this is given in double quotation marks (e.g., "tree"). Physiognomic or life-form terms used are after National Vegetation Working Group (1990), with the addition of the term "algae." Canadian species nomenclature after Scoggan (1978). See Scoggan (1978) for authorities on Canadian species.

Binomial	*Local name*
Abies amabilis	amabilis or Pacific silver fir
Abies balsamea	balsam fir
Abies concolor	white fir
Abies faxoniana	'fir'
Abies fraseri	Fraser fir
Abies grandis	grand fir
Abies homolepis	Nikko fir
Abies lasiocarpa	subalpine fir
Abies magnifica	red fir
Abies mariesii	Marie's fir
Abies sachalinensis	'fir'
Abies sibirica	Siberian fir
Abies veitchii	Veitch's silver fir
Acacia	'acacia'
Acaena magellanica	"shrub"
Acer campestre	field maple

Acer circinatum	vine maple
Acer glabrum	false boxwood
Acer macrophyllum	broadleaf maple
Acer mono var. *glabrum*	'maple'
Acer negundo	Manitoba maple
Acer pensylvanicum	striped maple
Acer pseudoplatanus	sycamore
Acer rubrum	red maple
Acer saccharinum	silver maple
Acer saccharum	sugar maple
Acer skutchii	'maple'
Acer spicatum	mountain-maple
Acer tataricum	Tatarian maple
Achillea millefolium	common yarrow
Aciphylla	"tree"
Acorus calamus	sweet flag
Adiantum pedatum	maidenhair fern
Aesculus glabra	Ohio buckeye
Aesculus hippocastanum	horse chestnut
Aesculus octandra	sweet buckeye
Agropyron cristatum	crested wheat-grass
Agropyron dasystachyum	northern wheat-grass or blue joint grass
Agropyron michnoi	"graminoid"
Agropyron repens var. *glaucum*	'wheat-grass'
Agropyron smithii	western wheat-grass
Agropyron spicatum	bluebunch wheat-grass
Agropyron trachycaulum	slender wheat-grass
Albizia kalkora	"tree"
Alectoria ochroleuca	"lichen"
Alectoria pubescens	"lichen"
Alnus crispa	mountain-alder
Alnus glutinosa	common or black alder
Alnus incana	grey alder
Alnus japonica	'alder'
Alnus kamtschatica	'alder'
Alnus maximowiezii	'alder'
Alnus rubra	red alder
Alnus rugosa	speckled alder
Alopecurus alpinus	alpine foxtail
Amelanchier alnifolia	saskatoon
Amorpha canescens	lead plant
Anaphalis margaritacea	pearly everlasting
Andreanea	"bryophyte"

Andromeda polifolia	bog rosemary
Andropogon gerardii	big bluestem
Andropogon scoparius	little bluestem
Anemone canadensis	Canada anemone
Aneurolepidium chinense	sheepgrass
Anisotome	"forb"
Aralia nudicaulis	wild sarsaparilla
Araucaria araucana	Chile pine or monkey-puzzle
Arbutus menziesii	Pacific madrone
Arceuthobium pusillum	dwarf mistletoe
Arctagrostis latifolia	polar grass
Arctophila fulva	pendent grass
Arctostaphylos alpina	alpine bearberry
Arctostaphylos uva-ursi	common bearberry or kinnikinnick
Argyroxiphium sandwicense	silversword
Aristida dichotoma	poverty-grass
Aristida longiseta	"graminoid"
Aristida purpurascens	"graminoid"
Artemisia arbuscula	low sagebrush
Artemisia cana	hoary sagebrush
Artemisia communata	'sagebrush'
Artemisia frigida	prairie sagewort
Artemisia larscheana	'sagebrush'
Artemisia maritima	'sagebrush'
Artemisia rupestris	'sagebrush'
Artemisia tridentata	big sagebrush
Artemisia tridentata var. *vaseyana*	big sagebrush
Artemisia tripartita	three-tip sagebrush
Asimina triloba	paw-paw
Astelia	"forb"
Aster cordifolius	'aster'
Aster laevis	smooth aster
Aster praealtus	willow aster
Astragalus alpinus	"forb"
Athenasia acerosa	"shrub"
Atriplex confertifolia	saltbush
Aulacomnium stricta	"bryophyte"
Aulacomnium turgidum	"bryophyte"
Austrocedrus	Chilean cedar
Avena fatua	wild oats
Bactris	'palm'
Betula albosienensis	'birch'
Betula alleghaniensis	yellow birch

Betula dahurica	'birch'
Betula ermanii	Russian rock birch
Betula glandulosa	scrub birch
Betula lenta	sweet birch
Betula lutea	yellow birch
Betula nana	dwarf birch
Betula papyrifera	white birch
Betula pendula	European birch
Betula platyphylla	'birch'
Betula populifolia	grey birch
Betula pubescens	downy birch
Betula pubescens ssp. *tortuosa*	mountain birch
Betula pumila	swamp birch
Betula tortuosa	'birch'
Betula utilis	Himalayan birch
Biatorella	"lichen"
Bombax	"evergreen tree"
Bothriochloa lagurioides	"graminoid"
Bouteloua curtipendula	sideoats grama
Bouteloua gracilis	blue grama
Brachyelytrum erectum	bearded short-husk
Brachythecium austrostamineum	"bryophyte"
Bromus japonicus	Japanese brome
Bromus pubescens	Canada brome
Bromus riparius	'brome'
Bromus tectorum	downy brome
Bryocaulon divergens	"lichen"
Buchloë dactyloides	buffalo grass
Buellia	"lichen"
Butomus umbellatus	flowering rush
Calamagrostis antonianus	"graminoid"
Calamagrostis arundinacea	"graminoid"
Calamagrostis canadensis	bluejoint
Calamagrostis densiflora	"graminoid"
Calamagrostis epigejos	feathertop
Calamagrostis inexpansa	northern reedgrass
Calamagrostis rubescens	pinegrass
Calamovilfa longifolia	big sandgrass
Calluna vulgaris	ling or Scotch heather
Calocedrus decurrens	incense cedar
Caloplaca	"lichen"
Caltha leptosepala	marsh marigold
Caltha palustris	cowslip

Campylopus atrovirens	"bryophyte"
Campylopus flexuosis	"bryophyte"
Caragana	pea shrubs
Carduus nyssanus	thistle
Carex aquatilis	aquatic sedge
Carex atherodes	hair sedge
Carex capillaris	hair sedge
Carex chordorrhiza	creeping sedge
Carex curvula	"graminoid"
Carex eleocharis	"graminoid"
Carex elynoides	"graminoid"
Carex ensifolia	"graminoid"
Carex filifolia	"graminoid"
Carex humilis	"graminoid"
Carex lanuginosa	"graminoid"
Carex lasiocarpa	"graminoid"
Carex limosa	mud sedge
Carex lyngbyei	"graminoid"
Carex membranacea	fragile sedge
Carex meyeriana	"graminoid"
Carex misandra	"graminoid"
Carex nardina	"graminoid"
Carex nigricans	"graminoid"
Carex obtusata	"graminoid"
Carex pensylvanica	"graminoid"
Carex pluriflora	"graminoid"
Carex praecox	"graminoid"
Carex rariflora	scant sedge
Carex rostrata	"graminoid"
Carex rupestris	rock sedge
Carex secta	"graminoid"
Carex spectabilis	"graminoid"
Carex stans	"graminoid"
Carex supina	weak Arctic sedge
Carpha	"graminoid"
Carpinus betulus	common hornbeam
Carpinus caroliniana	ironwood
Carpinus fargesii	"tree"
Carpinus orientalis	oriental hornbeam
Carya arkansana	'hickory'
Carya cordiformis	bitternut hickory
Carya glabra	pignut hickory
Carya laciniosa	big shellbark hickory

Carya ovata	shellbark hickory
Carya tomentosa	mockernut
Carya villosa	'hickory'
Cassiope mertensiana	mountain heather
Cassiope tetragona	Arctic white heather
Castanea crenata	Japanese chestnut
Castanea dentata	American sweet chestnut
Castanea sativa	sweet chestnut
Castanea sequinii	'chestnut'
Casuarina	"tree"
Cecropia	umbrella tree
Celtis spp.	hackberries
Celtis sinensis	Chinese hackberry
Centrolepis	"forb"
Cephalanthus occidentalis	buttonbush
Ceratoides lanata	greasewood
Cercis canadensis	redbud
Cercocarpus montanus	birchleaf mahogany
Cetraria cucullata	"lichen"
Cetraria delisei	"lichen"
Cetraria icelandica	"lichen"
Cetraria nivalis	"lichen"
Chamaecyparis nootkatensis	Alaska yellow cedar
Chamaecyparis thyoides	Atlantic white cedar
Chamaedaphne calyculata	leather leaf
Chinochlora flavescens	broad-leaved snow tussock
Chlamydomonas	"algae"
Chlorosphaera	"algae"
Chodatella	"algae"
Chorisodontium aciphyllum	"bryophyte"
Chrysothamnus viscidiflorus	"shrub"
Chusquea quila	'bamboo'
Cinclidium arcticum	"bryophyte"
Cladina alpestris	"lichen"
Cladina mitis	yellow reindeer lichen
Cladina rangiferina	reindeer lichen
Cladina stellaris	"lichen"
Cladium jamaicense	saw grass
Cladium mariscoides	twig-rush
Cladonia	"lichen"
Cleistogenes squarrosa	cleistogenes
Cleistopholis patens	"tree"
Clintonia borealis	bluebead-lily

Colobanthus crassifolius	"forb"
Colobanthus quitensis	"forb"
Combretocarpus	"tree"
Cornicularia divergens	"lichen"
Cornus canadensis	bunchberry
Cornus drummondii	roughleaf dogwood
Cornus florida	eastern flowering dogwood
Cornus sanguinea	'dogwood'
Cornus stolonifera	red-osier dogwood
Corylus americana	American hazel
Corylus avellana	'hazel'
Corylus cornuta	beaked hazelnut
Coryphantha vivipara	pin cushion cactus
Crataegus mollis	hawthorn
Cyathea	'tree fern'
Cymbopogon	"graminoid"
Cyperus papyrus	papyrus
Dacrycarpus dacrydioides	"tree"
Dactylocladus	"tree"
Danthonia archiboldii	'oat-grass'
Danthonia intermedia	wild oat-grass
Danthonia parryii	Parry's oat-grass
Danthonia schneideri	'oat-grass'
Danthonia semiannularis	'oat-grass'
Danthonia vestita	'oat-grass'
Dendrosenecio brassica	cabbage groundsel
Deschampsia antarctica	'hairgrass'
Deschampsia caespitosa	tufted hairgrass
Deschampsia klossii	'hairgrass'
Diapensia lapponica	"dwarf shrub"
Dicranopteris	"fern"
Dicranowesia antarctica	"bryophyte"
Dicranum elongatum	"bryophyte"
Diospyros	"tree"
Distichium capillaceum	"bryophyte"
Distichlis stricta	alkali grass
Draba arbuscula	"herb"
Draba chinophila	"herb"
Drepanocladus aduncus	"bryophyte"
Drepanocladus fluitans	"bryophyte"
Drepanocladus revolvens	"bryophyte"
Drepanocladus uncinatus	"bryophyte"
Drosera rotundifolia	round-leaved sundew

Dryas integrifolia	white mountain avens
Dryas integrifolia ssp. *crenulata*	white mountain avens
Dryas octopetala	dryas
Dryas octopetala var. *hookeriana*	mountain avens
Dryas octopetala ssp. *punctata*	mountain avens
Echinochloa polystachya	"graminoid"
Echinochloa pyromidalis	"graminoid"
Elaeagnus commutata	silverberry or wolf willow
Eleocharis palustris	spike-rush
Eleocharis smallii	spike-rush
Eleusine	"graminoid"
Elymus arenarius	sea lime grass
Elymus canadensis	Canada wild rye
Elymus condensatus	giant wild rye
Empetrum nigrum	black crowberry
Encalypta rhaptocarpa	"bryophyte"
Endothia parasitica	fungus blight
Epilobium angustifolium	fireweed
Equisetum arvense	common horsetail
Equisetum fluviatile	water horsetail
Erica tetralix	bell heather
Erigeron	"graminoid"
Eriogonum	umbrella plant
Eriophorum angustifolium	tall cotton-grass
Eriophorum medium	'cotton-grass'
Eriophorum scheuchzeri	'cotton-grass'
Eriophorum triste	'cotton-grass'
Eriophorum vaginatum	sheathead cotton-grass
Eriophorum vaginatum ssp. *spissum*	'cotton-grass'
Erodium cicutarium	filaree
Erysimum asperum	western wallflower
Erythronium grandiflorum	yellow avalanche lily
Espeletia schultzii	"forb"
Eucalyptus rudis	'eucalyptus'
Fagus crenata	'beech'
Fagus grandifolia	American beech
Fagus japonica	'beech'
Fagus mexicana	'beech'
Fagus orientalis	oriental beech
Fagus sylvatica	common beech
Fauria crista-galli	deer-cabbage
Ferula	"forb"
Festuca campestris	rough fescue

Festuca contracta	'fescue'
Festuca dahurica	'fescue'
Festuca dolybergetum	'fescue'
Festuca halleri	'fescue'
Festuca hallii	'fescue'
Festuca heterophylla	'fescue'
Festuca idahoensis	Idaho fescue
Festuca ovina	sheep's-fescue
Festuca pilgeri	'fescue'
Festuca pilgeri var. *supina*	'fescue'
Festuca pseudovina	'fescue'
Festuca scabrella	rough fescue
Festuca scabrella var. *major*	'fescue'
Festuca sulcata	'fescue'
Ficus congensis	'fig'
Ficus verruculosa	'fig'
Filofolium sibiricum	"forb"
Fimbristylis	"graminoid"
Fragaria viridis	"herb"
Fraxinus americana	white ash
Fraxinus excelsior	European ash
Fraxinus nigra	black ash
Fraxinus ornis	'ash'
Fraxinus pennsylvanica	green ash
Fraxinus profunda	red ash
Fraxinus quadrangulata	blue ash
Fuchsia excorticata	fuchsia
Gagea	"forb"
Gaultheria procumbens	checkerberry
Gaultheria shallon	salal
Gaylussacia baccata	black huckleberry
Gaylussacia dumosa	dwarf huckleberry
Gentiana propinqua	Arctic gentian
Gingidium	"forb"
Ginkgo biloba	maidenhair tree
Gleichenia vulcanica	false staghorn fern
Gonystylus	"tree"
Grimmia affinis	"bryophyte"
Gutierrezia sarothrae	broom-snakeroot
Gymnocladus dioica	Kentuchy coffee-tree
Helichrysum	everlastings
Hierochloe alpina	holy grass
Hierochloe pauciflora	"graminoid"

Hippuris vulgaris	common mare's-tail
Hoheria glabrata	ribbonwood
Honckenya peploides	sea-purslane
Hordeum jubatum	squirrel-tail grass
Hydrocotyle	water-pennywort
Hylocomium splendens	"bryophyte"
Hyparrhenia rufa	elephant grass
Hyphaene	'palm'
Ilex aquifolium	holly
Ilex verticillata	winterberry
Inga	"evergreen tree"
Iris	fleur-de-lis
Juglans nigra	black walnut
Juncus acutiflorus	sharp-flowered rush
Juncus balticus	Baltic rush
Juncus biglumis	"graminoid"
Juncus scheuchzerioides	"graminoid"
Juniperus communis	common juniper
Juniperus deppeana	alligator-bark juniper
Juniperus monosperma	one-seed juniper
Juniperus osteosperma	Utah juniper
Juniperus scopulorum	Rocky Mountain juniper
Juniperus sibirica	'juniper'
Juniperus virginiana	red cedar
Kalmia angustifolia	lambkill
Kalmia latifolia	mountain-laurel
Kalmia polifolia	bog-laurel
Kniphofia thommonii	red-hot poker
Kobresia bellardii	cobresia
Koeleria cristata	June grass
Koeleria gracilis	"graminoid"
Koeleria pyramidata	"graminoid"
Larix cajanderi	Asian larch
Larix dahurica	Asian larch
Larix decidua	European larch
Larix gmelinii	Asian larch
Larix laricina	tamarack
Larix leptolepis	'larch'
Larix lyallii	alpine larch
Larix occidentalis	western larch
Larix russica	Siberian larch
Lecanora	"lichen"
Lecidia	"lichen"

Ledum glandulosum	trapper's tea
Ledum groenlandicum	Labrador tea
Ledum palustre	dwarf Labrador tea
Ledum palustre var. *decumbens*	dwarf Labrador tea
Leopoldinia	'palm'
Leucophyllum	"tree"
Lilium philadelphicum	western red lily
Linosyris	"shrub"
Liquidambar macrophylla	gumtree
Liquidambar styraciflua	sweetgum
Liriodendron grandifolia	tulip-tree
Liriodendron tulipifera	tulip-tree
Lithospermum ruderale	puccoon
Lobelia keniensis	'lobelia'
Loiseleuria procumbens	alpine azalea
Loudetia phragmitoides	"graminoid"
Luetkea pectinata	partridge foot
Lupinus lepidus	lupine
Luzula confusa	woodrush
Luzula nivalis	'rush'
Luzula wahlenbergii	woodrush
Lycopodium annotium	stiff club-moss
Lycopodium lucidulum	shining club-moss
Lycopodium obscurum	ground-pine
Lysichitum americanum	western skunk cabbage
Lythrum salicaria	purple loosestrife
Magnolia acuminata	cucumber-tree
Magnolia grandiflora	laurel magnolia
Magnolia virginiana	swamp bay
Matteuccia struthiopteris	ostrich fern
Mauritia	aguaje palm
Medeola virginiana	Indian cucumber root
Melaleuca	paperbark eucalyptus
Melaleuca quinquenervia	tea tree
Metasequoia	water larch
Metrosideris polymorpha	ohia
Metroxylon sagu	sago palm
Minuartia obtusiloba	sandwort
Miscanthidium violaceum	"graminoid"
Mitragyna stipulosa	"tree"
Molinia caerulea	purple moor grass
Monostachya	"graminoid"
Morus rubra	red mulberry

Muhlenbergia asperifolia	scratchgrass
Musanga cecropioides	umbrella tree
Myosotis	"forb"
Myrica gale	sweet gale
Nardus stricta	matgrass
Nectandra	sweetwood
Neoscortechinia	"tree"
Nothofagus alpina	raulí
Nothofagus antarctica	ñirre
Nothofagus dombeyi	coigue
Nothofagus glauca	'southern hemisphere beech'
Nothofagus obliqua	roble
Nothofagus pumilio	lenga
Nothofagus solandri	black beech
Nymphaea odorata	water lily
Nyssa aquatica	water tupelo
Nyssa sylvatica	black gum
Nyssa sylvatica var. *biflora*	swamp tupelo
Oncophorus wahlenbergii	"bryophyte"
Onoclea sensibilis	sensitive fern
Opuntia fragilis	'prickly pear'
Opuntia polyacantha	prickly pear
Oryza longistaminata	wild rice
Oryzopsis asperifolia	white-grained mountain rice grass
Oryzopsis canadensis	'mountain-rice'
Oryzopsis pungens	northern rice grass
Osmunda trispera	"fern"
Ostrya carpinifolia	'hornbeam'
Ostrya virginiana	hop-hornbeam
Oxycoccus microcarpus	cranberry
Oxycoccus palustris	swamp cranberry
Oxydendrum arboreum	sourwood
Oxytropis arctobia	"forb"
Oxytropis lambertii	purple locoweed
Oxytropis maydelliana	"forb"
Pandanus	pandan
Panicum hemitomum	"graminoid"
Panicum virgatum	switchgrass
Parmelia	"lichen"
Paronychia pulvinata	"forb"
Parrya nudicaulis	"forb"
Paspalum fasciculatum	"graminoid"
Paspalum plicatulum	"graminoid"

Paspalum quadrifarium	"graminoid"
Pentaschistris minor	"graminoid"
Persea borbonia	'evergreen broadleaf'
Petasites palmatus	palmated-leaved colt's foot
Phleum alpinum	alpine timothy
Phleum phleoides	"graminoid"
Phoenix reclinata	'palm'
Phragmites australis	reed
Phragmites communis	'reed'
Phragmites karka	'reed'
Phyllocladus alpinus	mountain toatoa
Phyllodoce empetriformis	pink mountain heather
Picea abies	Norway spruce
Picea ajanensis	Yezo spruce
Picea engelmannii	Engelmann spruce
Picea glauca	white spruce
Picea jezoensis	yeso spruce
Picea mariana	black spruce
Picea obovata	Siberian spruce
Picea pungens	Colorado spruce
Picea rubens	red spruce
Picea sitchensis	Sitka spruce
Pilgerondendrum uviferum	'cypress'
Pinus albicaulis	whitebark pine
Pinus aristata	Rocky Mountain bristlecone pine
Pinus balfouriana	foxtail pine
Pinus banksiana	Jack pine
Pinus cembra	Aroila pine
Pinus contorta	lodgepole pine
Pinus contorta ssp. *latifolia*	lodgepole pine
Pinus densiflora	Japanese red pine
Pinus divaricata	'pine'
Pinus echinata	short-leaf pine
Pinus edulis	Colorado pinyon pine
Pinus flexilis	limber pine
Pinus jeffreyi	Jeffrey pine
Pinus lambertiana	sugar pine
Pinus longaeva	bristlecone pine
Pinus monophylla	singleleaf or single-needle pinyon pine
Pinus monticola	western white pine
Pinus nigra	Austrian pine
Pinus ponderosa	ponderosa or western yellow pine
Pinus pumila	dwarf Siberian pine

Pinus resinosa	red (or Norway) pine
Pinus rigida	pitch pine
Pinus serotinia	'pine'
Pinus sibirica	cedar pine
Pinus strobus	white pine
Pinus sylvestris	Scots pine
Pinus sylvestris var. *mongolica*	'pine'
Pinus taeda	loblolly pine
Pinus virginia	Virginia pine
Plagianthus betulinus	"tree"
Platanus occidentalis	silver sycamore
Pleurophyllum hookeri	"forb"
Pleurozia purpurea	"bryophyte"
Pleurozium schreberi	feathermoss
Poa alpigena	"graminoid"
Poa alpina	"graminoid"
Poa angustifolia	"graminoid"
Poa annua	annual bluegrass
Poa cusickii	Cusick's bluegrass
Poa flabellata	"graminoid"
Poa foliosa	"graminoid"
Poa keyessii ssp. *sarawagetica*	"graminoid"
Poa lingularis	"graminoid"
Poa palustris	fowl meadow-grass
Poa pratensis	Kentucky bluegrass
Poa ruwenzoriense	"graminoid"
Poa secunda	Sandberg's bluegrass
Podocarpus nivalis	snow totara
Podocarpus usambarensis	"tree"
Polygonum viviparum	alpine bistort
Polylepis incana	quenua
Polystichum munitum	western sword fern
Polytrichum alpestre	"bryophyte"
Polytrichum alpinum	"bryophyte"
Polytrichum norvegicum	"bryophyte"
Polytrichum strictum	"bryophyte"
Populus balsamifera	balsam poplar
Populus deltoides	cottonwood
Populus grandidentata	largetooth aspen
Populus tremuloides	trembling aspen
Potamogeton gramineus	pondweed
Potamogeton pusillus	pondweed
Potentilla	cinquefoil

Prenanthes altissima	rattlesnake-root
Prosopis calderina	mesquite
Prosopis glandulosa	honey mesquite
Prunus emarginata	bitter cherry
Prunus pensylvanica	pin cherry
Prunus pumila	low sand cherry
Prunus serotina	black cherry
Prunus spinosa	blackthorn bush
Prunus virginiana	choke-cherry
Pseudotsuga macrocarpa	bigcone Douglas-fir
Pseudotsuga menziesii	Douglas-fir
Pseudotsuga menziesii var. *glauca*	blue Douglas-fir
Psoralea agrophylla	silverleaf psoralea
Pteridium aquilinum	bracken
Ptilidium ciliare	"bryophyte"
Ptilium crista-castrensis	'feathermoss'
Puccinellia distans	alkali-grass
Puccinellia nuttalliana	Nuttall's alkali grass
Puccinellia phryganodes	alkali-grass, goose grass
Pulsatilla patens	"forb"
Purshia tridentata	bitter brush
Pyrola	wintergreen
Pyrus	"deciduous tree"
Quercus acutissima	'oak'
Quercus alba	white oak
Quercus aliena	'oak'
Quercus bicolor	swamp white oak
Quercus cerris	Turkey oak
Quercus coccinea	scarlet oak
Quercus densiflora	'oak'
Quercus dentata	Daimyo oak
Quercus ellipsoidalis	northern pin-oak
Quercus faginea	'oak'
Quercus frainetto	Hungarian oak
Quercus garryana	Garry or Oregon white oak
Quercus ilicifolia	bear oak
Quercus macrocarpa	bur oak
Quercus marilandica	blackjack oak
Quercus mongolica	'oak'
Quercus mongolia var. *grosseserrata*	'oak'
Quercus muehlenbergii	'oak'
Quercus palustris	pin-oak

Quercus petraea	sessile oak
Quercus prinoides	willow oak
Quercus prinus	chestnut oak
Quercus pubescens	white oak
Quercus pyrenaica	Pyrenean oak
Quercus robur	pedunculate oak
Quercus rubra	Northern red oak
Quercus shumardii	Shumard red oak
Quercus stellata	post oak
Quercus variabilis	oriental cork oak
Quercus velutina	black oak
Racomitrium lanuginosum	"bryophyte"
Racomitrium sudeticum	"bryophyte"
Ranunculus	buttercup
Raphia farinifera	'palm'
Rhamnus catharticus	buckthorn
Rhizocarpon geographicum	"lichen"
Rhododendron canadense	rhodora
Rhododendron lapponicum	Lapland rose-bay
Rhus aromatica	fragrant sumac
Rhus radicans	poison ivy
Rhus vernix	poison sumac
Rorippa islandica	marsh yellow cress
Rosa acicularis	prickly rose
Rosa arkansana	wild praire rose
Rostkovia magellanica	"graminoid"
Rubus chamaemorus	cloudberry
Rubus hispidus	"shrub"
Rubus parviflorus	thimbleberry
Ruppia occidentalis	widgeon-grass
Saccharum robustum	"graminoid"
Sagittaria latifolia	arrowhead
Sagittaria rigida	"forb"
Salicornia rubra	glasswort
Salix alaxensis	'willow'
Salix amygdaloides	peach-leaved willow
Salix arbusculoides	'willow'
Salix arctica	Arctic willow
Salix babylonica	weeping willow
Salix brachycarpa	short-capsuled willow
Salix candida	silver willow
Salix glauca	blue-green willow
Salix hastata	'willow'

Salix interior	sandbar willow
Salix lanata	lime willow
Salix matsudana	corkscrew willow
Salix pedicellaris	bog willow
Salix petiolaris	'willow'
Salix phylicifolia	'willow'
Salix planifolia	flat-leaved willow
Salix polaris	'willow'
Salix pulchra	'willow'
Salix reptans	'willow'
Salix reticulata	snow willow
Salix reticulata var. *nivalis*	snow willow
Salvia herzogii	water hyacinth
Salvinia auriculata	water hyacinth
Sambucus canadensis	common elder
Sanionia uncinata	"bryophyte"
Sasa kurilensis	dwarf bamboo
Sassafras albidum	white sassafras
Saxifraga hirculus	yellow marsh-saxifrage
Saxifraga oppositifolia	purple saxifrage
Saxifraga tolmiei	'saxifrage'
Saxifraga tricuspidata	three-toothed saxifrage
Schizachyrium scoparium	little bluestem
Schoenus nigricans	black bog-rush
Scirpus acutus	'bulrush'
Scirpus americanus	'bulrush'
Scirpus californicus	'bulrush'
Scirpus cespitosus	deergrass
Scirpus fluviatilis	river-bulrush
Scirpus hudsonianus	'bulrush'
Scirpus lacustris var. *glaucus*	'bulrush'
Scirpus lacustris var. *validus*	'bulrush'
Scirpus maritimus	'bulrush'
Scirpus olneyi	'bulrush'
Scirpus paludosus	prairie bulrush
Scirpus validus	'bulrush'
Scolochloa festucacea	spangle-top
Scorpidium scorpioides	"bryophyte"
Scotiella	"algae"
Sedum roseum	roseroot
Selaginella	spikemoss
Senecio	groundsel
Senecio congestus	marsh ragwort

Sequoia sempervirens	coastal redwood
Sequoiadendron giganteum	giant sequoia
Sesleria coerulea	"graminoid"
Setaria	"graminoid"
Shepherdia argentea	buffaloberry
Shepherdia canadensis	soapberry
Shorea albida	"evergreen tree"
Sibbaldia	"forb"
Silene acaulis	moss-campion
Sinarundinaria fangiana	'bamboo'
Solarina saccata	"lichen"
Solidago canadensis	tall goldenrod
Solidago missouriensis	low goldenrod
Sorbus americana var. *japonica*	mountain ash
Sorbus aria	whitebeam
Sorbus aucuparia	rowan
Sorghastrum nutans	Indian grass
Sparganium eurycarpum	bur-reed
Spartina alterniflora	salt-water cord-grass
Spartina gracilis	alkali cord-grass
Spartina pectinata	slough or cord grass
Sphaeralcea coccinea	cowboy's delight
Sphaerophorus fragilis	"lichen"
Sphagnum angustifolium	'sphagnum'
Sphagnum balticum	'sphagnum'
Sphagnum capillifolium	'sphagnum'
Sphagnum cuspidatum	'sphagnum'
Sphagnum flavicomans	'sphagnum'
Sphagnum fuscum	'sphagnum'
Sphagnum girgensohnii	'sphagnum'
Sphagnum imbricatum	'sphagnum'
Sphagnum lenense	'sphagnum'
Sphagnum magellanicum	'sphagnum'
Sphagnum majus	'sphagnum'
Sphagnum nemoreum	'sphagnum'
Sphagnum papillosum	'sphagnum'
Sphagnum pulchrum	'sphagnum'
Sphagnum recurvum	'sphagnum'
Sphagnum rubellum	'sphagnum'
Sphagnum russowii	'sphagnum'
Sphagnum tenellum	'sphagnum'
Sphagnum warnstorfii	'sphagnum'
Sporobolus heterolepis	prairie dropseed

Stereocaulon arenarium	"lichen"
Stereocaulon paschale	"lichen"
Stereocaulon rivulorum	"lichen"
Stilbocarpa polaris	"forb"
Stipa baicalensis	'feathergrass'
Stipa brachychaeta	'feathergrass'
Stipa breviflora	'feathergrass'
Stipa bungeana	'feathergrass'
Stipa capillata	'feathergrass'
Stipa comata	needle-and-thread or speargrass
Stipa grandis	'feathergrass'
Stipa humilis	'feathergrass'
Stipa ichu	ichu
Stipa ioannis	'feathergrass'
Stipa krylovii	Krylov's feathergrass
Stipa obtusa	'feathergrass'
Stipa spartea	porcupine grass
Stipa speciosa	'feathergrass'
Stipa trichotoma	'feathergrass'
Stipa viridula	green needlegrass
Symphoricarpos occidentalis	wolfberry
Symphoricarpos orbiculatus	buckbrush
Symphoricarpos oreophilus	western snowberry
Syzygium guineense	"evergreen tree"
Taxodium ascendens	pond cypress
Taxodium distichum	bald cypress
Taxus baccata	yew
Thelypteris noveboracensis	New York fern
Themeda triandra	rooigrass
Thuja occidentalis	eastern white cedar
Thuja plicata	western red cedar
Thymus serpyllum	creeping thyme
Tilia americana	basswood or linden
Tilia amurensis	'basswood'
Tilia caroliniana	'basswood'
Tilia cordata	small-leaved lime
Tilia floridana	'basswood'
Tilia heterophylla	white basswood
Tilia japonica	'basswood'
Tilia neglecta	'basswood'
Tilia tomentosa	silver lime
Tomenthypnum nitens	"bryophyte"
Tortella arctica	"bryophyte"

Tortula robusta	"bryophyte"
Tortula ruralis	"bryophyte"
Trachypogon spicatus	"graminoid"
Trichophorum caespitosum	"graminoid"
Trientalis borealis	American star-flower
Trifolium cryptopodium	clover
Trifolium nanum	"forb"
Triglochin maritima	arrow-grass
Trillium undulatum	painted trillium
Tsuga canadensis	eastern hemlock
Tsuga diversiflora	northern Japanese hemlock
Tsuga heterophylla	western hemlock
Tsuga mertensiana	mountain hemlock
Tulipa	tulip
Typha angustifolia	narrow-leaved cat-tail
Typha domingensis	'cat-tail'
Typha glauca	'cat-tail'
Typha latifolia	common cat-tail
Typha orientalis	'cat-tail'
Ulex europaeus	gorse
Ulmus alata	winged elm
Ulmus americana	American white elm
Ulmus davidiana var. *japonica*	'elm'
Ulmus fulva	slippery elm
Ulmus rubra	slippery elm
Ulmus thomasii	rock elm
Utricularia vulgaris	common bladderwort
Vaccinium angustifolium	low sweet blueberry
Vaccinium membranaceum	mountain huckleberry
Vaccinium myrtillus	dwarf bilberry
Vaccinium oxycoccus	cranberry
Vaccinium stamineum	deerberry
Vaccinium uliginosum	alpine bilberry
Vaccinium vitis-idaea	rock-cranberry
Verrucaria	"lichen"
Viburnum cassinoides	wild-raisin
Viburnum lentago	sweet viburnum
Vitis aestivalis	summer-grape
Xanthoria elegans	"lichen"
Zanthoxylum americanum	northern prickly ash
Zizania aquatica	wild rice

Bibliography

Aarssen, L.W. (1989). "Competitive ability and species coexistence: a 'plant's-eye' view." *Oikos*, 56:386–401.

Abate, T. (1992). "Which bird is the better indicator species for old-growth forest?" *BioScience*, 42:8–9.

Abrams, M.D. (1992). "Fire and the development of oak forests." *BioScience*, 42:346–53.

– (1985). "Fire history of oak gallery forests in a northeast Kansas tallgrass prairie." *American Midland Naturalist*, 114:188–91.

– and Nowacki, G.J. (1992). "Historical variation in fire, oak recruitment, and post-logging accelerated succession in central Pennsylvania." *Bulletin of the Torrey Botanical Club*, 119:19–28.

– and Scott, L.M. (1989). "Disturbance-mediated accelerated succession in two Michigan forest types." *Forest Science*, 35:42–9.

Achuff, P.L., and La Roi, G.H. (1977). "*Picea-Abies* forests in the highlands of northern Alberta." *Vegetatio*, 33:127–146.

Acton, D.F. (1992). "Grassland soils." Pp. 25–54 in Coupland, R.T. (ed.), *Natural Grasslands: Introduction and Western Hemisphere*. Ecosystems of the World 8A. Elsevier, New York.

Adams, G.D. (1988). "Wetlands of the Prairies of Canada." Pp. 157–98 in National Wetlands Working Group, *Wetlands of Canada*. Ecological Land Classification Series. no. 24. Sustainable Development Branch, Environment Canada, Ottawa, and Polyscience Publications Inc., Montreal.

Adams, H.S., and Stephenson, S.L. (1989). "Old-growth red spruce communities in the mid-Appalachians." *Vegetatio*, 85:45–56.

Aerts, R., Wallen, B., and Malmer, N. (1992). "Growth-limiting nutrients in *Sphagnum*-dominated bogs subject to low and high atmospheric nitrogen supply." *Journal of Ecology*, 80:131–40.

– Berendse, F., de Caluwe, H., and Schmitz, M. (1990). "Competition in heathland along an experimental gradient of nutrient availability." *Oikos*, 57:310–18.

Aleksandrova, V.D. (1988). *Vegetation of the Soviet Polar Deserts*. Cambridge University Press, Cambridge.

– (1980). *The Arctic and Antarctic: Their Division into Geobotanical Areas.* Cambridge University Press, Cambridge.

Allen, G.M., Eagles, P.F.J., and Price, S.D. (eds.) (1990). *Conserving Carolinian Canada: Conservation Biology in the Deciduous Forest Region.* University of Waterloo Press, University of Waterloo, Waterloo, Ont.

Ambrose, J.D., and Kevan, P.G. (1990). "Reproductive biology of rare Carolinian plants with regard to conservation management." Pp. 57–64 in Allen, G.M., Eagles, P.F.J., and Price, S.D. (eds.), *Conserving Carolinian Canada: Conservation Biology in the Deciduous Forest Region.* University of Waterloo Press, University of Waterloo, Waterloo, Ont.

Amiro, B.D., and Courtin, G.M. (1981). "Patterns of vegetation in the vicinity of an industrially disturbed ecosystem, Sudbury, Ontario." *Canadian Journal of Botany,* 59:1623–981.

Anderson, J.A.R. (1983). "The tropical peat swamps of western Malesia," Pp. 181–99 in Gore, A.J.P. (ed.), *Mires: Swamps, Bog, Fen and Moor.* Regional Studies. Ecosystems of the World 4B. Elsevier, New York.

Anderson, R.C. (1992). "The historic role of fire in the North American grassland." Pp. 8–18 in Collins, S.L., and Wallace, L.L. (eds.), *Fire in North American Tallgrass Prairie.* University of Oklahoma Press, Norman.

Andreas, B.K., and Knoop, J.D. (1992). "100 years of changes in Ohio peatlands." *Ohio Journal of Science,* 92:130–8.

– and Bryan, G.R. (1990). "The vegetation of three *Sphagnum*-dominated basin-type bogs in northeastern Ohio." *Ohio Journal of Science,* 90(3):54–66.

Archambault, S., and Bergeron, Y. (1992). "Discovery of a living 900 year-old northern white cedar, *Thuja occidentalis,* in northwestern Québec." *Canadian Field-Naturalist,* 106:192–5.

Archibold, O.W., and M.R. Wilson (1980). "The natural vegetation of Saskatchewan prior to agricultural settlement." *Canadian Journal of Botany,* 58:2031–42.

Arnalds, A. (1987). "Ecosystem disturbance in Iceland." *Arctic and Alpine Research,* 19:508–13.

Arnborg, T. (1990). "Forest types of northern Sweden." *Vegetatio,* 90:1–13.

Arndt, J.L., and Richardson, J.L. (1989). "A comparison of soils to wetland hydrology in North Dakota." Pp. 76–90 in *Proceedings, Manitoba Society of Soil Science,* 10–11 January 1989. Manitoba Agriculture, Winnipeg.

Arno, S.F., and Hammerly, R.P. (1984). *Timberline: Mountain and Arctic Forest Frontiers.* The Mountaineers, Seattle.

Arseneault, D., and Payette, S. (1992). "A postfire shift from lichen-spruce to lichen-tundra vegetation at tree line." *Ecology,* 73:1067–81.

Art, H.W., and Marks, P.L. (1971). *A Summary Table of Biomass and Net Annual Primary Production in Forest Ecosystems of the World.* Working Group on Forest Biomass Studies. XVth IUFRO Congress, University of Florida, Gainsville.

Atkinson, K. (1981). "Vegetation zonation in the Canadian Subarctic." *Area,* 15:13–17.

Auclair, A.N.D., and Carter, T.B. (1993). "Forest wildfires as a recent source of CO_2 at northern latitudes." *Canadian Journal of Forest Research,* 23:1528–36.

Austin, D.D. (1988). "Plant community changes within a mature pinyon-juniper woodland." *Great Basin Naturalist,* 47:96–9.

Axelrod, D.I. (1985). "Rise of the grassland biome, central North America." *The Botanical Review*, 51:163–201.

Baker, W.L. (1992). "Structure, disturbance, and change in the bristlecone pine forests of Colorado, U.S.A." *Arctic and Alpine Research*, 24:17–26.

Bakowsky, W.D. (1988). "The phytosociology of midwestern savanna in the Carolinian region of southern Ontario." Botany MSc thesis, University of Toronto.

Ball, T.F. (1986). "Historical evidence and climatic implications of a shift in the boreal tundra transition in central Canada." *Climatic Change*, 8:121–34.

– and Scott, G.A.J. (1987). "Climate change and the geographic limits of forest in central Canada." Pp. 164–5 in Marsh, W.M., *Earthscapes: A Physical Geography*. John Wiley and Sons, New York.

Ball, W.J., and Kolabinski, V.S. (1979). *An Aerial Reconnaissance of Softwood Regeneration on Mixedwood Sites in Saskatchewan*. Information Report NOR-X-216. Northern Forest Research Centre, Edmonton.

Banner, A., Hebda, R.J., Oswald, E.T., Pojar, J., and Trowbridge, R. (1988). "Wetlands of Pacific Canada." Pp. 307–46 in National Wetlands Working Group, *Wetlands of Canada*. Ecological Land Classification Series. no. 24. Sustainable Development Branch, Environment Canada, Ottawa, and Polyscience Publications Inc., Montreal.

– Pojar, J., and Kimmins, J.P. (1987). "The bog-forest complex of north-coastal British Columbia." Pp. 483–91 in Rubec, C.D.A., and Overend, R.P. (compilers), *Proceedings Symposium '87: Wetlands/Peatlands*. Edmonton, Alta.

Barclay-Estrup, P. (1970). "The description and interpretation of cyclical processes in a heath community. II. Changes in biomass and shoot production during the *Calluna* cycle." *Journal of Ecology*, 58:243–9.

Barden, L.C. (1981). "Forest development in canopy gaps of a diverse hardwood forest in the southern Appalachian Mountains." *Oikos*, 37:205–9.

Barnes, B.V. (1991). "Deciduous forests of North America." Pp. 219–344 in Röhrig, E., and Ulrich, B. (eds.), *Temperate Deciduous Forests*. Ecosystems of the World 7. Elsevier, Amsterdam.

Barr, B.M. (1989). "The forest sector of the Soviet Far East: a review and summary." *Soviet Geography*, 30:283–302.

Barrett, S.W., and Arno, S.F. (1982). "Indian fires as an ecological influence in the northern Rockies." *Journal of Forestry*, 80:647–51.

Bartlett, R.M., Reader, R.J., and Larson, D.W. (1991). "Multiple controls of cliff-edge distribution patterns of *Thuja occidentalis* and *Acer saccharum* at the stage of seedling recruitment." *Journal of Ecology*, 79:183–97.

Basile, R.M. (1971). *A Geography of Soils*. Brown, Dubuque.

Baskerville, G.L. (1988). "Redevelopment of a degrading forest system." *Ambio*, 17:314–22.

Battles, J.J., Johnson, A.H., Siccama, T.G., Friedland, A.J., and Miller, E.K. (1992). "Red spruce death: Effects on forest composition and structure on Whiteface Mountain, New York." *Bulletin of the Torrey Botanical Club*, 119:418–30.

Bay, C. (1992). "A phytogeographical study of the vascular plants of northern Greenland – north of 74° northern latitude." *Meddelelser om Gronland, Bioscience*, 36:1–102.

Beadle, L.C. (1974). *The Inland Waters of Tropical Africa: An Introduction to Tropical Limnology.* Longman, London.

Bellamy, D. (1987). *The Wild Boglands: Bellamy's Ireland.* Facts on File Publications, New York.

Bernard, J.M., and Hankinson, G. (1979). "Seasonal changes in standing crop, primary production, and nutrient levels in *Carex rostrata* wetland." *Oikos*, 32:328–36.

Berger, A.L. (1992). "Astronomical theory of paleoclimates and the last glacial-interglacial cycle." *Quaternary Science Reviews*, 11:571–81.

Bergeron, Y., and Bouchard, A. (1983). "Use of ecological groups in analysis and classification of plant communities in a section of western Québec." *Vegetatio*, 56:45–63.

– and Brisson, J. (1990). "Fire regime and red pine stands at the northern edge of the species' range." *Ecology*, 71:1352–64.

– and Dansereau, P.-R. (1993). "Predicting the composition of Canadian southern boreal forest in different fire cycles." *Journal of Vegetation Science*, 4:827–32.

Bernsohn, K. (1983). "Is sustained yield just another myth?" *Forest Talk*, 7:14–19.

Billings, W.D. (1988). "Alpine vegetation." Pp. 391–420 in Barbour, M.G., and Billings, W.D. (eds.), *North American Terrestrial Vegetation.* Cambridge University Press, New York.

– (1987). "Constraints to plant growth, reproduction, and establishment in Arctic environments." *Arctic and Alpine Research*, 19:357–65.

– (1974). "Arctic and Alpine vegetation: plant adaptations to cold summer climates." Pp. 403–43 in Ives, J.D., and Barry, R.G. (eds.), *Arctic and Alpine Environments.* Methuen, London.

Biondini, M.E., Steuter, A.A., and Grygiel, C.E. (1989). "Seasonal fire effects on the diversity patterns, spatial distribution and community structure of forbs in the Northern Mixed Prairie, USA." *Vegetatio*, 85:21–31.

Bird, G.A., and Chatarpaul, L. (1988). "Effect of forest harvest on decomposition and colonization of maple litter by soil microarthropods." *Canadian Journal of Soil Science*, 68:29–40.

Blauel, R.A., and Hocking, D. (1974). *Air Pollution and Forest Decline near a Nickel Smelter: The Thompson, Manitoba Smoke Easement Survey, 1972–74.* Information Report NOR-X-115. Northern Forest Research Centre, Edmonton.

Bliss, L.C. (1990). "High Arctic ecosystems: how they develop and are maintained." Pp. 350–80 in Harington, C.R. (ed.), *Canada's Missing Dimension. Science and History in the Canadian Arctic Islands,* vol. 1. Canadian Museum of Nature, Ottawa.

– (1988). "Arctic tundra and polar desert biome." Pp. 1–32 in Barbour, M.G., and Billings, W.D. (eds.), *North American Terrestrial Vegetation.* Cambridge University Press, New York.

– (1985). "Alpine." Pp. 41–65 in Chabot, B.F., and Mooney, H.A. (eds.), *Physiological Ecology of North American Plant Communities.* Chapman and Hall, New York.

– (ed.) (1977). *Truelove Lowland, Devon Island, Canada: A High Arctic Ecosystem.* University of Alberta Press, Edmonton.

– Heal, O.W., and Moore, J.J. (eds.) (1981). *Tundra Ecosystems: A Comparative Analysis.* The International Biological Program no.25, Cambridge University, Cambridge.

Bliss, M., and Vogelmann, H.W. (1982). "Decline of red spruce in the Green Mountains of Vermont." *Bulletin of the Torrey Botanical Club*, 109:162–8.

Boerner, R.E.J. (1983). "Nutrient dynamics of vegetation and detritus following two intensities of fire in the New Jersey pine barrens." *Oecologia*, 59:129–34.

– and Cho, D.-S. (1987). "Structure and composition of Goll Woods, an old-growth forest remnant in northeastern Ohio." *Bulletin of the Torrey Botanical Club*, 114:173–9.

Bolgiano, C. (1993). "The aliens." *American Forests*, 99(7&8):17–19, 53–4.

– (1989). "Yellowstone and the let-burn policy." *American Forests*, 95 (1&2):21–5, 74.

Bonan, G.B., Pollard, D., and Thompson, S.L. (1992). "Effects of boreal forest vegetation on global climate." *Nature*, 359:716–18.

Bonetto, A.A. (1975). "Hydrologic regime of the Parana River and its influence on ecosystems." Pp. 175–98 in Hasler, A.D. (ed.), *Coupling of Land and Water Systems*. Springer-Verlag, New York.

Bonner, W.N. (1985). "Impact of fur seals on the terrestrial environment at South Georgia." Pp. 641–6 in Siegfried, W.R., Condy, P.R., and Laws, R.M. (eds.), *Antarctic Nutrient Cycles and Food Webs*. Springer-Verlag, Berlin.

Boone, R.D., Sollins, P., and Cromack, K., Jr. (1988). "Stand and soil changes along a mountain hemlock death and regrowth sequence." *Ecology*, 69:714–22.

Bord na Móna. (1985). *Fuel Peat in Developing Countries*. World Bank Technical Paper no. 41. The World Bank, Washington, DC.

Borisova, I.V., and Popova, T.A. (1990). "Vertical structure of biogeocenoses in the steppe zone of Mongolia." *Soviet Journal of Ecology*, 20(3):147–53.

Bormann, F.H., and Likens, G.E. (1981). *Pattern and Process in a Forest Ecosystem*. Springer-Verlag, New York.

– Siccama, T.G., Likens, G.E., and Whittaker, R.H. (1970). "The Hubbard Brook Ecosystem Study: composition and dynamics of the tree stratum." *Ecological Monographs*, 40: 373–88.

Botch, M.S. and Masing, V.V. (1983). "Mire ecosystems in the U.S.S.R." Pp. 95–152 in Gore, A.J.P. (ed.), *Mires: Swamps, Bog, Fen and Moor: Regional Studies*. Ecosystems of the World 4B. Elsevier, New York.

Bouchard, A., Hay, S., Bergeron, Y., and Leduc, A. (1987). "Phytogeographical and life-form analysis of the vascular flora of Gros Morne National Park, Newfoundland, Canada." *Journal of Biogeography*, 14:343–58.

Box, E.O. (1981). *Macroclimate and Plant Forms: An Introduction to Predictive Modelling in Phytogeography*. Junk, The Hague.

Bradbury, I.K., and Grace, J. (1983) "Primary production in wetlands." Pp. 285–310 in Gore, A.J.P. (ed.), *Mires: Swamp, Bog, Fen and Moor: General Studies*. Ecosystems of the World 4A. Elsevier, New York.

Bradley, R.S. (1990). "Holocene paleoclimatology of the Queen Elizabeth Islands, Canadian high Arctic." *Quaternary Science Reviews*, 9:365–84.

Bradshaw, R.H.W., and Miller, N.G. (1988). "Recent successional processes investigated by pollen analysis of closed-canopy forest sites." *Vegetatio*, 76:45–54.

Braun, E.L. (1967). *Deciduous Forests of Eastern North America*. Hafner Publishing Co., New York.

Brayshaw, T.C. (1965). "The dry forest of southern British Columbia," Pp. 65–75 in Krajina, V.J. (ed.), *Ecology of Western North America*, vol. 1. Department of Botany, University of British Columbia, Vancouver.

Brejda, J.J., Yocom, D.H., Moser, L.E., and Waller, S.S. (1993). "Dependence of 3 Nebraska Sandhills warm-season grasses on vesicular-arbuscular mycorrhizae." *Journal of Range Management*, 46:14–20.

Briggs, J.M., and Gibson, D.J. (1992). "Effect of fire on tree spatial patterns in a tallgrass prairie landscape." *Bulletin of the Torrey Botanical Club*, 119:300–7.

Brinson, M.M. (1990). "Riverine forests." Pp. 87–141 in Lugo, A.E., Brinson, M., and Brown, S. (eds.), *Forested Wetlands*. Ecosystems of the World 15. Elscvier, Oxford.

Brisson, J., Bergeron, Y., and Bouchard, A. (1988). "Les successions secondaires sur sites mesiques dans le Haut-Saint-Laurent, Québec, Canada." *Canadian Journal of Botany*, 66:1192–203.

Bronaugh, W. (1992a). "The fall of the Dyerville Giant." *American Forests*, 98(1&2):18–21.

– (1992b). "The biggest western red cedar." *American Forests*, 98(3&4):51.

– (1991). "The biggest bristlecone pine." *American Forests*, 97(7&8):32.

Brooke, R.C., Peterson, E.B., and Krajina, V.J. (1969). "The subalpine mountain hemlock zone." Pp. 148–349 in Krajina, V.J. (ed.), *Ecology of Western North America*, vol. 2. Department of Botany, University of British Columbia, Vancouver.

Brotherson, J.D., Evenson, W.E., Rushforth, S.R., Fairchild, J., and Johansen, J.R. (1985). "Spatial patterns of plant communities and differential weathering in Navajo National Monument, Arizona." *Great Basin Naturalist*, 45:1–13.

Brown, L. (1985). *The Audubon Society Nature Guides: Grasslands*. Random House Canada, Toronto.

Brown, L.F., and Trlica, M.J. (1977). "Interacting effects of soil water, temperature, and irradiance on CO_2 exchange rates of two dominant grasses of the shortgrass prairie." *Journal of Applied Ecology*, 14:197–204.

Bruenig, E.F. (1990). "Oligotrophic forested wetlands in Borneo." Pp. 299–334 in Lugo, A.E., Brinson, M., and Brown, S. (eds.), *Forested Wetlands*. Ecosystems of the World 15. Elsevier, Oxford.

Brunstein, F.C., and Yamaguchi, D.K. (1992). "The oldest known Rocky Mountain bristlecone pines (*Pinus aristata* Engelm.)." *Arctic and Alpine Research*, 24:253–6.

Bryson, R.A., and Wendland, W.M. (1967). "Tentative climatic patterns for some late-glacial and postglacial episodes in central North America. Pp. 271–98 in Mayer-Oakes, W.J. (ed.), *Life, Land and Water*. University of Manitoba Press, Winnipeg.

Bubier, J., Costello, A., Moore, T.R., Roulet, N.T., and Savage, K. (1993). "Microtopography and methane flux in boreal peatlands, northern Ontario, Canada." *Canadian Journal of Botany*, 71:1056–63.

Buell, M.F., and Buell, H.F. (1959). "Aspen invasion of prairie." *Bulletin of the Torrey Botanical Club*, 86:264–5.

Burchill, C.A., and Kenkel, N.C. (1991). "Vegetation-environment relationships of an inland boreal salt pan." *Canadian Journal of Botany*, 69:722–32.

Burgess, R.L., Johnson, W.C., and Keammerer, W.R. (1973). *Vegetation of the Missouri River*

Floodplain in North Dakota. North Dakota Water Resources Research Institute, University of North Dakota, Grand Forks.

Buringh, P. (1984). "Organic carbon in soils of the world." Pp. 91–109 in Woodwell G.M. (ed.), *The Role of Terrestrial Vegetation in the Global Carbon Cycle: Measurement by Remote Sensing.* J. Wiley and Sons, New York.

Burrows, C.J., McQueen, D.R., Esler, A.E., and Wardle, P. (1979). "New Zealand heathlands." Pp. 339–64 in Specht, R.L. (ed.), *Heathlands and Related Shrublands: Descriptive Studies.* Ecosystems of the World 9A. Elsevier Scientific, Amsterdam.

Busch, D.E., and Smith, S.D. (1993). "Effects of fire and salinity relations of riparian woody taxa." *Oecologia*, 94:186–94.

Bush, J.K., and Van Auken, O.W. (1989). "Soil resource levels and competition between a woody and herbaceous species." *Bulletin of the Torrey Botanical Club*, 116:22–30.

Busing, R.T., and Clebsch, E.E.C. (1988). "Two decades of change in a Great Smoky Mountains spruce-fir forest." *Bulletin of the Torrey Botanical Club*, 115:25–31.

Cajander, A.K. (1913). "Studien uber die Moore Finnlands." *Acta For. Fenn.*, 2(3):1–208.

Cameron, C.C. (1987). "Comparison of two tropical and north temperate peatlands in Sumatra and Maine." Pp. 355–61 in Rubec, C.D.A., and Overend, R.P. (compilers), *Proceedings Symposium '87: Wetlands/Peatlands.* Edmonton, Alta.

Cammermeyer, J. (1993). "Life's a beech – and then you die." *American Forests*, 99(7&8): 20–1, 46.

Campbell, E.O. (1983). "Mires in Australasia," Pp. 153–80 in Gore, A.J.P. (ed.), *Mires: Swamps, Bog, Fen and Moor: Regional Studies.* Ecosystems of the World 4B, Elsevier, New York.

Campbell, J.B., Lodge, R.W., Johnson, A., and Smoliak, S. (1962). *Range Management of Grasslands and Adjacent Parklands in the Prairie Provinces.* Publication #1133. Department of Agriculture, Research Branch, Ottawa.

Campbell, J.J.N. (1982). "Pears and persimmons: a comparison of temperate forests in Europe and eastern North America." *Vegetation*, 49:85–101.

Canada, Department of Energy, Mines and Resources (1973). *The National Atlas of Canada.* Vegetation Map 45–46. Survey and Mapping Branch, Ottawa.

Canada Soil Survey Committee (1978). *The Canadian System of Soil Classification.* Publication #1646. Subcommittee on Soil Classification. Research Branch, Department of Agriculture, Ottawa.

Canadian Parks Service (1989). *Keepers of the Flame: Implementing Fire Management in the Canadian Parks Service.* Natural Resources Branch, Canadian Parks Service, Ottawa.

Canadian Wildlife Service (1992). *A Landowner's Guide: Conservation of Canadian Prairie Grasslands.* Environment Canada, Ottawa.

Candolle, A.I. de (1855). *Géographique botanique raisonée.* Masson, Paris.

Carey, A.B., Horton, S.P., and Biswell, B.L. (1992). "Northern spotted owls: influence of prey base and landscape character." *Ecological Monographs*, 62:223–50.

Carleton, T.J. (1990). "Variation in terricolous bryophyte and macrolichen vegetation along primary gradients in Canadian boreal forests." *Journal of Vegetation Science*, 1:585–94.

Carter, A. (1978). "Some aspects of the fungal flora in nickel contaminated and noncontaminated soils near Sudbury, Ontario, Canada." Botany MSc thesis, University of Toronto.

Catling, P.M., Catling, V.R., and McKay-Kuja, S.M. (1992). "The extent, floristic composition and maintenance of the Rice Lake Plains, Ontario, based on historical records." *Canadian Field-Naturalist*, 106:73–86.

Cawker, K.B. (1983). "Fire history and grassland vegetation change: three pollen diagrams from southern British Columbia." *Canadian Journal of Botany*, 61:1126–39.

CEAC (1991). *A Protected Area Visions for Canada*. Canadian Environmental Advisory Council, Environment Canada, Ottawa.

Chadwick, D.H. (1989). "Sagebrush country: America's outback." *National Geographic*, 175(1):52–83.

Chapin, F.S., III (1983). "Direct and indirect effects of temperature on Arctic plants." *Polar Biology*, 2:47–52.

– Fetcher, N., Kielland, K., Everett, K.R., and Linkens, A.E. (1988). "Productivity and nutrient cycling in Alaskan tundra: enhancement by flowing soil water." *Ecology*, 69: 693–702.

– Jefferies, R., Reynolds, J., Shaver, G., Svoboda, J., and Chu, E. (eds.) (1992). *Arctic Ecosystems in a Changing Climate: An Ecophysiological Perspective*. Academic Press Inc., Toronto.

– McGraw, J.B., and Shaver, G.R. (1989). "Competition causes regular spacing of alder in Alaskan shrub tundra." *Oecologia*, 79:412–16.

Charman, D.J. (1993). "Patterned fens in Scotland: evidence from vegetation and water chemistry." *Journal of Vegetation Science*, 4:543–52.

– (1992). "Blanket mire formation at the Cross Lochs, Sutherland, northern Scotland." *Boreas*, 21:53–72.

Chen, J., Franklin, J.F., and Spies, T.A. (1993). "Contrasting microclimates among clearcut, edge, and interior of old-growth Douglas-fir forest." *Agricultural and Forest Meteorology*, 63:219–37.

Chernov, Y.I. (1989). "Thermal conditions and Arctic biota." *Soviet Geography*, 30:102–9.

– (1985). *The Living Tundra*. Cambridge University Press, Cambridge.

Chernov, Y.I., et al. (1975). "Tareya, USSR." Pp. 159–81 in Rosswall, T., and Heal, O.W. (eds.), *Structure and Function of Tundra Ecosystems*. Ecol. Bull. 20. Swedish Natural Science Research Council. Stockholm.

Ching, K.K. (1991). "Temperate deciduous forests in East Asia." Pp. 539–55 in Röhrig, E., and Ulrich, B. (eds.), *Temperate Deciduous Forests*. Ecosystems of the World 7. Elsevier, Amsterdam.

Christenson, L.M. (1990). "Reductions of SO_2 concentrations and the apparent changes in the vegeation and soils of the Flin Flon Manitoba region." Geography BSc thesis, University of Winnipeg.

Cincotta, R.P., Van Soest, P.J., and Robertson, J.B. (1991). "Foraging ecology of livestock on the Tibetan *Changtang*: a comparison of three adjacent grazing areas." *Arctic and Alpine Research*, 23:149–61.

Clark, J.S., Merkt, J., and Muller, H. (1989). "Post-glacial fire, vegetation, and human history on the northern alpine forelands, southwestern Germany." *Journal of Ecology*, 77:897–925.

Clarke, H.A., Rubec, C.D.A., McKechnie, M.R., and McCuaig, J.M. (1989) "Wetland conservation leadership in Canada." Pp. 743–50 in Marchand, M., and de Haes, U. (eds.), *Proceedings: Conference on Peoples Role in Wetland Management*. Leiden University Press. Leiden, Netherlands.

Cochran, M.F. (1990). "Back from the brink: chestnuts." *National Geographic*, 177(2): 128–40.

Cody, W.J., Scotter, G.W., and Zoltai, S.C. (1992). "Vascular plant flora of the Melville Hills Region, Northwest Territories." *Canadian Field-Naturalist*, 106:87–99.

Cogbill, C. Van H. (1982). "Analysis of vegetation, environment and dynamics in the boreal forests of the Laurentian Highlands, Québec." Botany PHD thesis, University of Toronto.

Cole, D.W., and Rapp, M. (1981). Elemental cycling in forest Ecosystems." Pp. 341–409 in Reichle, D.E. (ed.), *Dynamic Properties of Forest Ecosystems*. Cambridge University Press, New York.

Collins, N.J., Baker, J.H., and Tilbrook, P.J. (1975). "Signy Island, Maritime Antarctic." Pp. 345–74 in Rosswall, T., and Heal, O.W. (eds.), *Structure and Function of Tundra Ecosystems*. Ecol. Bull. 20. Swedish Natural Science Research Council, Stockholm.

Collins, S.L. (1992a). "Introduction: Fire as a natural disturbance in tallgrass prairie ecosystems." Pp. 3–7 in Collins, S.C., and Wallace, L.L. (eds.), *Fire in North American Tallgrass Prairie*. University of Oklahoma Press, Norman.

– (1992b). "Fire frequency and community heterogeneity in tallgrass prairie vegetation." *Ecology*, 73:2001–6.

– and Adams, D.E. (1983). "Succession in grasslands: thirty-two years of change in a central Oklahoma tallgrass prairie." *Vegetatio*, 51:181–90.

– and Gibson, D.J. (1992). "Effects of fire on community structure in tallgrass and mixed-grass prairie." Pp. 81–98 in Collins, S.C., and Wallace, L.L. (eds.), *Fire in North American Tallgrass Prairie*. University of Oklahoma Press, Norman.

Cook, B.D., Jastrow, J.D., and Miller, R.M. (1988). "Root and mycorrhizal endophyte development in a chronosequence of restored tallgrass prairie." *New Phytologist*, 110:355–62.

Cortlett, R.T. (1984). "Human impact on the subalpine vegetation of Mt. Wilhelm, Papua New Guinea." *Journal of Ecology*, 72:841–54.

Coupland, R.T. (1992a). "Overview of the grasslands of North America." Pp. 147–50 in Coupland, R.T. (ed.), *Natural Grasslands: Introduction and Western Hemisphere*. Ecosystems of the World 8A. Elsevier, New York.

– (1992b). "Fescue prairie." Pp. 291–5 in Coupland, R.T. (ed.), *Natural Grasslands: Introduction and Western Hemisphere*. Ecosystems of the World 8A. Elsevier, New York.

– (1961). "A reconsideration of grassland classification in the northern Great Plains of North America." *Journal of Ecology*, 49:135–67.

– and Brayshaw, T.C. (1953). "The fescue grassland in Saskatchewan." *Ecology*, 34: 386–405.

Crete, M., Morneau, C., and Nault, R. (1990). "Biomasse et espèces de lichens terrestres disponibles pour le caribou dans le nord du Québec." *Canadian Journal of Botany*, 68: 2047–53.

Critchfield, W.B., and Little, E.L. (1966). *Geographic Distribution of the Pines of the World*. Miscellaneous Publication 991. U.S. Department of Agriculture, Washington.

Crow, T.R. (1988). "Reproductive mode and mechanisms for self-replacement of northern red oak (*Quercus rubra*) – a review." *Forest Science,* 34:19–40.

Curtis, J.T. (1971). *The Vegetation of Wisconsin: An Ordination of Plant Communities.* University of Wisconsin Press, Madison.

Cwynar, L.C. (1987). "Fire and the history of the north Cascade Range." *Ecology,* 68: 791–802.

– (1975). "The fire history of Barron Township, Algonquin Park." Botany MSc thesis, University of Toronto.

– and Spear, R.W. (1991). "Reversion of forest to tundra in the central Yukon." *Ecology,* 72:202–12.

Dansereau, P. (1957). *Biogeography: An Ecological Perspective.* Roland Press, New York.

Dansereau, P.-R., and Bergeron, Y. (1993). "Fire history in the southern boreal forest of north-western Québec." *Canadian Journal of Forest Research,* 23:25–32.

Daubenmire, R. (1992). "Palouse prairie." Pp. 297–312 in Coupland, R.T. (ed.), *Natural Grasslands: Introduction and Western Hemisphere.* Elsevier, London.

– (1978). *Plant Geography: with Special Reference to North America.* Academic Press, New York.

Dawson, T.E., and Bliss, L.C. (1989). "Patterns of water use and tissue water relations in the dioecious shrub, *Salix arctica*: the physiological basis for habitat partitioning between the sexes." *Oecologia,* 79:332–43.

Day, R.T., Keddy, P.A., and McNeill, J. (1988). "Fertility and disturbance gradients: a summary model for riverine marsh vegetation." *Ecology,* 69:1044–54.

Dearden, P. (1989). "Wilderness and Our Common Future." *Natural Resources Journal,* 29:205–21.

– and Rollins, R. (1993). *Parks and Protected Areas in Canada. Planning and Management.* Oxford University Press, Toronto.

de Laubenfels, D.J. (1975). *Mapping the World's Vegetation: Regionalization of Formations and Flora.* Syracuse University Press, New York.

Delcourt, H.R., and Delcourt, P.A. (1991). *Quaternary Ecology. A Paleoecological Perspective.* Chapman and Hall, London.

del Moral, R., and Fleming, R.S. (1979). "Structure of coniferous forests in western Washington: Diversity and ecotope properties." *Vegetatio,* 41:143–54.

– and Wood, D.M. (1993). "Early primary succession on the volcano Mount St. Helens." *Journal of Vegetation Science,* 4:223–34.

– and Wood, D.M. (1988a). "Dynamics of herbaceous vegetation recovery on Mount St. Helens, Washington, USA, after a volcanic eruption." *Vegetatio,* 74:11–27.

– and Wood, D.M. (1988b). "The high elevation flora of Mount St. Helens, Washington." *Madroño,* 35:309–19.

Denny, P. (1993). "Wetlands of Africa: Introduction." Pp. 1–31 in Whigham, D., Dykyjová, D., and Hejuý, S. (eds.), *Wetlands of the World: Inventory, Ecology and Management.* Kluwer, London.

Deshmukh, I. (1986). *Ecology and Tropical Biology.* Blackwell Scientific Publications, Boston.

Development Report No. 73–46. Information Canada, Ottawa.

Dhillion, S.S., and Anderson, R.C. (1993). "Growth dynamics and associated mycorrhizal fungi of little bluestem grass [*Schizachyrium scoparium* (Michz.) Nash] on burned and unburned sand prairies." *New Phytologist*, 123:77–91.

Diamond, D.D., and Smeins, F.E. (1988). "Gradient analysis of remnant true and upper coastal prairie grasslands of North America." *Canadian Journal of Botany*, 66:2152–61.

Diotte, M., and Bergeron, Y. (1989). "Fire and the distribution of *Juniperus communis* L. in the boreal forest of Québec, Canada." *Journal of Biogeography*, 16:91–6.

Dixon, R., and Stewart, J. (1984). *Peatland Inventory of Manitoba I – The Pas*. (Using LANDSAT.) Manitoba Remote Sensing Centre, Surveys and Mapping Branch, Department of Natural Resources, Winnipeg.

Dixon, R.K., Winjum, J.K., and Schroeder, P.E. (1993). "Conservation and sequestration of carbon." *Global Environmental Change*, 3:159–73.

Dormaar, J.F., and Willms, W.D. (1990). "Sustainable production from the rough fescue prairie." *Journal of Soil and Water Conservation*, 45:137–40.

Douglas, L.A., and Bilgin, A. (1975). "Nutrient regimes of soils, landscapes, and streams, Prudhoe Bay, Alaska." Pp. 61–70 in Brown, J. (ed.), *Ecological Investigations of the Tundra Biome in the Prudhoe Bay Region, Alaska*. Biological papers of the University of Alaska, Special Report #2. Fairbanks.

Dzwonko, Z., and Loster, S. (1988). "Species richness of small woodlands on the western Carpathian foothills." *Vegetatio*, 76:15–27.

Eagles, P.F.J. (1993). "Environmental management in parks." Pp. 154–84 in Dearden, P., and Rollins, R. (eds.), *Parks and Protected Areas in Canada. Planning and Management*. Oxford University Press, Toronto.

– (1990). "Implications of the Carolinian Canada research." Pp. 333–9 in Allen, G.M., Eagles, P.F.J., and Price, S.D. (eds.), *Conserving Carolinian Canada: Conservation Biology in the Deciduous Forest Region*. University of Waterloo Press, University of Waterloo, Waterloo, Ont.

Earl, J.C., and Kershaw, K.A. (1989). "Vegetation patterns in James Bay coastal marshes: III. Salinity and elevation as factors influencing plant zonations." *Canadian Journal of Botany*, 67:2967–74.

Ecoregions Working Group. (1989). *Ecoclimatic Regions of Canada, First Approximation*. Ecoregions Working Group of the Canada Committee on Ecological Land Classification. Ecological Land Classification Service, Conservation and Protection, Environment Canada, Ottawa. Map at scale of 1:7,500,000.

Eddleman, L.E., and Ward, R.T. (1984). "Phytoedaphic relationships in alpine tundra of north-central Colorado, U.S.A." *Arctic and Alpine Research*, 16:343–59.

Editorial. (1985). "The soils of Canada." *Bridges*, May/June, 29–30.

Edlund, S.A. (1990). "Bioclimatic zones in the Canadian Arctic Archipelago." Pp. 421–40 in Harington, C.R. (ed.), *Canada's Missing Dimension. Science and History in the Canadian Arctic Islands*, vol. 1. Canadian Museum of Nature, Ottawa.

– (1983). "Bioclimatic zonation in a high Arctic region: central Queen Elizabeth Islands." Pp. 381–90 in *Current Research*, part A., Geological Survey of Canada, Paper 83–1A. Ottawa.

– and Alt, B.T. (1989). "Regional congruence of vegetation and summer climate patterns in the Queen Elizabeth Islands, Northwest Territories, Canada." *Arctic*, 42:3–23.

Edmonds, R.L. (ed.) (1982). *Analysis of Coniferous Forest Ecosystems in the Western United States*. Hutchinson Ross Pub. Co., Strasbourg, Pa.

– and Vogt, D.J. (1986). "Decomposition of Douglas-fir and red alder wood in clear cuttings." *Canadian Journal of Forest Research,* 16:822–31.

Edwards, C.A., Reichle, D.E., and Crossley, D.A. (1973). "The role of soil invertebrates in turnover of organic matter and nutrients." Pp. 147–72 in Reichle, D.E. (ed.), *Analysis of Temperate Forest Ecosystems*. Springer-Verlag, New York.

Ehrendorfer, F. (1980). "Polyploidy and distribution." Pp. 45–58 in Lewis, W.H. (ed.), *Polyploidy: Biological Relevance*. Plenum Press, New York.

Ellenberg, H. (1988). *Vegetation Ecology of Central Europe*. 4th ed. Cambridge University Press, Cambridge.

Ellenbroek, G.A. (1987). *Ecology and Productivity of an African Wetland System*. W. Junk, Boston.

Elliott, W.P. (1978). *Artificial land drainage in Maintoba: History – Administration – Law*. Practicum, Natural Resource Institute, University of Manitoba, Winnipeg.

Elliott-Fisk, D.H. (1988). "The boreal forest." Pp. 33–62 in Barbour, M.G., and Billings, W.D. (eds.), *North American Terrestrial Vegetation*. Cambridge University Press, New York.

– (1983). "The stability of the northern Canadian tree line." *Annals, Association of American Geographers*, 73:560–76.

Ellis, J. (1992). *Grasslands and Grassland Sciences in Northern China*. A report of the Committee on Scholarly Communication with the People's Republic of China. Office of International Affairs, National Research Council, National Academy Press, Washington, DC.

Energy, Mines and Resources Canada (1993). *Canada: Vegetation Cover*. Map MCR 4182, 1:7 500 000. The National Atlas of Canada, 5th ed. Produced by the National Atlas Information Service and Petawawa Forestry Institute, Canada Map Office, Ottawa.

– (1989). *Canada: Land Cover Associations*. Map MCR 4113, 1:7 500 000. The National Atlas of Canada, 5th ed. Produced by the Geographic Services Division, Canada Centre for Mapping, Ottawa.

Environment Canada (1982). *Canadian Climatic Normals: Temperature and Precipitation 1951–1980*. Canadian Climatic Program, Ottawa.

Evans, E.E. (1957). *Irish Folk Ways*. Routledge and Kegan Paul, London.

Everett, R.L., and Thran, D.F. (1992). "Above-ground biomass and N, P and S capital in singleleaf pinyon pine woodlands." *Journal of Environmental Management*, 34:137–47.

Ewing, A.L., and Engle, D.M. (1988). "Effects of late summer fire on tallgrass prairie microclimate and community composition." *American Midland Naturalist*, 120:212–23.

Eyre, S.R. (1968). *Vegetation and Soils: A World Picture*. Aldine Press, Chicago.

Faber-Langendoen, D. (1984). "The ecology of tallgrass prairie in southern Ontario." Botany MSC thesis, University of Toronto.

Famelis, T.V., and Nikonova, N.N. (1989). "Differentiation of high altitude vegetation on the basis of aerial photographs." *Soviet Journal of Ecology*, 19:146–51.

Fang, J.-Y., and Yoda, K. (1990). "Climate and vegetation in China: III. Water balance and distribution of vegetation." *Ecological Research*, 5:9–23.

Farrell, J.D., and Ware, S. (1991). "Edaphic factors and forest vegetation in the Piedmont of Virginia." *Bulletin of the Torrey Botanical Club*, 118:161–9.

Filion, L., and Payette, S. (1989). "Subarctic lichen polygons and soil development along a colonization gradient on eolian sands." *Arctic and Alpine Research*, 21:175–84.

Finkelstein, M. (1992). "National Park Dreams." *Borealis*, 3(2):32–42.

– (1982). "Fire in our national parks – an ecological perspective." *Parks News*, 18(3):4–11.

Finlayson, M., and Moser, M. (eds.) (1991). *Wetlands*. International Waterfowl and Wetland Research Bureau (IWRB). Facts on File Ltd., Oxford, England.

Fittkau, E. J., and Klinge, H. (1973). "On biomass and trophic structure of the central Amazonian rain forest ecosystem" *BioTropica*, 5:2–14.

Flint, P.S., and Gersper, P.L. (1974). "Nitrogen nutrient levels in Arctic tundra soils." Pp. 375–87 in Holding, A.J., Heal, O.W., MacLean, S.F., Jr., and Flanagan, P.W. (eds.), *Soil Organisms and Decomposition in Tundra*. Tundra Biome Steering Committee, Stockholm.

Food and Agricultural Organization of the United Nations (1982). *Forestry in China*. FAO Forestry Paper. Rome.

Forest Management Institute (1975). *Vegetation Types of the Mackenzie and Yukon Corridor*. Canadian Forestry Service, Environment Canada, for the Environmental-Social Program, Northern Pipelines. Ottawa.

Forest Planning Canada (1991). "Rate of change data 1975–1989." *Forestry Planning Canada*, 7(2):46–8.

Forest Resource Development Agreement (1986). *Mixedwood Section in an Ecological Perspective, Saskatchewan*. Technical Bulletin no. 8. 2nd ed. Canada-Saskatchewan Development Agreement, Canadian Forestry Service [Ottawa].

Forrest, G.I. (1971). "Structure and production of North Pennine blanket bog vegetation." *Journal of Ecology*, 59:453–79.

Foster, D.R., Zebryk, T., Schoonmaker, P., and Lezberg, A. (1992). "Post-settlement history of human land-use and vegetation dynamics of a *Tsuga canadensis* (hemlock) woodlot in central New England." *Journal of Ecology*, 80:773–86.

– Wright, H.E., Jr., Thelaus, M., and King, G.A. (1988). "Bog development and landform dynamics in central Sweden and south-eastern Labrador, Canada." *Journal of Ecology*, 76:1164–85.

Foster, J.R., and Reiners, W.A. (1983). "Vegetation patterns in a virgin subalpine forest at Crawford Notch, White Mountains, New Hampshire." *Bulletin of the Torrey Botanical Club*, 110:141–53.

Fox, C.A., Trowbridge, R., and Tarnocai, C. (1987). "Classification, macromorphology and chemical characteristics of Folisols from British Columbia." *Canadian Journal of Soil Science*, 67:765–78.

Fox, J.F. (1992). "Response of diversity and growth-form dominance to fertility in Alaskan tundra fellfield communities." *Arctic and Alpine Research*, 24:233–7.

FPC Research Staff (1991). "Does British Columbia grow more than it cuts?" *Forest Planning Canada*, 7(1):46–8.

Franklin, J.F. (1988). "Pacific Northwest forests." Pp. 103–30 in Barbour, M.G., and Billings, W.D. (eds.), *North American Terrestrial Vegetation*. Cambridge University Press, New York.

Franklin, R.T. (1973). "Insect influences on the forest canopy." Pp. 86–99 in Reichle, D.E. (ed.), *Analysis of Temperate Forest Ecosystems*. Ecological Studies, vol. 1. Springer-Verlag, New York.

Franklin, S.B., Robertson, P.A., Fralish, J.S., and Kettler, S.M. (1993). "Overstory vegetation and successional trends of Land between the Lakes, USA." *Journal of Vegetation Science*, 4:509–20.

Fraser, W.P., and Russell, R.C. (1944). *A Revised, Annotated List of the Plants of Saskatchewan*. University of Saskatoon, Saskatchewan.

Frego, K.A., and Staniforth., R.J. (1985). "Factors determining the distribution of *Opuntia fragilis* in the boreal forest of southeastern Manitoba." *Canadian Journal of Botany*, 63: 2377–82.

Froning, K. (1980). *Logging Hardwoods to Reduce Damage to White Spruce Understory*. Information Report NOR-X-229. Northern Forest Research Centre, Edmonton.

Fuller, L.G., and Anderson, D.W. (1990). "Changes in black soils following forest invasion of prairie." Pp. 47–59 in *Papers Presented at the 33rd Manitoba Society of Soil Science Meeting, January 9–10, 1990*. Manitoba Agriculture, Winnipeg.

Furley, P.A., Newey, W.W., Kirby, R.P., and Hotson, J. (1983). *Geography of the Biosphere: An Introduction to the Nature, Distribution and Evolution of the Word's Life Zones*. Butterworths, London.

Gafni, A., and Brooks, K.N. (1990). "Hydraulic characteristics of four peatlands in Minnesota." *Canadian Journal of Soil Science,* 70:2239–53.

Gajewski, K., Payette, S., and Ritchie, J.C. (1993). "Holocene vegetation history at the boreal-forest–strub-tundra transition in north-western Québec." *Journal of Ecology*, 81:433–43.

Garcia-Gonzalez, A. (1988). "Cantabrian mountain beechwoods: a survey and the case for their preservation." *Biological Conservation*, 45:121–34.

Gee, H. (1992). "The objective case for conservation." *Nature*, 357:639.

Gensac, P. (1990). "Plant and soil groups in the Alpine grasslands of the Vanoise Massif, French Alps." *Arctic and Alpine Research*, 22:195–201.

Gentry, A.H. (1992). "Tropical forest biodiversity: distributional patterns and their conservational significance." *Oikos*, 63:19–28.

Geographical Survey Institute. (1977). *The National Atlas of Japan*. The Japan Map Center, Tokyo.

Geological Survey of Canada. (1969). *Permafrost in Canada*. Map 1246A, 1st ed. Ottawa.

Gerardin, V., and Grondin, P. (1987). "For a review of the classification of ombotrophic peatland species and vegetation types." Pp. 411–21 in Rubec, C.D.A., and Overend, R.P. (compilers), *Proceedings Symposium '87: Wetlands/Peatlands*. Edmonton, Alta.

Gibson, D.J. (1989). "Effects of animal disturbance on tallgrass prairie vegetation." *American Midland Naturalist*, 121:144–54.

– (1988). "Regeneration and fluctuation of tallgrass prairie vegetation in response to burning frequency." *Bulletin of the Torrey Botanical Club*, 115:1–12.

Gigon, A. (1987). "A hierarchic approach in causal ecosystem analysis: the calcifuge-calcicole problem in Alpine grasslands." Pp. 228–44 in Schulze, E.-D., and Zwolfer, H. (eds.), *Potentials and Limitations of Ecosystem Analysis*. Springer-Verlag, New York.

Gill, D. (1973). "Floristics of a plant succession sequence in the Mackenzie Delta, Northwest Territories." *Polarforschung*, 43:55–65.

Gillespie, D.I., Boyd, H., and Logan, P. (1991). *Wetlands for the World: Canada's Ramsar Sites*. Canadian Wildlife Service, Environment Canada, Ottawa.

Gimingham, C.H. (1988). "A reappraisal of cyclical processes in *Calluna* heath." *Vegetatio*, 77:61–4.

– (1972). *Ecology of Heathlands*. Chapman and Hall, London.

– Chapman, S.B., and Webb, N.R. (1979). "European heathlands." Pp. 365–414 in Specht, R.L. (ed.), *Heathlands and Related Shrublands: Descriptive Studies*. Ecosystems of the World 9A. Elsevier Scientific, Amsterdam.

– and Lewis Smith, R.I. (1970). "Bryophyte and lichen communities in the Maritime Antarctic." Pp. 752–85 in Holgate, M.W. (ed.), *Antarctic Ecology*. Academic Press, London.

Glaser, P.H. (1992). "Raised bogs in eastern North America – regional controls for species richness and floristic assemblages." *Journal of Ecology*, 80:535–54.

Gleason, H.A., and Cronquist, A. (1964). *The Natural Geography of Plants*. Columbia University Press, New York.

Glebov, F.Z., and Korzukhin, M.D. (1992). "Transitions between boreal forest and wetland." Pp. 241–66 in Shugart, H.H., Leemans, R., and Bonan, G.B. (eds.), *A Systems Analysis of the Global Boreal Forest*. Cambridge University Press, New York.

Glooschenko, V., and Grondin, P. (1988). "Wetlands of eastern Temperate Canada." Pp. 199–248 in National Wetlands Working Group, *Wetlands of Canada*. Ecological Land Classification Series, no. 24. Sustainable Development Branch, Environment Canada, Ottawa, and Polyscience Publications Inc., Montreal.

Glooschenko, W.A., Tarnocai, C., Zoltai, S., and Glooschenko, V. (1993). "Wetlands of Canada and Greenland." Pp. 415–514 in Whigham, D., Dykyjová, D., and Hejuý, S. (eds.), *Wetlands of the World: Inventory, Ecology and Management*. Kluwer, London.

Godfread, C.S., and Barker, W.T. (1975). "Vascular flora of Barnes and Stutsman counties, North Dakota." Pp. 333–9 in Wali, M.K. (ed.), *Prairie: A Multiple View*. University of North Dakota Press, Grand Forks.

Godzic, B. (1991). "Heavy metals and macroelements in the tundra of southern Spitsbergen: the effect of little auk *Alle alle* (L.) colonies." *Polar Research*, 9:121–31.

Golley, F.B., McGinnis, J.T., and Clements, R.G. (1971). "La biomasa y la estructura mineral de algunos bosques de Darien." *Turrialba*, 21:189–96.

Goodwillie, R.N. (1987). "Coniferous woods." Pp. 121–3 in Mitchell, F., (ed.), *The Book of the Irish Countryside*. Blackstaff Press, Belfast.

– (1975). "Growth studies in field layer of *Acer saccharum* forest in southern Ontario." Botany MSc thesis, University of Toronto.

Gorchakovskii, P.L. (1987). "Ecology of boreal relics in forest oases of Central Kazakhstan." *Soviet Journal of Ecology*, 18:85–90.

Gordon, B.H.C. (1979). *Of Men and Herds in Canadian Plains Prehistory*. National Museum of Canada, Ottawa.

Gosz, J.R. (1991). "Fundamental ecological characteristics of landscape boundaries." Pp. 8–30 in Holland, M.M., Risser, P.G., and Naiman, R.J. (eds.), *Ecotones: The Role of Landscape Boundaries in the Management and Restoration of Changing Environments*. Chapman and Hall, New York.

Gower, S.T., and Richards, J.H. (1990). "Larches: deciduous conifers in an evergreen world." *BioScience*, 40:818–26.

Green, L. (1990). "The U.S. Forest Service's darkest secret: they've been raping the giant sequoias." *Audubon*, 92(3):112–24.

Greene, S.E., Harcombe, P.A., Harmon, M.E., and Spycher, G. (1992). "Patterns of growth, mortality and biomass change in a coastal *Picea sitchensis–Tsuga heterophylla* forest." *Journal of Vegetation Science*, 3:697–706.

Greenlee, J.M., and Langenheim, J.H. (1990). "Historic fire regimes and their relation to vegetation patterns in the Monterey Bay area of California." *American Midland Naturalist*, 124:239–53.

Greller, A.M. (1989). "Correlation of warmth and temperateness with the distributional limits of zonal forests in eastern North America." *Bulletin of the Torrey Botanical Club*, 116: 145–63.

– (1988). "Deciduous forest." Pp. 288–316 in Barbour, M.G., and Billings, W.D. (eds.), *North American Terrestrial Vegetation*. Cambridge University Press, New York.

Grigal, D.F. (1991). "Elemental dynamics in forested bogs in northern Minnesota." *Canadian Journal of Botany*, 69:539–46.

Griggs (1934). "The problem of Arctic vegetation." *Journal of the Washington Academy of Science*, 24(4):153–75.

Grillas, P. (1990). "Distribution of submerged macrophytes in the Camargue in relation to environmental factors." *Journal of Vegetation Science*, 1:393–402.

Haavisto, V.F., and Wearn, V.H. (1987). "Intensifying forestry uses of peatlands with drainage in Ontario." Pp. 249–56 in Rubec, C.D.A., and Overend, R.P. (compilers), *Proceedings Symposium '87: Wetlands/Peatlands*. Edmonton, Alta.

Hadley, J.L., and Smith, W.K. (1988). "Influence of krummholz mat microclimate on needle physiology and survival." *Oecologia*, 73:82–90.

– and Smith, W.K. (1986). "Wind effects on needles of timberline conifers: Seasonal influence on mortality." *Ecology*, 67:12–19.

Hadley, K.S. (1987). "Vascular alpine plant distributions within the central and southern Rocky Mountains, U.S.A." *Arctic and Alpine Research*, 19:242–51.

Hanley, D.P. (1976). *Tree Biomass and Productivity Estimated for Three Habitat Types of Northern Idaho*. Bulletin no. 14. College of Forestry, Wildlife and Range Sciences, University of Idaho, Moscow, Idaho.

Hannon, G.E., and Bradshaw, R.H.W. (1989). "Recent vegetation dynamics on two Connemara lake islands, western Ireland." *Journal of Biogeography*, 16:75–81.

Hardin, E.D. (1988). "Succession in Buffalo Beats prairie and surrounding forest." *Bulletin of the Torrey Botanical Club*, 115:13–24.

Hare, F.K., and Ritchie, J.C. (1972). "The boreal bioclimates." *Geographical Review*, 62:333–65.

Harmon, M.E., and Franklin, J.F. (1989). "Tree seedlings on logs in *Picea-Tsuga* forests of Oregon and Washington." *Ecology*, 70:48–59.

Harper, H.J. (1957). "Effects of fertilization and climate conditions on prairie hay." *Oklahoma Agricultural Experiment Station Bulletin*, no. 492.

Harshberger, J.W. (1970). *The Vegetation of the New Jersey Pine-Barrens: An Ecological Investigation*. Dover Pub. Co., New York.

– (1958). *Phytogeographic Survey of North America*. Hafner Pub. Co., New York.

Hartman, E.L., and Rottman, M.L. (1988). "The vegetation and alpine vascular flora of the Sawatch Range, Colorado." *Madroño*, 35:202–25.

Hayashi, I., Shu, J., and Nakamura, T. (1988). "Phytomass production of grasslands in Xilin River Basin, Zilingol, Inner Mongolia, China." *Bulletin Sugadaira Montane Research Center*, no. 9:19–31.

Heal, O.W., Jones, H.E., and Whittaker, J.B. (1975). "Moor House, UK." Pp. 159–81 in Rosswall, T. and Heal, O.W. (eds.), *Structure and Function of Tundra Ecosystems*. Ecol. Bull. 20. Swedish Natural Science Research Council, Stockholm.

Heath, G.W., Arnold, M.K., and Edwards, C.A. (1966). "Studies in leaf litter breakdown: I. Breakdown rates among leaves of different species." *Pedobiolgy*, 6:1–12.

Helle, T., and Aspi, J. (1983). "Effects of winter grazing by reindeer on vegetation." *Oikos*, 40:337–43.

Heinselman, M.L. (1975). "Boreal peatlands in relation to environment." Pp. 93–103 in Hasler, A.D. (ed.), *Coupling of Land and Water Systems*. Springer-Verlag, New York.

Henry, G.H.R. (1987). "Ecology of sedge meadow communities of a polar oasis: Alexandra Fjord, Ellesmere Island, Canada." Botany PhD thesis, University of Toronto.

– and Gunn, A. (1991). "Recovery of tundra vegetation after overgrazing by caribou in Arctic Canada." *Arctic*, 44:38–42.

Hetrick, B.A.D., Wilson, G.W.T., and Hartnett, D.C. (1989). "Relationship between mycorrhizal dependence and competitive ability of two tallgrass prairie grasses." *Canadian Journal of Botany*, 67:2608–15.

Hicks, D.J., and Chabot, B.F. (1985). "Deciduous forest." Pp. 257–77 in Chabot, B.F., and Mooney, H.A. (eds.), *Physiological Plant Ecology of North American Plant Communities*. Chapman and Hall, New York.

Hiebert, R.D., and Hamrick, J.L. (1984). "An ecological study of bristlecone pine (*Pinus longaeva*) in Utah and eastern Nevada." *Great Basin Naturalist*, 44:487–94.

Hildebrand, D.V., and Scott, G.A.J. (1987). "Relationships between moisture deficiency and amount of tree cover on the pre-agricultural Canadian prairies." *Prairie Forum*, 12:203–16.

Hill, G.R., and Platt, W.J. (1975). "Some effects of fire upon a tall grass prairie plant community in nonrthwestern Iowa." Pp. 103–14 in Wali, M.K. (ed.), *Prairie: A Multiple View*. University of North Dakota Press, Grand Forks.

Hillman, G.R. (1988). *Improving Wetlands for Forestry in Canada*. Project #2760–30,

Canada-Alberta Forest Resource Development Agreement. Supply and Services Canada, Ottawa.

Hills, T.L. (1965). "Savannas: a review of a major research problem in tropical geography." *Canadian Geographer*, 9:216–26.

Hilts, S.G., and Cook, F.S. (eds.) (1982). *Significant Natural Areas of Middlesex County.* McIlwraith Field Naturalists Inc., London, Ont.

Hobbie, J.E. (1980). "Introduction and Site Description". Pp. 19–50 in Hobbie, J.E.(ed.), *Limnology in Tundra Ponds: Barrow, Alaska.* Dowden, Hutchinson and Ross Inc., Strasbourg.

Hobbs, R.J. (1984). "Length of burning rotation and community composition in high-level *Calluna-Eriophorum* bog in N. England." *Vegetatio*: 57:129–36.

Hofgaard, A. (1993). "Structure and regeneration patterns in a virgin *Picea abies* forest in northern Sweden." *Journal of Vegetation Science*, 4:601–8.

Hofstetter, R.H. (1983). "Wetlands in the United States." Pp. 201–44 in Gore, A.J.P. (ed.), *Mires: Swamps, Bog, Fen and Moor: Regional Studies.* Ecosystems of the World 4B. Elsevier, New York.

Hogg, E.H. (1993). "Decay potential of hummock and hollow *Sphagnum* peats at different depths in Swedish raised bog." *Oikos*, 66:269–78.

Holdridge, L.R. (1967). *Life Zone Ecology.* Tropical Science Centre, San Jose, Costa Rica.

Holland, W.D., and Coen, G.M. (1982). *Ecological (Biophysical) Land Classification of Banff and Jasper National Parks.* Vol. 2, *Soil and Vegetation Resources.* Pub. no. SS–82–44. Alberta Institute of Pedology, Edmonton.

Holtmeier, F.-K., and Broll, G. (1992). "The influence of tree islands and microtopography on pedoecological conditions in the forest-alpine tundra ecotone on Niwot Ridge, Colorado Front Range, U.S.A." *Arctic and Alpine Research*, 24:216–28.

Hukusima, T., and Kershaw, K.A. (1988). "The ecology of a beech forest on Mt. Sanpoiwadake, Hakusan National Park, Japan. I. Braun-Blanquet, TWINSPAN and DCA analysis." *Ecological Research*, 3:89–100.

Hulbert, L.C. (1989). "Causes of fire effects in tallgrass prairie." *Ecology*, 69:46–58.

Humboldt, A. von, and Bonpland, A. (1805). *Essai sur la géographie des plantes.* Accompagné d'un tableau physique des regions equinoxiales. Paris.

Humphrey, L.D. (1984). "Patterns and mechanisms of plant succession after fire on *Artemisia*-grass sites in southeastern Idaho." *Vegetatio*, 57:91–101.

Hustich, I. (1983). "Tree-line and tree growth studies during 50 years: some subjective observations." Pp. 181–8 in Morisset, P., and Payette, S. (eds.), *Tree-Line Ecology: Proceedings of the Northern Québec Tree-Line Conference.* Centre d'études nordiques, Université Laval, Québec.

– (1966). "On the forest-tundra and the northern tree-lines." *Rep. Kevo Subarctic Res.*, 3: 7–47.

Hutchinson, T.C. (1973). The impact of pollution."Pp. 25–8 in Morton, J.K. (ed.), *Man's Impact on the Canadian Flora.* Canadian Botanical Association, Supplement to CBA/ABC *Bulletin*, 9(1).

– and Scott, M.G. (1988). "The response of feather moss, *Pleurozium schreberi*, to 5 years of simulated acid precipitation in the Canadian boreal forest." *Canadian Journal of Botany*, 66:82–8.

Innes, J.L. (1992). "Structure of evergreen temperate rain forest on the Taitao Peninsula, southern Chile." *Journal of Biogeography*, 19:555–62.

Ivanov, K.E. (1981). *Water Movement in Mirelands*. Translated from the Russian by Thomson, A., and Ingram, H.A.P. Academic Press, New York.

Jackson, L.E., and Bliss, L.C. (1984). "Phenology and water relations of three plant life-forms in a dry treeline meadow." *Ecology*, 65:1302–14.

Jacoby, G.C. (1983). "A dendroclimatic study in the forest-tundra ecotone on the east shore of Hudson Bay." Pp. 95–100 in Morisset, P., and Payette, S. (eds.), *Tree-Line Ecology: Proceedings of the Northern Québec Tree-Line Conference*. Centre d'études nordiques, Université Laval, Québec.

Jahn, G. (1991). "Temperate deciduous forests of Europe." Pp. 377–502 in Röhrig, E., and Ulrich, B. (eds.), *Temperate Deciduous Forests*. Ecosystems of the World 7. Elsevier, Amsterdam.

James, S.W. (1991). "Soil, nitrogen, phosphorus, and organic matter processing by earthworms in tallgrass prairie." *Ecology*, 72:2101–9.

Jean, M., and Bouchard, A. (1993). "Riverine wetland vegetation: importance of small-scale and large-scale environmental variation." *Journal of Vegetation Science*, 4:609–20.

Jeffery, W.W. (1961). "A prairie to forest succession in Wood Buffalo Park, Alberta." *Ecology*, 42:442–4.

Jenkin, J.F. (1975). "Macquarie Island, Subantarctic." Pp. 375–97 in Rosswall, T., and Heal, O.W. (eds.), *Structure and Function of Tundra Ecosystems*. Ecol. Bull. 20. Swedish Natural Science Research Council, Stockholm.

– and Ashton, D.H. (1970). "Productivity studies on Macquarie Island vegetation." Pp. 851–63 Holgate, M.W. (ed.), *Antarctic Ecology*. Academic Press, London.

Jensen, M.E. (1989). "Soil characteristics of mountainous northeastern Nevada sagebrush community types." *The Great Basin Naturalist*, 49:469–81.

Jessee, C.M. (1990). "Water retention and drainage in the forest floor organic layer under upland black spruce." Botany MSc thesis, University of Toronto.

Johnson, E.A. (1992). *Fire and Vegetation Dynamics: Studies from the North American Boreal Forest*. Cambridge University Press, Cambridge.

– and Larsen, P.S. (1991). "Climatically induced change in fire frequency in the southern Canadian Rockies." *Ecology*, 72:194–201.

Johnson, J.E., Smith, D.W., and Burger, J.A. (1985). "Effects on the forest floor of whole-tree harvesting in an Appalachian oak forest." *American Midland Naturalist*, 114:51–61.

Johnson, L.C. (1987). Macrostructure of *Sphagnum* peat as an indication of bog processes." Pp. 61–9 in Rubec, C.D.A., and Overend, R.P. (compilers), *Proceedings Symposium '87: Wetlands/Peatlands*. Edmonton, Alta.

Jonasson, S., and Skold, S.V. (1983). "Influences of frost heaving on vegetation and nutrient regime of polygon-patterned ground." *Vegetatio*, 53:97–112.

Jones, D.C., Wotton, D.L., McEachern, D.B., and Phillips, S.F. (1985). *The Growth of Jack Pine and Black Spruce Seedlings in Heavy Metal Contaminated Soils Collected Near the*

Inco Smelter at Thompson, Manitoba. Terrestrial Standards and Studies, Report #84–4. Manitoba Environment and Workplace Safety and Health, Winnipeg.

Jones, M.B. (1988). "Photosynthetic responses of c_3 and c_4 wetland species in a tropical swamp." *Journal of Ecology,* 76:253–62.

Junk, W.J. (1993). "Wetlands of tropical South America." Pp. 679–739 in Whigham, D., Dykyjová, D., and Hejuý, S. (eds.), *Wetlands of the World: Inventory, Ecology and Management.* Kluwer, London.

– (1983). "Ecology of swamps on the Middle Amazon." Pp. 269–94 in Gore, A.J.P. (ed.), *Mires: Swamps, Bog, Fen and Moor: Regional Studies.* Ecosystems of the World 4B. Elsevier, New York.

Kading, M.M. (1989). "An investigation of the nucleation of pines around isolated red oaks during sand dune succession at Wasaga Beach, Ontario." Geography MSC thesis, York University, Toronto.

Kahn, F., and de Granville, J.-J. (1992). *Palms and Forest Ecosystems of Amazonia.* Springer-Verlag, Berlin.

Kallio, P. (1975). "Kevo, Finland." Pp. 193–233 in Rosswall, T., and Heal, O.W. (eds.), *Structure and Function of Tundra Ecosystems.* Ecol. Bull. 20. Swedish Natural Science Research Council, Stockholm.

Kantrud, H.A., Millar, J.B., and van der Valk, A.G. (1989). "Vegetation in wetlands of the prairie pothole region." Pp. 132–87 in van der Valk, A.G. (ed.), *Northern Prairie Wetlands.* Iowa State University Press, Ames.

Karagatzides, J.D., and Hutchinson, I. (1991). "Intraspecific comparisons of biomass dynamics in *Scirpus americanus* and *Scirpus maritimus* on the Fraser River Delta." *Journal of Ecology,* 79:459–76.

– Lewis, M.C., and Schulman, H.M. (1985). "Nitrogen fixation in the High Arctic tundra at Sarcpa Lake, Northwest territories." *Canadian Journal of Botany,* 63:974–9.

Kauppi, P.E., Mielikainen, K., and Kuusela, K. (1992). "Biomass and carbon budget of European forests, 1971 to 1990." *Science,* 256: 70–4.

– and Posch, M. (1985). "Sensitivity of boreal forests to possible climatic warming." *Climatic Change,* 7:45–54.

Kavanagh, T.A. (1987). "Species composition, productivity and nutrition of three feather mosses under upland *Picea mariana.*" Botany MSC thesis, University of Toronto.

Kearney, M S., and Luckman, B H. (1983). "Holocene timberline fluctuations in Jasper National Park." *Science,* 221:261–3.

Keller, B.A. (1927). "The distribution of vegetation on the plains of European Russia." *Journal of Ecology,* 15:209–18, 227–9.

Kemp, D. (1991). "The greenhouse effect and global warming: a Canadian perspective." *Geography,* 76:121–30.

Kenkel, N.C. (1987). "Trends and interrelationships in boreal wetland vegetation." *Canadian Journal of Botany,* 65:12–22.

Kennedy, A.D. (1993). "Water as a limiting factor in the Antarctic terrestrial environment: a biogeographical analysis." *Arctic and Alpine Research,* 25:308–15.

Kershaw, G.P., and Kershaw, L.J. (1987). "Successful plant colonizers on disturbances in tundra areas of northwestern Canada." *Arctic and Alpine Research,* 19:451–60.

Kershaw, K.A. (1977). "Studies on lichen dominated systems: an examination of some aspects of the northern boreal lichen woodlands in Canada." *Canadian Journal of Botany*, 55: 393–410.

Keyser, J.M. (1988). "The AT trailblazing for tomorrow." *American Forests*, 94(9 & 10): 26–9.

Khanna, P.K., and Ulrich, B. (1991). "Ecochemistry of temperate deciduous forest." Pp. 121–63 in Röhrig, E., and Ulrich, B. (eds.), *Temperate Deciduous Forests*. Ecosystems of the World 7. Elsevier, Amsterdam.

Kilgore, B.M., and Taylor, D. (1979). "Fire history of a sequoia-mixed conifer forest." *Ecology*, 60:129–42.

– and Biswell, H.H. (1971). "Seedling germination following fire in a giant sequoia forest." *California Agriculture*, 25:8–10.

Kimerling, A.J., and Jackson, P.L. (1985). *Atlas of the Pacific Northwest*. 7th ed. Oregon State University Press.

King, R.H. (1991). "Paleolimnology of a polar oasis, Truelove Lowland, Devon Island, N.W.T., Canada." *Hydrobiologia*, 214:317–25.

Klokk, T., and Ronning, O.I. (1987). "Revegetation experiments at Ny-Alesund, Spitsbergen, Svarlbad." *Arctic and Alpine Research*, 19:549–53.

Knapp, A.K. (1985). "Effect of fire and drought on the ecophysiology of *Andropogon gerardi* and *Panicum virgatum* in a tallgrass prairie." *Ecology*, 66:1309–20.

Knystautas, A. (1987). *The Natural History of the USSR*. McGraw-Hill, New York.

Kohyama, T. (1984). "Regeneration and coexistence of two *Abies* species dominating subalpine forests in central Japan." *Oecologia*, 62:1556–61.

Kolchugina, T.P., and Vinson, T.S. (1993). "Carbon sources and sinks in forest biomes of the former Soviet Union." *Global Biogeochemical Cycles*, 7:291–304.

Koniac, S. (1985). "Succession in pinyon-juniper woodlands following wildfire in the Great Basin." *Great Basin Naturalist*, 45:556–66.

Kononova, M.M. (1966). *Soil Organic Matter*. 2nd ed. Pergamon Press, Oxford.

Köppen, W. (1918). "Klassifikation der Klimate nach Temperatur, Niederschlag und Jahreslauf." *Petermanns Geogr. Mitt.*, 64:193–203, 243–8.

Krajina, V.J. (1969). "Ecology of forest trees in British Columbia". Pp. 1–146 in Krajina, V.J. (ed.), *Ecology of Western North America,* vol. 2. Department of Botany, University of British Columbia, Vancouver.

– (1965). "Biogeoclimatic zones and classification of British Columbia." Pp. 1–17 in Krajina, V.J. (ed.), *Ecology of Western North America,* vol. 1. Department of Botany, University of British Columbia, Vancouver.

Krantz, W.B. (1990). "Self-organization manifest as patterned ground in recurrently frozen soils." *Earth-Science Reviews*, 29:117–30.

Kucera, C.L. (1992). "Tall-grass prairie." Pp. 227–68 in Coupland, R.T. (ed.), *Natural Grasslands: Introduction and Western Hemisphere*. Ecosystems of the World 8A. Elsevier, New York.

Küchler, A.W. (1964). *Potential Natural Vegetation of the Coterminous United States*. American Geographical Society Special Publication, #36. Washington, DC.

Kuhry, P., Nicholson, B.J., Gignac, L.D., Vitt, D.H., and Bayley, S.E. (1993). "Development of *Sphagnum*-dominated peatlands in boreal continental Canada." *Canadian Journal of Botany*, 71:10–22.

Kullman, L. (1989). "Recent regression of the forest-alpine tundra ecotone (*Betula pubescens* Ehrh. ssp. *tortuosa* (Ledeb) Nyman) in the Scandes Mountains, Sweden." *Journal of Biogeography*, 16:83–90.

Kumler, M.L. (1969). "Plant succession on the sand dunes of the Oregon coast." *Ecology*, 50:695–704.

Kurz, W.A., Apps, M.J., Webb, T.M., and McNamee, P.J. (1991). "The contribution of biomass burning to the carbon budget of the Canadian forest sector: a conceptual model." Pp. 339–44 in Levine, J.S. (ed.), *Global Biomass Burning: Atmospheric, Climatic, and Biospheric Implications*. MIT Press, Cambridge, Mass.

LaFarge-England, C., Vitt, D.H., and England, J. (1991). "Holocene soligenous fens on a high Arctic fault block, northern Ellesmere Island (82°N), N.W.T., Canada." *Arctic and Alpine Research*, 23:80–90.

Lafleur, P.M., Renzetti, A.V., and Bello, R. (1993). "Seasonal changes in the radiation balance of subarctic forest and tundra." *Arctic and Alpine Research*, 25:32–6.

Lamb, R. (1979). *World without Trees*. Wildwood House, London.

Lamontagne, L., Camire, C., and Ansseau, C. (1991). "La végétation forestière du delta de Lanoraie, Québec." *Canadian Journal of Botany*, 69:1839–52.

Lands Directorate (1967). *Canada Land Inventory: Land Capability Maps*. Environment Canada, Ottawa.

Lanner, R.M. (1981). *The Piñon Pine: A Natural and Cultural History*. University of Nevada Press, Reno.

Laprise, D., and Payette, S. (1988). "Evolution récente d'une tourbiere à palses (Québec subarctique): analyse cartographique et dendrochronologique." *Canadian Journal of Botany*, 66:2217–27.

La Roi, G.H. (1992). "Classification and ordination of southern boreal forests from the Hondo–Slave Lake area of central Alberta." *Canadian Journal of Botany*, 70:614–28.

Larsen, J.A. (1989). *The Northern Forest Boundary in Canada and Alaska: Biotic Communities and Ecological Relationships*. Springer-Verlag, New York.

– (1980). *The Boreal Ecosystem*. Academic Press, New York.

Larson, A.M. (1972). *History of the White Pine Industry in Minnesota*. Arno Press, New York.

Larson, D.W., and Kelly, P.E. (1991). "The extent of old-growth *Thuja occidentalis* on cliffs of the Niagara Escarpment." *Canadian Journal of Botany*, 69:1628–36.

– Spring, S.H., Matthes-Sears, U., and Bartlett, R.M. (1989). "Organization of the Niagara Escarpment cliff community." *Canadian Journal of Botany*, 67:2731–42.

Lauenroth, W.K. (1979). "Grassland primary production: North American grasslands in perspective." Pp. 3–24 in French, N. R. (ed.), *Perspectives in Grassland Ecology: Results and Applications of the US/IBP Grassland Biome Study*. Springer-Verlag, New York.

– and Milchunas, D.G. (1992). "Short-grass steppe." Pp. 183–226 in Coupland, R.T. (ed.), *Natural Grasslands: Introduction and Western Hemisphere*. Ecosystems of the World 8A. Elsevier, New York.

Lavoie, C., and Payette, S. (1992). "Black spruce growth forms as a record of a changing winter environment at treeline, Québec, Canada." *Arctic and Alpine Research*, 24:40–9.

Lavoie, G. (1992). "Vascular plants likely to be listed as vulnerable or threatened in Québec." *Canadian Biodiversity*, 2(1):6–13.

Laws, R.M. (ed.) (1984). *Antarctic Ecology*. Vol. 1. Academic Press, New York.

Leader-Williams, N., and Ricketts, C. (1981). "Seasonal and sexual patterns of growth and condition of reindeer introduced into South Georgia." *Oikos*, 38:27–39.

Leak, W.B. (1991). "Secondary forest succession in New Hampshire." *Forest Ecology and Management*, 43:69–86.

Lee, T.D., and La Roi, G.H. (1979). "Bryophyte and understory vascular plant beta diversity in relation to moisture and elevation gradients." *Vegetatio*, 40:29–38.

Lemieux, G.J. (1963). "Soil-vegetation relationships in the Northern Hardwoods of Québec." Chapter 13 in Youngberg, C.T. (ed.), *Forest-Soil Relationships in North America*. Oregon State University Press, Corvallis.

Lemmen, D.S. (1989). "The last glaciation of Marvin Peninsula northern Ellesmere Island, high Arctic Canada." *Canadian Journal of Earth Sciences*, 26:2578–90.

Lenihan, J.M. (1993). "Ecological response surfaces for North American boreal tree species and their use in forest classification." *Journal of Vegetation Science*, 4:667–80.

Lertzman, K.P. (1992). "Patterns of gap-phase replacement in a subalpine, old-growth forest." *Ecology*, 73:657–69.

Lewis Smith, R.I. (1988). "Destruction of Antarctic terrestrial ecosystems by a rapidly increasing fur seal population." *Biological Conservation*, 45:55–72.

– and Walton, D.W.H. (1975). "South Georgia, Subantarctic." Pp. 399–423 in Rosswall, T., and Heal, O.W. (eds.), *Structure and Function of Tundra Ecosystems*. Ecol. Bull. 20. Swedish Natural Science Research Council, Stockholm.

Lieth, H. (1975). "Primary production of the major vegetation units of the world." Pp. 203–15 in Lieth, H., and Whittaker, R.H. (eds.), *Primary Productivity of the Biosphere*. Springer-Verlag, New York.

Likens, G.E., Bormann, F.H., Pierce, R.S., and Reiners, W.A. (1978). "Recovery of a deforested ecosystem." *Science*, 199:492–6.

– Bormann, F.H., Pierce, R.S., Eaton, J.S., and Johnson, N.M. (1977). *Biogeochemistry of a Forested Ecosystem*. Springer-Verlag, New York.

Lind, E.M., and Morrison, M.E.S. (1974). *East African Vegetation*. Longman, London.

Linder, S. (1987). "Responses to water and nutrients in coniferous forest." Pp. 180–201 in Schulze, E.-D., and Zwolfer, H. (eds.), *Potentials and Limitations in Ecosystem Analysis*. Ecological Studies, vol. 61. Springer-Verlag, Berlin.

Lisetskii, F.N. (1987). "Evaluation of changes in humus formation conditions in Holocene steppe ecosystems of Prichernomor'e." *Soviet Journal of Ecology*, 18:134–9.

Little, S. (1979). "The Pine Barrens of New Jersey." Pp. 451–64 in Specht, R.L. (ed.), *Heathlands and Related Shrublands: Descriptive Studies*. Ecosystems of the World 9A. Elsevier, Amsterdam.

Liu, C., and Wa, K. (1985). "The effects of forest on water and soil conservation in the Loess Plateau of China." Pp. 25–38 in Pannell, C.W., and Salter, C.L. (eds.), *China Geographer*, no. 12. Westview Press, Boulder.

Liu, K.-B. (1988). "Quaternary history of the temperate forests of China." *Quaternary Science Review*, 7:1–20.

Liu, X., and Zhao, K. (1987). "Site types of mire-wetlands and their utilization and conservation in China." Pp. 451–8 in Rubec, C.D.A., and Overend, R.P. (compilers), *Proceedings Symposium '87: Wetlands/Peatlands*. Edmonton, Alta.

Loidi, J., and Herrera, M. (1990). "The *Quercus pubescens* and *Quercus faginea* forests in the Basque Country (Spain): distribution and typology in relation to climatic factors." *Vegetatio*, 90:81–92.

Loneragan, W.A., and del Moral, R. (1984). "The influence of microlief on community structure of subalpine meadows." *Bulletin of the Torrey Botanical Club*, 111:209–16.

Longton, R.E. (1988). "Adaptations and strategies of polar bryophytes." *Botanical Journal of the Linnean Society*, 98:253–68.

– (1985). "Reproductive biology and susceptibility to air pollution in *Pleurozium schreberi* (Brid.)Mitt. (Musci) with particular reference to Manitoba, Canada." *Monogr. Syst. Bot. Missouri Bot. Gard.*, 11:51–69.

Looman, J. (1987a). "The vegetation of the Canadian Prairie provinces IV. The woody vegetation, part 4." *Phytocoenologia*, 15:289–327.

– (1987b). "The vegetation of the Canadian Prairie provinces IV. The woody vegetation, part 3." *Phytocoenologia*, 15:51–84.

– (1984). "The biological flora of Canada. 4. *Shepherdia argentea* (Pursh) Nutt., Buffaloberry." *Canadian Field Naturalist*, 98:231–44.

– (1983a). "Distribution of plant species and vegetation types in relation to climate." *Vegetatio*, 54:17–25.

– (1983b). "Grassland as natural or semi-natural vegetation." Pp. 173–84 in Holzner, W., Werger, M.J.A., and Ikusima, I. (eds.), *Man's Impact on Vegetation*. W. Junk, The Hague.

– (1983c). "The vegetation of the Canadian Prairie provinces IV. The woody vegetation, part I." *Phytocoenologia*, 11:297–330.

– (1982). "Grasslands of western North America: Fescue grasslands." Pp. 209–22 in Nicholson, A.C., McLean, A., and Baker, T.E. (eds.), *Grassland Ecology and Classification Symposium Proceedings*. Ministry of Forests, Province of British Columbia.

– (1981). "The vegetation of the Canadian Prairie provinces II. The grasslands, part 2. Mesic grasslands and meadows." *Phytocoenologia*, 9:1–26.

– (1980). "The vegetation of the Canadian Prairie provinces II. The grasslands, part I." *Phytocoenologia*, 8:153–90.

– (1979). "The vegetation of the Canadian prairies I. An overview." *Phytocoenologia*, 5: 347–66.

Love, D. (1970). "Subarctic and subalpine: where and what?" *Arctic and Alpine Research*, 2:63–73.

– (1959). "The postglacial development of the flora of Manitoba: a discussion." *Canadian Journal of Botany*, 37:547–85.

Love, M. (1991). "The Red River tall grass prairie revival." *Borealis*, 3(1):10–12.

Luckman, B.H., Jozsa, L.A., and Murphy, P.J. (1984). "Living seven-hundred-year old *Picea engelmannii* and *Pinus albicaulis* in the Canadian Rockies." *Arctic and Alpine Research*, 16:419–22.

Lugo, A.E., Brinson, M., and Brown, S. (eds.) (1990a). *Forested Wetlands*. Ecosystems of the World 15. Elsevier, Oxford.

– Brown, S., and Brinson, M. (1990b). "Concepts in wetland ecology." Pp. 53–85 in Lugo, A.E., Brinson, M., and Brown, S. (eds.), *Forested Wetlands*. Ecosystems of the World 15. Elsevier, Oxford.

McAllister, D.E. (1991). "Estimating the pharmaceutical values of forests, Canadian and tropical." *Canadian Biodiversity*, 1(3):16–25.

– and Dalton, K.W. (1992). "How global warming affects species survival." *Canadian Biodiversity*, 2(2):7–14.

McCarthy, B.C., Hammer, C.A., Kauffman, G.L., and Cantina, P.D. (1987). "Vegetation patterns and structure of an old-growth forest is southeastern Ohio." *Bulletin of the Torrey Botanical Club*, 114:33–45.

McCaw, P.E. (1985). "The status of black gum *(Nyssa sylvatica)* in Bachus Woods, southern Ontario." Botany MSC thesis, University of Toronto.

McComb, A.L., and Loomis, W.E. (1944). "Subclimax prairie." *Bulletin of the Torrey Botanical Club*, 71:46–76.

MacDonald, G.M., Edwards, T.W.D., Moser, K.A., Pienitz, R., and Smol, J.P. (1993). "Rapid response of treeline vegetation and lakes to past climatic warming." *Nature*, 361:243–6.

MacDonald, P.O., Frayer, W.E., and Clauser, J.K. (1979). *Documentation, Chronology, and Future Projections of Bottomwood Hardwood Habitat Loss in the Lower Mississippi Alluvial Plain*. Vol. 1. Basic Report. US Fish and Wildlife Service, Ecological Services, Vicksburg, Miss.

McInnis, P.F., Naiman, R.J., Pastor, J., and Cohen, Y. (1992). "Effects of moose browsing on vegetation and litter of the boreal forest, Isle Royal, Michigan, USA." *Ecology*, 73:2059–75.

Mackay, J.R. (1970) "Disturbances to the tundra and forest tundra environment of the western Arctic." *Can. Geotech. J.*, 7:420–32.

Mackenzie Lamb, I. (1970). "Antarctic terrestrial plants and their ecology", Pp. 733–51 in Holgate, M.W. (ed.), *Antarctic Environments*. Academic Press, London.

MacLellan, P. (1991). "Dynamics of post-logged plant communities in the boreal forest of northern Ontario, Canada." Botany PHD thesis, University of Toronto.

McNabb, D.H., Gaweda, F., and Froehlich, H.A. (1989). "Infiltration, water repellency, and soil moisture content after broadcast burning a forest site in southwest Oregon." *Journal of Soil and Water Conservation*, 44(1):87–90.

Macy, H. (1991). "Ecostroika." *Forest Planning Canada*, 7(2):44.

Mahendrappa, M.K., Foster, N.W., Weetman, G.F., and Krause, H.H. (1986). "Nutrient cycling and availability in forest soils." *Canadian Journal of Forestry*, 66:547–72.

Makarevskii, M.F. (1992). "Stocks and balance of organic carbon in forest and marsh biogeocenoses in Karelia." *Soviet Journal of Ecology*, 22:133–9.

Malecki, R., Blossey, B., Hight, S., Schroeder, D., Kok, L., and Coulson, J. (1993). "Biological controls of purple loosestrife." *BioScience*, 43:680–9.

Malhotra, S.S., and Khan, A.A. (1983). "Sensitivity to SO_2 of various processes in an epiphytic lichen, *Evernia mesomorpha*." *Biochem. Physiol. Pflanzen.*, 178:121–30.

Malmer, N. (1988). "Patterns in the growth and the accumulation or inorganic constituents in the *Spahagnum* cover on ombrotrophic bogs in Scandinavia." *Oikos*, 53:105–20.

Marcuzzi, G. (1979). *European Ecosystems*. W. Junk, The Hague.

Mark, A.F., and Adams, N.M. (1973). *New Zealand: Alpine Plants*. A.H. and A.W. Reed, Wellington.

Martikainen, P., Nykänen, H., Crill, P., and Silvola, J. (1993). "Effect of a lowered water table on nitrous oxide fluxes from northern peatlands." *Nature*, 366:51–3.

Martinsen, G.D., Cushman, J.H., and Whitham, T.G. (1990). "Impact of pocket gopher disturbance on plant species diversity in a shortgrass prairie community." *Oecologia*, 83: 132–8.

Mathiasen, R.L., Blake, E.A., and Edminster, C.B. (1987). "Estimates of site potential for ponderosa pine based on site index for several southwestern habitat types." *Great Basin Naturalist*, 47:467–72.

Matthes-Sears, U., Neeser, C., and Larson, D.W. (1992). "Mycorrhizal colonization and macronutrient status of cliff-edge *Thuja occidentalis* and *Acer saccharum*." *Ecography*, 15: 262–6.

Matveyeva, N.V., Parinkina, O.M., and Chernov, Y.I. (1975). "Maria Pronchitsheva Bay, USSR." Pp. 61–72 in Rosswall, T., and Heal, O.W. (eds.), *Structure and Function of Tundra Ecosystems*. Ecol. Bull. 20. Swedish Natural Science Research Council, Stockholm.

Maycock, P.F., and Fahselt, D. (1992). "Vegetation on stressed screes and slopes in Sveerdrup Pass, Ellesmere Island, Canada." *Canadian Journal of Botany*, 70:2357–77.

Mazurski, K.R. (1990). "Industrial pollution: the threat to Polish forests." *Ambio*, 19(2):70–4.

Mcdnis, R.J. (1981). "Indigenous plants and animals of Newfoundland: their geographic affinities and distribution." Pp. 218–49 in Macpherson, A.G., and Macpherson J.B. (eds.), *The Natural Environment of Newfoundland, Past and Present*. Department of Geography, Memorial University of Newfoundland, St John's.

Messier, C., and Kimmins, J.P. (1991). "Above- and below-ground vegetation recovery in recently clearcut and burned sites dominated by *Gaultheria shallon* in coastal British Columbia." *Forest Ecology and Management*, 46:275–94.

Mikola, P. (1982). "Application of vegetation science to forestry in Finland." Pp. 199–224 in Jahn, G. (ed.), *Application of Vegetation Science to Forestry*. W. Junk, The Hague.

Miller, N.G., and Alpert, P. (1984). "Plant associations and edaphic features of a high Arctic mesotopographic setting." *Arctic and Alpine Research*, 16:11–24.

Mills, G.F., Hopkins, L.A., and Smith, R.E. (1977). *Organic Soils of the Roseau River Watershed in Manitoba*. Monograph no. 17. Canada Department of Agriculture, Winnipeg, Man.

Milne, B.T. (1985). "Upland vegetational gradients and post-fire succession in the Albany Pine Bush, New York." *Bulletin of the Torrey Botanical Club*, 112:21–34.

Ministry of Natural Resources (1990). *A Biological Inventory and Evaluation of the Sudden Bog Area of Natural and Scientific Interest*. Ontario Ministry of Natural Resources, Toronto.

Mitsch, J.M. (1988). "Productivity-hydrology-nutrient models of forested wetlands." Pp. 115–32 in Mitsch, J.M., Straskraba, M., and Jorgansen, S.E. (eds.), *Wetland Modelling*. Elsevier, Amsterdam.

Moise, B. (1989). "Temagami wilderness under siege." *Canadian Geographic*, 109:28–39.

Moore, J.J., Dowding, P., and Healy, B. (1975). "Glenamoy, Ireland." Pp. 159–81 in Rosswall, T., and Heal, O.W. (eds.), *Structure and Function of Tundra Ecosystems.* Ecol. Bull. 20. Swedish Natural Science Research Council, Stockholm.

Moore, T.R. (1981). "Controls on the decomposition of organic matter in subarctic spruce-lichen woodland soils." *Soil Science*, 131:107–13.

– (1978). "Soil formation in northeastern Canada." *Annals of the Association of American Geographers*, 68:518–34.

Morisset, P., Payette, S., and Deshaye, J. (1983). "The vascular flora of the Northern Québec-Labrador Peninsula: Phytogeographical structure with respect to the tree line." Pp. 141–151 in Morisset, P., and Payette, S. (eds.), *Tree-Line Ecology: Proceedings of the Northern Québec Tree-Line Conference.* Centre d'études nordiques, Université Laval, Québec.

Morneau, C., and Payette, S. (1989). "Postfire lichen-spruce woodland recovery at the limit of the boreal forest in northern Québec." *Canadian Journal of Botany*, 67:2770–82.

Mörs, I. von, and Bégin, Y. (1993). "Shoreline scrub population extension in response to recent isostatic rebound in eastern Hudson Bay, Québec, Canada." *Arctic and Alpine Research*, 25:15–23.

Moser, K.A., and MacDonald, G.M. (1990). "Holocene vegetation change at treeline north of Yellowknife, Northwest Territories, Canada." *Quaternary Research,* 34:227–39.

Moss, E.H. (1932). "The vegetation of Alberta: IV. The poplar association and related vegetation of central Alberta." *Journal of Ecology*, 20:401–7.

Muc, M., and Bliss, L.C. (1977). "Plant communities of Truelove Lowland." Pp. 143–54 in Bliss, L.C. (ed.), *Truelove Lowland, Devon Island, Canada: A High Arctic Ecosystem.* University of Alberta Press, Edmonton.

Mueller-Dombois, D. (1965). "Initial stages of succession in the Coastal Douglas-Fir and Western Hemlock Zones." Pp. 38–41 in Krajina, V.J. (ed.), *Ecology of Western North America,* vol. 1. Department of Botany, University of British Columbia, Vancouver.

– (1964). "The forest habitat types in southeastern Manitoba and their application to forest management." *Canadian Journal of Botany*, 42:1417–44.

– and Ellenberg, H. (1974). *Aims and Methods of Vegetation Ecology.* John Wiley, New York.

Muir, P.S., and Lotan, J.E. (1985). "Disturbance history and serotiny of *Pinus contorta* in western Montana." *Ecology*, 66:1658–68.

Murphy, P.J., Rousseau, A., and Stewart, D. (1993). "Sustainable forests: a Canadian commitment. National Forest Strategy and Canada Forest Accord process and results." *Forestry Chronicle*, 69:278–84.

Mutch, R.W. (1970). "Wildland fires and ecosystems – a hypothesis." *Ecology*, 51:1046–52.

Myers, R.L. (1990). "Palm swamps." Pp. 267–86 in Lugo, A.E., Brinson, M., and Brown, S. (eds.), *Forested Wetlands.* Ecosystems of the World 15. Elsevier, Oxford.

Nadelhoffer, K.J, Giblin, A.E., Shaver, G.R., and Laundre, J.A. (1991). "Effects of temperature and substrate quality on element mineralization in six Arctic soils." *Ecology*, 72: 242–53.

Naiman, R.J., Melillo, J.M., and Hobbie, J.E. (1986). "Ecosystem alteration of boreal forest streams by beaver (*Castor canadensis*)." *Ecology*, 67:1254–69.

Nakashizuka, T. (1991). "Population dynamics of coniferous and broad-leaved trees in a Japanese temperate mixed forest." *Journal of Vegetation Science*, 2:413–18.

National Vegetation Working Group (1990). *The Canadian Vegetation Classification System*. National Working Group of the Canadian Committee on Ecological Land Classification. Edited by Strong, W.L., Oswald, E.T., and Downing, D.J. Ecological Land Classification Series, no. 25. Sustainable Development, Corporate Policy Group, Canadian Wildlife Service, Environment Canada, Ottawa.

National Wetlands Working Group (1988). *Wetlands of Canada*. Ecological Land Classification Series, no. 24. Sustainable Development Branch, Canadian Wildlife Service, Environment Canada, Ottawa, and Polyscience Publications Inc., Montreal.

– (1987). *The Canadian Wetland Classification System*. Ecological Land Classification Series, no. 21. Provisional ed. Lands Conservation Branch, Canadian Wildlife Service, Environment Canada, Ottawa.

– (1986). *Canada's Wetlands*. Maps of wetland regions and distribution in Canada, prepared by the National Wetlands Working Group of the Canada Committee on Ecological Land Classification and forming part of the 5th ed. of the National Atlas of Canada. Canada Map Office, Energy, Mines and Resources Canada, Ottawa.

Nelson, J.G. (1989). "Wilderness in Canada: past, present, future." *Natural Resources Journal*, 29:83–102.

– and England, R.E. (1971). "Some comments on the causes and effects of fire in the northern grassland area of Canada and the nearby United States, 1750–1900." *Canadian Geographer*, 15:295–306.

Nelson, T.C. (1979). "Fire management policy in the national forests – a new era." *Journal of Forestry*, 77:723–5.

Nettleship, D.N., and Smith, P.A. (eds.) (1975). *Ecological Sites in Northern Canada*. Canadian Committee for the International Biological Programme Conservation Terrestrial – Panel 9. Canadian Wildlife Service, Ottawa.

Nichols, G.E. (1935). "The hemlock–white pine northern hardwood region of eastern North America." *Ecology*, 16:403–22.

Nichols, H. (1967). "Pollen diagrams from sub-arctic central Canada." *Science*, 155:1665–8.

Nicholson, A.C., McClean, A., and Baker, T.E. (eds.) (1982). *Grassland Ecology and Classification Symposium Proceedings*. Ministry of Forests, Province of British Columbia.

Nieppola, J. (1992). "Long-term vegetation changes in stands of *Pinus sylvestris* in southern Finland." *Journal of Vegetation Science*, 3:475–84.

Nihlgard, B., and Lindgren, L. (1977). "Plant biomass, primary production and bioelements of three mature beech forests in South Sweden." *Oikos*, 28:95–104.

Nikolov, N., and Helmisaari, H. (1992). "Silvics of the circumpolar boreal forest tree species." Pp. 13–84 in Shugart, H.H., Leemans, R., and Bonan, G.B. (eds.), *A Systems Analysis of the Global Boreal Forest*. Cambridge University Press, New York.

Norin, B.N., and Ignatenko, I.V. (1975). "Ary-Mas, USSR." Pp. 183–91 in Rosswall, T. and Heal, O.W. (eds.), *Structure and Function of Tundra Ecosystems*. Ecol. Bull. 20. Swedish Natural Science Research Council, Stockholm.

North, M.E.A. (1976). *A Plant Geography of Alberta*. Studies in Geography, Monograph 2. University of Alberta, Edmonton.

Norton, D.A., and Schonenberger, W. (1984). "The growth forms and ecology of *Nothofagus solandri* at alpine timberline, Craigieburn Range, New Zealand." *Arctic and Alpine Research*, 16:361–70.

Norton, P. (1988). "Seeing the forest. Documenting forest decline: a national survey." *Harrowsmith*, 13:48–53.

Noy-Meir, I. (1973). "Desert ecosystems: environment and producers. *Annual Review of Ecology and Systematics*, 4:23–51.

Numata, M. (ed.) (1979). *Ecology of Grasslands and Bamboolands in the World*. Gustav Fischer Verlag, Jena.

O'Connell, M. (1990). "Origins of Irish lowland blanket bog." Pp. 49–71 in Doyle, G. L. (ed.), *Ecology and Conservation of Irish Peatlands*. Royal Irish Academy, Dublin.

Oechel, W.C., Hastings, S.J., Vourlitis, G., Jenkins, M., Riechers, G., and Grulke, N. (1993). "Recent change in Arctic tundra ecosystems from a net carbon dioxide sink to a source." *Nature*, 361:520–3.

Ohsawa, M. (1984). "Differentiation of vegetation zones and species strategies in the subalpine region of Mt. Fuji." *Vegetatio*, 57:15–52.

Olson, C.G., and Hupp, C.R. (1986). "Coincidence and spatial variability of geology, soils, and vegetation, Mill Run Watershed, Virginia." *Earth Surface Processes and Landforms*, 11:619–29.

Olson, J.S. (1975). "Productivity of forest ecosystems." Pp. 33–43 in *Proceedings of the Special Committee for the International Biological Program, Seattle, Washington*. National Acadamy of Sciences, Washington, DC.

Omernik, J.M. (1987). "Ecoregions of the conterminous United States." *Annals of the Association of American Geographers*, 77:118–25.

Onega, T.L., and Eickmeier, W.G. (1991). "Woody detritus inputs and decomposition kinetics in a southern temperate deciduous forest." *Bulletin of the Torrey Botanical Club*, 118: 52–7.

O'Neill, R.V., and DeAngelis, D.L. (1980). "Comparative productivity and biomass relations of crest ecosystems." Pp. 411–49 in Reichle, D.E. (ed.), *Dynamic Properties of Forest Ecosystems*. Cambridge University Press, New York.

Orloci, L. (1965). "The coastal western hemlock zone of the south-western British Columbia mainland." Pp. 18–34 in Krajina, V.J. (ed.), *Ecology of Western North America*, vol. 1. Department of Botany, University of British Columbia, Vancouver.

Ornduff, R. (1974). *An Introduction to California Plant Life*. University of California Press, Los Angeles.

Osborne, P.L. (1993). "Wetlands of Papua New Guinea." Pp. 305–44 in Whigham, D., Dykyjová, D., and Hejuý, S. (eds.), *Wetlands of the World: Inventory, Ecology and Management*. Kluwer, London.

Osvald, H. (1970). *Vegetation and Stratigraphy of Peatlands in North America*. Uppsala University, Sweden.

Ovenden, L.E. (1981). "Vegetation history of a polygonal peatland, Old Crow Flats, northern Yukon." Botany MSC thesis, University of Toronto.

Ovington, J.D. (1963). "Plant biomass and productivity of prairie, savanna, oakwood, and maize field ecosystems in central Minnesota." *Ecology*, 44:52–63.

Packer, J.G. (1969). "Polyploidy in the Canadian Arctic Archipelago." *Arctic and Alpine Research*, 1:15–28.

Paillet, F.L. (1984). "Growth-form and ecology of American chestnut sprout clones in northeastern Massachusetts." *Bulletin of the Torrey Botanical Club*, 111:316–28.

Paijmans, K. (1990). "Wooded swamps in New Guinea." Pp. 335–55 in Lugo, A.E., Brinson, M., and Brown, S. (eds.), *Forested Wetlands*. Ecosystems of the World 15. Elsevier, Oxford.

Parker, A.J. (1986). "Persistence of lodgepole pine forests in the central Sierra Nevada." *Ecology*, 67:1560–7.

Parker, G.G., Hill, S.M., and Kuehnel, L.A. (1993). "Decline of understory American chestnut (*Castanea dentata*) in a southern Appalachian forest." *Journal of Vegetation Science*, 23: 259–65.

Parker G.R., and Leopold, D.J. (1983). "Replacement of *Ulmus americana* L. in a mature east-central Indiana woods." *Bulletin of the Torrey Botanical Club*, 110:182–8.

Pastor, J., and Mladenoff, D.J. (1992). "The southern boreal-northern hardwood border." Pp. 216–40 in Shugart, H.H., Leemans, R., and Bonan, G.B. (eds.), *A Systems Analysis of the Global Boreal Forest*. Cambridge University Press, New York.

– Dewey, B., Naiman, R.J., McInnes, P.F., and Cohen, Y. (1993). "Moose browsing and soil fertility in the boreal forest of Isle Royale National Park." *Ecology*, 74:467–80.

Pavlick, L.E., and Looman, J. (1984). "Taxonomy and nomenclature of rough fescues, *Festuca altaica*, *F. campestris* (*F. scabrella* var. *major*), and *F. hallii*, in Canada and adjacent part of United States." *Canadian Journal of Botany*, 62:1739–49.

Payette, S. (1992). "Fire as a controlling process in the North American boreal forest." Pp. 144–69 in Shugart, H.H., Leemans, R., and Bonan, G.B. (eds.), *A Systems Analysis of the Global Boreal Forest*. Cambridge University Press, New York.

– (1988). "Late-Holocene development of subarctic ombotrophic peatlands: Allogenic and autogenic succession." *Ecology*, 69:516–31.

– (1983). "The forest tundra and present tree-lines of the Northern Québec-Labrador Peninsula." Pp. 3–23 in Morisset, P., and Payette, S. (eds.), *Tree-Line Ecology: Proceedings of the Northern Québec Tree-Line Conference*. Centre d'études nordiques, Université Laval, Québec.

– and Filion, L. (1993). "Origin and significance of subarctic patchy Podzolic soils and paleosols." *Arctic and Alpine Research*, 25:267–76.

– and Gagnon, R. (1979). "Tree-line dynamics in Ungava Peninsula, northern Québec." *Holarctic Ecology*, 2:239–48.

– and Morneau, C. (1993). "Holocene relict woodlands at the eastern Canadian treeline." *Quaternary Research*, 39:84–9.

Pearce, C.M. (1991). "Mapping muskox habitat in the Canadian High Arctic with SPOT satellite data." *Arctic*, 44, supp.1:49–57.

– (1987). "Vegetation dynamics in northern wetlands – examples from the Mackenzie Delta." Abstract in *Session Abstracts: Symposium '87, Wetlands/Peatlands*. Edmonton, Alta.

Pears, N. (1985). *Basic Biogeography*. Longman, New York.

Peart, M.R. (1988). "Forest fire in China." *Geography*, 73:152–4.

Peet, R.K. (1988). "Forests of the Rocky Mountains." Pp. 64–102 in Barbour, M.G., and

Billings, W.D. (eds.), *North American Terrestrial Vegetation*. Cambridge University Press, New York.

– (1984). "Twenty-six years of change in a *Pinus strobus, Acer saccharum* forest, Lake Itasca, Minnesota." *Bulletin of the Torrey Botanical Club*, 111:61–8.

– (1981). "Forest vegetation of the Colorado Front Range." *Vegetatio*, 45:3–75.

Pennington, W. (1969). *The History of British Vegetation*. English Universities Press Ltd., London.

Percy, M.B. (1986). *Forest Management and Economic Growth in British Columbia*. Economic Council of Canada, Ottawa.

Peters, R., and Ohkubo, T. (1990). "Architecture and development in *Fagus japonica–Fagus crenata* forest near Mount Takahara, Japan." *Journal of Vegetation Science*, 1:499–506.

Petzold, D.E., and Mulhern, T. (1987). "Vegetation distribution along lichen-dominated slopes of opposing aspect in the eastern Canadian Subarctic." *Arctic*, 40:221–4.

Pfitsch, W.A. (1988). "Microenvironment and the distribution of two species of *Draba* (Brassicaceae) in a Venezuelan *Paramo*." *Arctic and Alpine Research*, 20:333–41.

– and Bliss, L.C. (1988). "Recovery of net primary production in subalpine meadows of Mount St. Helens following the 1980 eruption." *Canadian Journal of Botany*, 66:989–97.

Phillips, D.E., Wild, A., and Jenkinson, D.S. (1990). "The soil's contribution to global warming." *Geographical Magazine*, 62:2, 36–8.

Phillips, W.S. (1963). *Vegetational Changes in the Northern Great Plains*. Agricultural Experimental Station, University of Arizona, Tucson.

Pidwirny, M.J. (1990). "Plant zonation in a brackish tidal marsh: descriptive verification of resource-based competition and community structure." *Canadian Journal of Botany*, 68: 1689–97.

Piedade, M.T.F., Junk, W.J., Long, S.P. (1991). "The productivity of the c_4 grass *Echinichloa polystachya* on the Amazon floodplain." *Ecology*, 72:1456–63.

Pielou, E.C. (1991). *After the Ice Age: The Return of Life to Glaciated North America*. University of Chicago Press, Chicago.

Pires, J.M., and Prance, G.T. (1985). "The vegetation types of the Brazilian Amazon" Pp. 109–45 in Prance, G.T., and Lovejoy, T.E. (eds.), *Key Environments: Amazonia*. Pergamon Press, New York.

Polunin, N. (1969). *Introduction to Plant Geography*. Longmans Green, London.

Polunin, O., and Walters, M. (1985). *A Guide to the Vegetation of Britain and Europe*. Oxford University Press, London.

Population Atlas of China (1987). Population Census Office of the State Council of the People's Republic of China and the Institute of Geography of the Chinese Acadamy of Sciences, Beijing.

Prentice, I.C., Cramer, W., Harrison, S.P., Leemans, R., Monserud, R.A., and Solomon, A.M. (1992). "A global biome model based on plant physiology and dominance, soil properties and climate." *Journal of Biogeography*, 19:117–34.

Preston, R.J. (1968). *Rocky Mountain Trees*. Dover Publications, Inc., New York.

Preston-Whyte, R.A. (series ed.) (1985). *Biogeography and Ecosystems of South Africa*. University of Natal, Durban.

Price, J.S. (1990). "Coastal salt marshes: hydrology and salinity." Pp. 83–5 in Roulet, N.

(guest ed.), "Focus: aspects of the physical geography of wetlands," *Canadian Geographer*, 34:79–88.

– Woo, K.-K., and Maxwell, B. (1989) "Salinity of marshes along the James Bay coast, Ontario, Canada." *Physical Geography*, 10:1–12.

– Ewing, K., Woo, M.-K., and Kershaw, K. A. (1988). "Vegetation patterns in James Bay coastal waters. II. Effects of hydrology on salinity and vegetation." *Canadian Journal of Botany*, 66:2586–94.

Pyne, S.J. (1984). *Introduction to Wildland Fire*. John Wiley and Sons, New York.

– (1983). "Indian fires." *Natural History*, 92(2):6–11.

Quinby, P.A. (1991). "Self-replacement in old-growth white pine forests of Temagami, Ontario." *Forest Ecology and Management*, 41:95–109.

Quintanilla, V. (1977). "A contribution to the phytogeographical study of temperate Chile." Pp. 31–41 in Muller, P. (ed.), *Ecosystem Research in South America*. W. Junk, The Hague.

Racine, C.H., Johnson, L.A., and Viereck, L.A. (1987). "Patterns of vegetation recovery after tundra fires in northwestern Alaska, U.S.A." *Arctic and Alpine Research*, 19:461–9.

Ram, J., Singh, S.P., and Singh, J.S. (1988). "Community level phenology of grassland above treeline in central Himalaya, India." *Arctic and Alpine Research*, 20:325–32.

Rannie, W.F. (1986). "Summer air temperature and number of vascular species in Arctic Canada." *Arctic*, 39:133–7.

Raunkiaer, C. (1934). *Life Forms of Plants and Statistical Plant Geography*. Clarendon Press, Oxford.

Rawat, G.S., and Pangtey, Y.P.S. (1987). "Floristic structure of snowline vegetation in central Himalaya, India." *Arctic and Alpine Research*, 19:195–201.

Rehder, H., Beck, E., and Kokwaro, J.O. (1988). "The Afroalpine plant communities of Mt. Kenya (Kenya)." *Phytocoenologia*, 16:433–63.

Reichle, D.E. (ed.) (1981). *Dynamic Properties of Forest Ecosystems*. Cambridge University Press, New York.

– (1969). *Analysis of Temperate Forest Ecosystems*. Ecological Studies, vol. 1. Springer-Verlag, New York.

– Dinger, B.E., Edwards, N.T., Harris, W.F., and Sollins, P. (1973). "Carbon flow and storage in a forest ecosystem." Pp. 345–65 in Woodwell, G.M., and Pecan, E.V. (eds.), *Carbon and the Biosphere, Proceedings of the 24th Brookhaven Symposium in Biology*. AEC–CONF–720510. NTIS, Springfield, Va.

Reichman, O.J. (1987). *Konza Prairie – A Tallgrass Natural History*. University Press of Kansas, Lawrence.

Reid, W.V., and Miller, K.R. (1989). *Keeping Options Alive: The Scientific Basis for Conserving Biodiversity*. World Resources Institute, Washington, DC.

Reimchen, T.E. (1991). "Biodiversity studies and forest management in British Columbia." *Canadian Biodiversity*, 1(3):34–5.

Reznicek, A.A., and Maycock, P.F. (1983). "Composition of an isolated prairie in central Ontario." *Canadian Journal of Botany*, 61:3107–16.

Richardson, D.H.S., and Finegan, E. (1977). "Studies on the lichens of Truelove Lowland."

Pp. 245–62 in Bliss, L.C. (ed.), *Truelove Lowland, Devon Island, Canada: A High Arctic Ecosystem.* University of Alberta Press, Edmonton.

Richardson, J.L., and Arndt, J.L. (1989). "What use prairie potholes?" *Journal of Soil and Water Conservation*, 44:196–8.

Risser, P.G. (1988). "Abiotic controls on primary productivity and nutrient cycling in North American grasslands." Pp. 115–29 in Pomeroy, L.R., and Alberts, J.J. (eds.), *Concepts of Ecosystem Ecology: A Comparative View.* Springer-Verlag, New York.

– Birney, E.C., Blocker, H.D., May, S.W., Parton, W.J., and Wiens, J.A. (1981). *The True Prairie Ecosystem.* Hutchinson Ross Publishing Company, Stroudsburg, Pa..

Ritchie, J.C. (1987). *Postglacial Vegetation of Canada.* Cambridge University Press, New York.

– (1984). *Past and Present Vegetation of the Far Northwest of Canada.* University of Toronto Press, Toronto.

– (1962). *A Geobotanical Survey of Northern Manitoba.* Technical Paper no. 9. Arctic Institute of North America, Montreal.

Robbins, R.G. (1961). "The montane vegetation of New Guinea." *Tuatara*, 8:124–33.

Roberts, M.R., and Christensen, N.L. (1988). "Vegetation variation among mesic successional forest stands in northern lower Michigan." *Canadian Journal of Botany*, 66:1080–90.

Rode, M.W. (1993). "Leaf-nutrient accumulation and turnover at three stages of succession from heathland to forest." *Journal of Vegetation Science*, 4:263–8.

Rodin, L.E., and Bazilevich, N.I. (1967). *Production and Mineral Cycling in Terrestrial Vegetation.* Oliver and Boyd, London.

Röhrig, E. (1991a). "Seasonality." Pp. 25–33 in Röhrig, E., and Ulrich, B. (eds.), *Temperate Deciduous Forests.* Ecosystems of the World 7. Elsevier, Amsterdam.

– (1991b). "Temperate deciduous forests in Mexico and Central America." Pp. 371–5 in Röhrig, E., and Ulrich, B. (eds.), *Temperate Deciduous Forests.* Ecosystems of the World 7. Elsevier, Amsterdam.

– (1991c). "Deciduous forests of the Near East." Pp. 527–37 in Röhrig, E., and Ulrich, B. (eds.), *Temperate Deciduous Forests.* Ecosystems of the World 7. Elsevier, Amsterdam.

– (1991d). "Biomass and productivity." Pp. 165–74 in Röhrig, E., and Ulrich, B. (eds.), *Temperate Deciduous Forests.* Ecosystems of the World 7. Elsevier, Amsterdam.

Rood, K.M. (1984). *An Aerial Photograph Inventory of the Frequency and Yield of Mass Wasting on the Queen Charlotte Islands, British Columbia.* British Columbia Minestry of Forests, Victoria.

Ross, M. (1991). "Environmental impact assessment and the forestry sector: lessons from the Alberta-Pacific review." *Newsletter of the Canadian Institute of Resources Law*, no. 31, Summer 1991.

Ross, R.L., and Hunter, H.E. (1976). *Climax Vegetation of Montana Based on Soils and Climate.* USDA Soil Conservation Service, Bozeman, Mont.

Rosswall, T., and Heal, O.W. (eds.) (1975). *Structure and Function of Tundra Ecosystems.* Ecol. Bull. 20. Swedish Natural Science Research Council, Stockholm.

Rostlund, E. (1957). "The myth of a natural prairie belt in Alabama: an interpretation of historical records." *Annals of the Association of American Geographers*, 47:392–411.

Rottman, M.L., and Hartman, E.L. (1985). "Tundra vegetation of three cirque basins in the northern San Juan Mountains, Colorado." *Great Basin Naturalist*, 45:87–93.

Rouse, W.R. (1991). "Impacts of Hudson Bay on the terrestrial climate of the Hudson Bay lowlands." *Arctic and Alpine Research*, 23:24–30.

– (1984a). "Microclimate at Arctic tree line: 1. Radiation balance of tundra and forest." *Water Resources Research*, 20:57–66.

– (1984b). "Microclimate of Arctic tree line: 2. Soil microclimate of tundra and forest." *Water Resources Research*, 20:67–73.

– (1984c). "Microclimate at Arctic tree line: 3. The effects of regional advection on the surface energy balance of upland tundra." *Water Resources Research*, 20:74–8.

Rowe, J.S. (1984). "Lichen woodlands in northern Canada." Pp. 225–37 in Olson, R., Hastings, R., and Geddes, F. (eds.), *Northern Ecology and Resource Management*. University of Albert Press, Edmonton.

– (1983). "Concepts of fire effects on plant individuals and species." Pp. 135–54 in Wein, R.W., and MacLean, D.A., (eds.), *The Role of Fire in Northern Circumpolar Ecosystems*. SCOPE 18, Fredricton, NB.

– (1972). *Forest Regions of Canada*. Canadian Forestry Publication no. 1300. Department of the Environment, Ottawa.

Rubec, C.D.A. (1992). "Assessing biodiversity risk in Canada." *National Round Table Review*, Fall 1992, 16–17.

Rump, P.C. (1987). "The state of Canada's wetlands." Pp. 259–66 in Rubec, C.D.A., and Overend, R.P. (compilers), *Proceedings Symposium '87: Wetlands/Peatlands*. Edmonton, Alta.

Russell, E.W.B. (1987). "Pre-blight distribution of *Castanea dentata* (Marsh.) Borkh." *Bulletin of the Torrey Botanical Club*, 114:183–90.

– (1983). "Indian-set fires in the forests of the northeastern United States." *Ecology*, 64:78–88.

Rutkowski, D.R., and Stottlemyer, R. (1993). "Composition, biomass and nutrient distribution in mature northern hardwood and boreal forest stands, Michigan." *American Midland Naturalist*, 130:13–30.

Ruuhijarvi, R. (1983). "The Finnish mire types and their regional distribution." Pp. 47–67 in Gore, A.J.P. (ed.), *Mires: Swamps, Bog, Fen and Moor: Regional Studies*. Ecosystems of the World 4B. Elsevier, New York.

Ryan, K.C., and Reinhardt, E.D. (1988). "Predicting postfire mortality of seven western conifers." *Canadian Journal of Forest Research*, 18:1291–7.

Sala, O.E., Parton, W.J., Joyce, L.A., and Lauenroth, W.K. (1988). "Primary production of the Central Grassland Region of the United States." *Ecology*, 69:40–5.

Sampson, N. (1990). "Updating the old-growth wars." *American Forests*, 96(11&12):17–20.

Samoil, J.K. (ed.) (1988). *Management and Utilization of Northern Mixedwoods*. Information Report NOR–X–296. Northern Forestry Centre, Edmonton.

Satoo, T. (1973). "A synthesis of studies by the harvest method: primary production relations in the temperate deciduous forests of Japan." Pp. 55–72 in Reichle, D.E. (ed.), *Analysis of Temperate Forest Ecosystems*. Springer-Verlag, New York.

Savage, M. (1991). "Structural dynamics of a southwestern pine forest under chronic human disturbance." *Annals of the Association of American Geographers*, 81:271–89.

– and Swetnam, T.W. (1990). "Early 19th-century fire decline following sheep pasturing in a Navajo ponderosa pine forest." *Ecology*, 71:2374–8.

Saville, D.B.O. (1972). *Arctic Adaptations in Plants*. Monograph no. 6. Research Branch, Canada Department of Agriculture, Ottawa.

Schaminée, J.H., Hennekens, S. M., and Thébaud, G. (1993). "A syntaxonomical study of sub-alpine heathland communities in West European low mountain ranges." *Journal of Vegetation Science*, 4:125–34.

Schimel, D., Stillwell, M.A., and Woodmansee, R.G. (1985). "Biogeochemistry of C, N and P in a soil catena of the shortgrass steppe." *Ecology*, 66:276–82.

Schimper, A.F.W. (1903). *Plant-Geography upon a Physiological Basis*. Clarendon Press, Oxford.

Schlesinger, W.H. (1984). "Soil organic matter: a source of atmospheric CO_2." Pp. 111–27 in Woodwell, G.M. (ed.), *The Role of Terrestrial Vegetation in the Global Carbon Cycle: Measurement by Remote Sensing*. John Wiley and Sons, New York.

– DeLucia, E.H., and Billings, W.D. (1989). "Nutrient-use efficiency of woody plants on contrasting soils in the western Great Basin, Nevada." *Ecology*, 70:105–13.

Schmaltz, J. (1991). "Deciduous forests of southern South America." Pp. 557–78 in Röhrig, E., and Ulrich, B. (eds.), *Temperate Deciduous Forests*. Ecosystems of the World 7. Elsevier, Amsterdam.

Schneider, R.L., and Sharitz, R.R. (1988). "Hydrochory and regeneration in a bald cypress–water tupelo swamp forest." *Ecology*, 69:1055–63.

Schoonmaker, P., and McKee, A. (1988). "Species composition and diversity during secondary succession of coniferous forests in the western Cascade forests of Oregon." *Forest Science*, 34:960–79.

Schueler, F.W. (1991). "Maps of the number of tree species in Canada: a pilot GIS study of tree biodiversity. Part 1." *Canadian Biodiversity*, 1(1):22–9.

Schwab, J.W. (1983). *Mass Wasting: October–November 1978 Storm, Rennell Sound, Queen Charlotte Islands, British Columbia*. British Columbia Minestry of Forests, Victoria.

Scoggan, H.J. (1978). *The Flora of Canada*. 4 vols. National Museum of Natural Sciences, Ottawa.

– (1957). *Flora of Manitoba*. Bulletin no. 140. National Museum of Canada, Ottawa.

Scott, G.A.J. (1987a). "The impact of agriculture on the soils and vegetation of the Canadian prairies." Pp. 276–8 in Marsh, W.M., *Earthscapes: A Physical Geography*. John Wiley and Sons, New York.

– (1987b). "Shifting cultivation where land is limited." Pp. 34–45 in Jordan, C. (ed.), *Amazonian Rain Forests: Ecosystem Disturbance and Recovery*. Ecological Studies #60. Springer-Verlag, Berlin.

– (1978). *Grassland Development in the Gran Pajonal of Eastern Peru: A Study of Soil-Vegetation Nutrient Systems*. Hawaii Monographs in Geography, no. 1. University Microfilms International, Ann Arbor.

– (1967). "Phytogeographical observations in North Inishowen." Geography BSC Hons. thesis, Queen's University of Belfast, Northern Ireland.

– and Crane, C.B. (1991). "Agroecosystem soils as a source of atmospheric CO_2: a case study from southern Manitoba." Pp. 46–54 in *The Dauphin Papers: Research by Prairie Geographers*. Brandon Geographical Studies # 1. Brandon, Man.

– and Scott, K. (1990). "Relationships between moisture availability and vegetation cover in the Canadian Prairies." Unpublished photocopy, Department of Geography, University of Winnipeg, Winnipeg.

Scott, J.T., Siccama, T.G., Johnson, A.H., and Breisch, A.R. (1984). "Decline of red spruce in the Adirondacks, New York." *Bulletin of the Torrey Botanical Club*, 111:438–44.

Scott, M. G., Hutchinson, T. C., and Feth, M. J. (1989). "Contrasting responses of lichens and *Vaccinium angustifolium* to long-term acidification of a boreal forest ecosystem." *Canadian Journal of Botany*, 67:579–88.

Scott, P.A., Hansell, R.I.C., and Erickson, W.R. (1993). "Influences of wind and snow on northern tree-line environments at Churchill, Manitoba, Canada." *Arctic*, 46:316–23.

– Bentley, C.V., Fayle, D.C., and Hansell, R.I.C. (1987a). "Crown forms and shoot elongation of white spruce at the treeline, Churchill, Manitoba, Canada." *Arctic and Alpine Research*, 19:175–86.

– Hansell, R.I.C., and Fayle, D.C. (1987b). "Establishment of white spruce populations and responses to climate change at the treeline, Churchill, Manitoba, Canada." *Arctic and Alpine Research*, 19:45–51.

Scottsberg, C. (1954). "Antarctic flowering plants." *Bot. Tidsskr*, 2:330–8.

Seastedt, T.R. (1989). "Mass, nitrogen, and phosphorus dynamics in foliage and root detritus of tallgrass prairie." *Ecology*, 69:59–65.

Sedjo, R. (1989) "Forests: a tool to moderate global warming?" *Environment*, 31(1):14–20.

Senevirantne, G. (1992). "Strategy to protect earth's biological variety offered by International Coalition." *Environmental Conservation*, 19:77–8, 84.

Shankman, D., and Daly, C. (1988). "Forest regeneration above tree limit depressed by fire in the Colorado Front Range." *Bulletin of the Torrey Botanical Club*, 115:272–9.

Sharma, N.P. (ed.) (1992). *Managing the World's Forests: Looking for Balance Between Conservation and Development*. Kendall/Hunt, Dubuque, Iowa.

Shaver, G.R. (1986). "Woody stem production in Alaskan tundra shrubs." *Ecology*, 67:660–9.

– and Chapin, F.S., III. (1991). "Production: biomass relationships and element cycling in contrasting Arctic vegetation types." *Ecological Monographs*, 61:1–31.

Sheail, J., and Wells, T.C.E. (1983). "The fenlands of Huntingdonshire, England: a case study in catastrophic change." Pp. 375–93 in Gore, A.J.P. (ed.), *Mires: Swamps, Bog, Fen and Moor: Regional Studies*. Ecosystems of the World 4B. Elsevier, New York.

Sherman, R.J., and Warren, R.K. (1988). "Factors in *Pinus ponderosa* and *Calocedrus decurrens* mortality in Yosemite Valley, USA." *Vegetatio*, 77:79–85.

Short, S.K., and Andrews, J.T. (1988). "A sixteen thousand year old organic deposit, northern Baffin Island, NWT, Canada: palynology and significance." *Géographie physique et Quaternaire*, 42:75–82.

Siccama, T.G., Bliss, M., and Vogelmann, H.W. (1982). "Decline of red spruce in the Green Mountains of Vermont." *Bulletin of the Torrey Botanical Club*, 109:162–8.

Sims, P.L. (1988). "Grasslands." Pp. 266–87 in Barbour, M.G., and Billings, W.D. (eds.), *North American Terrestrial Vegetation*. Cambridge University Press, New York.

– Singh, J.S., and Lauenroth, W.K. (1978). "The structure and function of ten western North American grasslands." *Journal of Ecology*, 66:251–85.

Singh, J.S., Lauenroth, W.K., Heitschmidt, R.K., and Dodd, J.L. (1983). "Structural and functional attributes of the vegetation of northern mixed prairie of North America." *Botanical Review*, 49:117–49.

Sioli, H. (1975). "Amazon tributaries and drainage basins." Pp. 199–214 in Hasler, A.D. (ed.), *Coupling of Land and Water Systems*. Springer-Verlag, New York.

Sirois, L. (1992). "The transition between boreal forest and tundra." Pp. 196–215 in Shugart, H.H., Leemans, R., and Bonan, G.B. (eds.), *A Systems Analysis of the Global Boreal Forest*. Cambridge University Press, New York.

– and Payette, S. (1991). "Reduced postfire tree regeneration along a boreal forest–tundra-tundra transect in northern Québec." *Ecology*, 72:619–27.

Sjörs, H. (1983). "Mires of Sweden." Pp. 69–94 in, Gore, A.J.P. (ed.), *Mires: Swamps, Bog, Fen and Moor: Regional Studies*. Ecosystems of the World 4B. Elsevier, New York.

– (1959). "Bogs and fens in the Hudson Bay lowlands." *Arctic*, 12:3–19.

Skye, E. (1989). "Changes to the climate and flora of Hopen Island during the last 110 years." *Arctic*, 42:323–32.

Smeins, F.E., Diamond, D.D., and Hanselka, C.W. (1992). "Coastal prairie." Pp. 269–90 in Coupland, R.T. (ed.), *Natural Grasslands: Introduction and Western Hemisphere*. Ecosystems of the World 8A. Elsevier, New York.

Smil, V. (1984). *The Bad Earth: Environmental Deterioration in China*. Sharpe, New York.

Smith, A.P. (1981). *Growth and Population Dynamics of Esplita (Compositae) in the Venezuelan Andes*. Smithsonian Contributions to Botany, no. 48. Smithsonian Institution Press, Washington, DC.

Smithers, L.A. (1961). *Lodgepole pine in Alberta*. Bulletin 127. Canada Department of Forestry, Ottawa.

Sochava, V.B. (1979). "Distribution of grasslands in the USSR." Pp. 103–10 in Numata, M. (ed.), *Ecology of Grasslands and Bamboolands in the World*. Gustav Fischer Verlag, Jena.

Soil Survey Staff (1975). *Soil Taxonomy: A Basic System of Soil Classification for Making and Interpreting Soil Surveys*. US Department of Agriculture, Washington, DC.

Soils Research Institute (1972). *Soils of Canada*. Map scale 1:5,000,000. Research Branch, Canada Department of Agriculture, Ottawa.

Solem, T. (1989) "Blanket mire formation at Haramsoy, More og Romsdal, western Norway." *Boreas*, 18:221–35.

Solomon, A.M. (1992). "The nature and distribution of past, present and future boreal forests: lessons for a research and modelling agenda." Pp. 291–307 in Shugart, H.H., Leemans, R., and Bonan, G.B. (eds.), *A Systems Analysis of the Global Boreal Forest*. Cambridge University Press, Cambridge.

Soriano, A. (1992). "Río de la Plata grasslands." Pp. 367–408 in Coupland, R.T. (ed.), *Natural Grasslands: Introduction and Western Hemisphere*. Ecosystems of the World 8A. Elsevier, New York.

– Golluscio, R.A., and Satorre, E. (1987). "Spatial heterogeneity of the root system of grasses in the Patagonian arid steppe." *Bulletin of the Torrey Botanical Club*, 114:103–8.

Specht, R.L. (1990). "Forested wetlands in Australia." Pp. 387–406 in Lugo, A.E., Brinson,

M., and Brown, S. (eds.). *Forested Wetlands*. Ecosystems of the World 15. Elsevier, Oxford.

– (1979). "Heathlands and related shrublands of the world." Pp. 1–18 in Specht, R.L. (ed.), *Heathlands and Related Shrublands: Descriptive Studies*. Ecosystems of the World 9A. Elsevier, Amsterdam.

Spies, T.A., and Franklin, J.F. (1988). "Coarse woody debris in Douglas-fir forests of western Oregon and Washington." *Ecology*, 69:1689–702.

Spira, T.P. (1987). "Alpine annual plant species in the White Mountains of eastern California." *Madroño*, 34:315–23.

Sprugel, D.G. (1991). "Disturbance, equilibrium, and environmental variability: what is 'natural' vegetation in a changing environment?" *Biological Conservation*, 58:1–18.

– (1984). "Density, biomass, productivity, and nutrient-cycling changes during stand development in wave-regenerated balsam fir forests." *Ecological Monographs*, 54:165–86.

Stall, P.D., Williams, S.E., and Christensen, M. (1988). "Efficacy of native vesicular-arbuscular mycorrhizal fungi after severe soil disturbance." *New Phytologist*, 110:347–54.

Stalter, R., Kincaid, D.T., and Lamont, E.E. (1991). "Life forms of the flora at Hampstead Plains, New York, and a comparison with four other sites." *Bulletin of the Torrey Botanical Club*, 118:191–4.

Stanek, W. (1977). "A list of terms and definitions." Pp. 367–87 in Radforth, N.W., and Brawner, C.O. (eds.), *Muskeg and the Northern Environment in Canada*. University of Toronto Press, Toronto.

Steere, W.C. (1967). "Phytogeography." Pp. 8–9 in Greene, S.W. et al. (eds.), *Terrestrial Life in Antarctica*. Antarctic Map Folio Series, folio 5. American Geographical Society, New York.

Steiguer, J.E. de., Pye, J.M., and Love, C.S. (1990). "Air pollution damage to U.S. forests." *Journal of Forestry*, 88(8):17–22.

Steinauer, E.M., and Bragg, T.B. (1987). "Ponderosa pine (*Pinus ponderosa*) invasion of Nebraska Sandhill Prairie." *American Midland Naturalist*, 118:358–65.

Stephenson, S.L., and Adams, H.S. (1986). "An ecological study of balsam fir communities in West Virginia." *Bulletin of the Torrey Botanical Club*, 113:372–81.

– and Adams, H.S. (1984). "The spruce-fir forest on the summit of Mount Rogers in southwestern Virginia." *Bulletin of the Torrey Botanical Club*, 111:69–75.

Stewart, C.N., and Nilson, E.T. (1993). "Association of edaphic factors and vegetation in several isolated Appalachian peat bogs." *Bulletin of the Torrey Botanical Club*, 120:128–35.

Stewart, G.H. (1988). "The influence of canopy cover on understory development in forests of the western Cascade Range, Oregon, USA." *Vegetatio*, 76:79–88.

– (1986). "Population dynamics of a montane conifer forest, western Cascade Range, Oregon, USA." *Ecology*, 67:534–44.

Stocks, B.J. (1993). "Global warming and forest fires in Canada." *Forestry Chronicle*, 69:290–3.

Stohlgren, T.J. (1993). "Spatial patterns of giant sequoia (*Sequoiadendron giganteum*) in two sequoia groves in Sequoia National Park, California." *Canadian Journal of Forest Research*, 23:120–32.

Stone, E.L., and Kalisz, P.J. (1991). "On the maximum extent of tree roots." *Forest Ecology and Management*, 46:59–102.

Streng, D.R., and Harcombe, P.A. (1982). "Why don't east Texas savannas grow up to be forest?" *American Midland Naturalist*, 108:278–94.

Strong, W.L., Pluth, D.J., LaRoi, G.H., and Corns, I.G.W. (1991). "Forest understory plants as predictors of lodgepole pine and white spruce site quality in west-central Alberta." *Canadian Journal of Forest Research*, 21:1675–83.

Stuart, J.D. (1987). "Fire history of an old growth forest of *Sequoia sempervirens* (Taxodiaceae) forest in Humboldt Redwoods State Park, California." *Madroño*, 34:128–41.

Sugden, D. (1982). *Arctic and Antarctic: A Modern Geographical Synthesis*. Barnes and Noble, Totowa, NJ.

Sukhachev, V.N. (1973). *Izbrannye Trudy* (Selected works), vol. 7:88–9. Nauka, Leningrad (in Russian).

Sundriyal, R.C. (1993). "Structure, productivity and energy flow in an alpine grassland in the Garhwal Himalaya." *Journal of Vegetation Science*, 4:125–34.

Svedarsky, W.D., and Buckley, P.E. (1975). "Some interactions of fire, prairie and aspen in northwestern Minnesota." Pp. 115–22 in Wali, M.K. (ed.), *Prairie: A Multiple View*. University of North Dakota Press, Grand Forks.

Svensson, G. (1988). "Bog development and environmental conditions as shown by the stratigraphy of Store Mosse mire in southern Sweden." *Boreas*, 17:89–111.

Szeicz, J.M., and MacDonald, G.M. (1991). "Postglacial vegetation history of oak savanna in southwestern Ontario." *Canadian Journal of Botany*, 69:1507–19.

Tainton, N.M., and Mentis, M.T. (1984). "Fire in grassland." Pp. 115–48 in Booysen, P. deV., and Tainton, N.M. (eds.), *Ecological Effects of Fire in South African Ecosystems*. Springer-Verlag, New York.

Tande, G.F. (1979). "Fire history and vegetation pattern of coniferous forests in Jasper National Park, Alberta." *Canadian Journal of Botany*, 57:1912–31.

Tardif, J., and Bergeron, Y. (1992). "Analyse écologique des peuplements de frêne noir (*Fraxinus nigra*) des rives du lac Duparquet, nord-ouest du Québec." *Canadian Journal of Botany*, 70:2294–302.

Tarnocai, C. (1984). *Peat Resources of Canada*. Publication no. 24140. National Research Council of Canada, Ottawa.

– (1980). "Canadian wetland registry." Pp. 9–39 in Rubec, C.D.A., and Pollett, F.C. (eds.), *Proceedings, Workshop on Canadian Wetlands*. Ecological Land Classification Series no. 12. Lands Directorate, Environment Canada. Ottawa.

– (1973). *Soils of the Mackenzie River Area*. Task Force on Northern Oil Development Report no. 73–26. Information Canada, Ottawa.

– (1972). "Some characteristics of cryic organic soils in northern Manitoba." *Canadian Journal of Soil Science*, 52:485–96.

– and Zoltai, S.C. (1988). "Wetlands of Arctic Canada." Pp. 27–53 in National Wetlands Working Group, *Wetlands of Canada*, Ecological Land Classification Series, no. 24. Sustainable Development Branch, Environment Canada, Ottawa, and Polyscience Publications Inc., Montreal.

Tausch, R.J., and West, N.W. (1988). "Differential establishment of pinyon and juniper following fire." *American Midland Naturalist,* 119:174–84.

Taylor, A.H. (1990). "Disturbance and persistence of sitka spruce (*Picea sitchensis* (Bong) Carr.) in coastal forests of the Pacific Northwest, North America." *Journal of Biogeography,* 17:47–58.

– and Zisheng, Q. (1988). "Regeneration patterns in old-growth *Abies-Betula* forests in the Wolong Natural Reserve, Sichuan, China." *Journal of Ecology,* 76:1204–18.

Taylor, J.A. (1983). "The peatlands of Great Britain and Ireland." Pp. 1–46 in Gore, A.J.P. (ed.), *Mires: Swamps, Bog, Fen and Moor: Regional Studies.* Ecosystems of the World 4B, Elsevier, New York.

Tester, J.R. (1989). "Effects of fire frequency on oak savanna in east-central Minnesota." *Bulletin of the Torrey Botanical Club,* 116:134–44.

Thaler, G.R. (1970). "The study of the tension zone between the Boreal and Carolinian floras in Ontario." Botany MSc thesis, University of Toronto.

Thomas, W. (1976). *The Swamp.* W.W. Norton and Co., New York.

Thompson, I.D., and Curran, W.J. (1993). "A reexamination of moose damage to balsam fir–white birch forests in central Newfoundland: 27 years later." *Canadian Journal of Forest Research,* 23:1388–95.

Thompson, K., and Hamilton, A.C. (1983). "Peatlands and swamps of the African continent." Pp. 331–74 in Gore, A.J.P. (ed.), *Mires: Swamps, Bog, Fen and Moor: Regional Studies.* Ecosystems of the World 4B. Elsevier, New York.

Thornthwaite, C.W. (1948). "An approach toward a rational classification of climate." *Geographical Review,* 38:55–94.

Tieszen, L.L. (ed.) (1978). *Vegetation and Production Ecology of the Alaskan Arctic Tundra.* Ecological Studies, vol. 29. Springer-Verlag, New York.

Tikhomirov, B.A. (ed.) (1969). *Vascular Plants of the Siberian North and the Northern Far East.* Israel Program for Scientific Translations, Jerusalem.

Timoney, K.P., La Roi, G.H., Zoltai, S.C., and Robinson, A.L. (1993a). "Vegetation communities and plant distributions and their relationships with parent materials in the forest-tundra of northwestern Canada." *Ecography,* 16:174–88.

– La Roi, G.H., and Dale, M.R.T. (1993b). "Subarctic forest-tundra vegetation gradients: the sigmoid wave hypothesis." *Journal of Vegetation Science,* 4:387–94.

– La Roi, G.H., Zoltai, S.C., and Robinson, A.L. (1992). "The high subArctic forest-tundra of northwestern Canada: position, width, and vegetation gradients in relation to climate." *Arctic,* 45:1–9.

– and Wein, R. W. (1991). "The areal pattern of burned tree vegetation in the subArctic region of northwestern Canada." *Arctic,* 44:223–30.

Titlyanova, A.A. (1982). "Ecosystems succession and biological turnover." *Vegetatio,* 50: 43–51.

Tivy, J. (1971). *Biogeography: A Study of Plants in the Ecosphere.* Oliver and Boyd, Edinburgh.

Tobey, R.C. (1981). *Saving the Prairies: The Life Cycle of the Founding School of American Plant Ecology, 1895–1955.* University of California Press, Berkeley.

Tonteri, T., Mikkola, K., and Lahti, T. (1990). "Compositional gradients in the forest vegetation of Finland." *Journal of Vegetation Science*, 1:691–8.

Tranquillini, W. (1979). *Physiological Ecology of the Alpine Timberline.* Springer-Verlag,
Berlin.

Trewartha, G.T. (1965). *Japan: A Geography.* University of Wisconsin Press, Madison.

Tritton, L.M., and Siccama, T.G. (1990). "What proportion of standing trees in forests of the
Northeast are dead?" *Bulletin of the Torrey Botanical Club*, 117(2):163–6.

Troll, C. (1966). *Geo-ecology of the Mountainous Regions of the Tropical Americas.*
Proceedings of the UNESCO Mexico Symposium, 1966. Ferd. Dummlers-Verlag, Bonn.

Trottier, G.C. (1986). "Disruption of rough fescue, *Festuca hallii*, grassland by livestock grazing in Riding Mountain National Park, Manitoba." *Canadian Field Naturalist*, 100:488–95.

Tsuyuzaki, S., and Tsujii, T. (1992). "Size and shape of *Carex meyeriana* tussocks in an alpine
wetland, northern Sichuan Province, China." *Canadian Journal Of Botany*, 70:2310–12.

Tuan, Y.-F. (1969). *China.* Aldine, Chicago.

Tueller, P.T., and Eckert, R.E. (1987). "Big sagebrush (*Artemisia tridentata vaseyana*) and
longleaf snowberry (*Symphoricarpos oreophilus*) plant associations in northeastern
Nevada." *Great Basin Naturalist*, 47:117–31.

Tuhkanen, S. (1984). "A circumboreal system of climate-phytogeographical regions." *Acta
Bot. Fennica*, 127:1–50.

Turner, D.P. (1985). "Successional relationships and a comparison of biological characteristics among six northwestern conifers." *Bulletin of the Torrey Botanical Club*, 112:421–8.

Tzedakis, P.C. (1992). "Effects of soils on the Holocene history of forest communities, Cape
Cod Massachusetts, U.S.A." *Géographie physique et Quaternaire*, 46:113–24.

Umbanhower, C.E., Jr (1992). "Reanalysis of the Wisconsin Prairie Continuum." *American
Midland Naturalist*, 127:268–75.

Valanne, T. (1983). "Fennoscandian birch and its evolution in the marginal forest zone."
Pp. 101–10 in Morisset, P., and Payette, S. (eds), *Tree-Line Ecology: Proceedings of the
Northern Québec Tree-Line Conference.* Centre d'études nordiques, Université Laval,
Québec.

Van Cleve, K., Chapin, F.S., III, Flanagan, P.W., Viereck, L.A., and Dyrness, C.T. (1986).
Forest Ecosystems in the Alaskan Taiga: A Synthesis of Structure and Function. Springer-
Verlag, New York.

van der Maarel, E. (1990). "Ecotones and ecoclines are different." *Journal of Vegetation
Science*, 1:135–8.

– and Titlyanova, A. (1989). "Above-ground and below-ground biomass relations in steppes
under different grazing conditions." *Oikos*, 56:364–70.

van der Valk, A.G. (ed.) (1989). *Northern Prairie Wetlands.* Iowa State University Press,
Ames.

Varga, S., Jalava, J., and Riley, J.L. (1991). *Ecological Survey of the Rouge Valley Park.*
Ontario Ministry of Natural Resources, Central Region, Aurora, Ont.

Vassiljevskaja, V.D. et al. (1975). "Agapa, USSR." Pp. 141–58 in Rosswall, T. and Heal, O.W.
(eds.), *Structure and Function of Tundra Ecosystems.* Ecol. Bull. 20. Swedish Natural
Science Research Council, Stockholm.

Veblen, T.T. (1986). "Age and size structure of subalpine forests in the Colorado Front Range." *Bulletin of the Torrey Botanical Club*, 113:225–40.

– Hadley, K.S., Reid, M.S., and Rebertus, A.J. (1991). "The response of subalpine forests to spruce beetle outbreak in Colorado." *Ecology*, 72:213–31.

– and Markgraf, V. (1988). "Steppe expansion in Patagonia?" *Quaternary Research*, 30: 331–8.

Viereck, L.A., Van Cleve, K., and Dryness, C.T. (1986). "Forest ecosystem distribution in the Taiga environment." Pp. 22–43 in Van Cleve, K., Chapin, F.S., III, Flanagan, P.W., Viereck, L.A., and Dryness, C.T. (eds.), *Forest Ecosystems in the Alaskan Taiga: A Synthesis of Structure and Function*. Springer-Verlag, New York.

Vil'chek, G.E. (1987). "Productivity of the tundras of the Taimyr." *Soviet Journal of Ecology*, 18:249–54.

Vinton, M.A., Hartnett, D.C., Finck, E.J., and Briggs, J.M. (1993). "Interactive effects of fire, bison (*Bison bison*) grazing and plant community composition in tallgrass prairie." *American Midland Naturalist*, 129:10–18.

Vitt, D.H., Marsh, J.E., and Bovey, R.B. (1988) *Mosses, Lichens and Ferns of Northwest North America*. University of Washington Press, Seattle.

Vogl, R.J. (1974). "Effects of fire on grasslands." Pp. 139–94 in Kozlowski, T.T., and Ahlgren, C.E. (eds.), *Fire and Ecosystems*. Academic Press, New York.

Vuilleumier, F., and Monasterio, M. (eds.) (1986). *High Altitude Tropical Biogeography*. Oxford University Press, New York.

Vukelić, J. (1991). "Synökologische Charakterisierung und syntaxonomische Einordnung von Carpinion-Gesellschaften Nordkroatiens." *Phytocoenologia*, 19:519–46.

Walker, D.A., Webber, P.J., Binnian, E.F., Everett, K.R., Lederer, N.D., Nordstrand, E.A., and Walker, M.D. (1987). "Cumulative impacts of oil fields on Northern Alaskan Landscapes." *Science*, 238:757–61.

Walker, M.D., Walker, D.A., Everett, K.R., and Short, S.K. (1991). "Steppe vegetation on south-facing slopes of pingos, central Arctic coastal plain, Alaska, U.S.A." *Arctic and Alpine Research*, 23:170–88.

Wallis, C.A. (1982). "An overview of the mixed grasslands of North America." Pp. 195–208 in Nicholson, A.C., McLean, A., and Baker, T.E. (eds.), *Grassland Ecology and Classification Symposium Proceedings*. Ministry of Forests, Province of British Columbia, Victoria.

Walter, H. (1979). *Vegetation of the Earth and Ecological Systems of the Geo-biosphere*. Springer-Verlag, New York.

– (1973). *Vegetation of the Earth in Relation to Climate and the Eco-Physiological Conditions*. Springer-Verlag, New York.

– Harnickell, E., and Mueller-Dombois, D. (1975). *Climate-Diagram Maps*. Springer-Verlag, New York.

Wang, C. (1961). *The Forests of China with a Survey of Grassland and Desert Vegetation*. Publication Series no. 5. Maria Moors Cabot Foundation, Harvard University, Cambridge, Mass.

Wang, C., and McKeague, J.A. (1986). "Short-range soil variability and classification of podzolic pedons along a transect in the Laurentian Highlands." *Canadian Journal of Soil Science*, 66:21–30.

Wardle, P., Bulfin, M.J.A., and Dugale, J. (1983). "Temperate broad-leaved evergreen forests of New Zealand." Pp. 33–72 in Ovington, J.D. (ed.), *Temperate Broad-leaved Evergreen Trees*. Ecosystems of the World 10. Elsevier, New York.

Waring, R.H., and Franklin, J.F. (1979). "Evergreen coniferous forests of the Pacific Northwest." *Science*, 204:1380–6.

Warkentin, J. (1968). *Canada: A Geographical Interpretation*. Methuen, Toronto.

Warner, B.G. (1992). "Peat: nature's compost." *Earth*, March 1992, 44–9.

– Clymo, R.S., and Tolonen, K. (1993). "Implications of peat accumulation at Point Escuminac, New Brunswick." *Quaternary Research*, 39:245–8.

– and Kubiw, H.J. (1987). "Origin of *Sphagnum* kettle bogs, southeastern Ontario." Pp. 543–50 in Rubec, C.D.A., and Overend, R.P. (compilers), *Proceedings Symposium '87: Wetlands/Peatlands*. Edmonton, Alta.

– Tolonen, K., and Tolonen, M. (1991). "A postglacial history of vegetation and bog formation at Point Escuminac, New Brunswick." *Canadian Journal of Earth Science*, 28: 1572–82.

Waters, I., and Shay, J.M. (1992). "Effect of water depth on population parameters of a *Typha glauca* stand." *Canadian Journal of Botany*, 70:349–51.

Watson, E.K., and Murtha, P.A. (1978). "A remote sensing rangeland classification for the Lac-du-Bois grassland, Kamloops. British Columbia." Pp. 16–26 in MacEwan, A. (ed.), *Proceedings of the 5th Canadian Symposium on Remote Sensing*, Canadian Aeronautics and Space Institute, Ottawa.

Watters, R.J., and Price, A.G. (1988). "Physical and chemical properties of a Podzol (upland series) with tonguing B horizon." *Canadian Journal of Soil Science*, 68:787–94.

Watts, F.B. (1969). "The natural vegetation of the Southern Great Plains of Canada." Pp. 93–112 in Nelson, J.G., and Chambers, M.J. (eds.), *Vegetation, Soils and Wildlife*. Methuen, Toronto.

Weaver, J.E. (1968). *Prairie Plants and Their Environment*. University of Nebraska Press, Lincoln.

– and Albertson, F.W. (1956). *Grasslands of the Great Plains: Their Nature and Use*. Johnsen Pub. Co., Lincoln.

Webber, P.J. (1974). "Tundra primary productivity." Pp. 445–73 in Ives, J.D., and Barry, R.G. (eds.), *Arctic and Alpine Research*. Methuen, London.

Wein, R.W., and MacLean, D.A. (1983). *The Role of Fire in Northern Circumpolar Ecosystems*. John Wiley and Sons, New York.

Wells, E.D., and Hirvonen, H.E. (1988). "Wetlands of Atlantic Canada." Pp. 249–303 in National Wetlands Working Group, *Wetlands of Canada*. Ecological Land Classification Series, no. 24. Sustainable Development Branch, Environment Canada, Ottawa, and Polyscience Publications Inc., Montreal.

Wells, P.V. (1970). "Postglacial vegetation history of the Great Plains." *Science*, 167: 1574–81.

Went, F.W., and Stark, N. (1968). "Mycorrhiza." *BioScience*, 18:1035–9.

Wentz, W.A. (1988) "Functional status of the nations wetlands." Pp. 50–9 in Hook, D.D. et al. (eds.), *The Ecology and Management of Wetlands,* vol. 2, *Management, Use and Value of Wetlands.* Timber Press, Portland.

West, N.E. (1988). "Intermountain deserts, shrub steppes, and woodlands." Pp. 210–31 in Barbour, M.G., and Billings, W.D. (eds.), *North American Terrestrial Vegetation.* Cambridge University Press, New York.

West, R. (1988). "Tolling the chestnut." *American Forests,* 94(9 & 10):10, 76.

Wheaton, E.E. (1992). "Using climatic change scenarios to estimate ecoclimatic impacts." *Canadian Geographer,* 36:67–9.

Whigham, D., Dykyjová, D., and Hejuý, S. (eds.), *Wetlands of the World: Inventory, Ecology and Management.* Kluwer, London.

White, F. (1983). *The Vegetation of Africa: A Descriptive Memoir to Accompany the UNESCO Vegetation Map of Africa.* UNESCO, Paris.

White, J.A., and Glenn-Lewin, D.C. (1984). "Regional and local variation in tallgrass prairie remnants of Iowa and eastern Nebraska." *Vegetatio,* 57:65–78.

White, J.H. (1977). *The Forest Trees of Ontario.* 6th ed. Ontario Ministry of Natural Resources, Toronto.

Whitman. W.C., and Wali, M.K. (1975). "Grasslands of North Dakota." Pp. 53–68 in Wali, M.K. (ed.), *Prairie: A Multiple View.* University of North Dakota Press, Grand Forks.

Whitney, G.G. (1990). "Multiple pattern analysis of an old-growth hemlock–white pine–northern hardwood stand." *Bulletin of the Torrey Botanical Club,* 117:39–47.

– (1986). "Relation of Michigan's presettlement pine forests to substrate and disturbance history." *Ecology,* 67:1548–59.

– and Runkle, J.R. (1981). "Edge versus age effects in the development of a beech-maple forest." *Oikos,* 37:377–81.

Whitney, S. (1985). *Western Forests.* Audubon Society Nature Guide. Random House, Inc., New York.

Whittaker, R.H. (1979). "Appalachian balds and other North American heathlands." Pp. 427–39 in Specht, R.L. (ed.), *Heathlands and Related Shrublands: Descriptive Studies.* Ecosystems of the World 9A. Elsevier Scientific, Amsterdam.

Whittaker, R.H. (1970). *Communities and Ecosystems.* Macmillan, New York.

– (1956). "Vegetation of the Great Smokey Mountains." *Ecological Monographs,* 26:1–80.

– and Likens, G.E. (1975). "The biosphere and man." Pp. 305–28 in Leith, H., and Whittaker, R.H. (eds.), *Primary Productivity of the Biosphere.* Springer-Verlag, New York.

– and Woodwell, G.M. (1969). "Structure, production and diversity of the oak-pine forest at Brookhaven, New York." *Ecology,* 57:155–74.

Whyte, R.O. (1968). *Grasslands of the Monsoon.* Faber and Faber, London.

Wickware, G.M., and Cowell, D.W. (1985). *Forest Ecosystem Classification of the Turkey Lakes Watershed, Ontario: A Research Contribution from the Federal LRTAP Calibration Watersheds Program.* Ecological Land Classification Series, no. 18. Lands Directorate, Environment Canada, Ottawa.

Wielgolaski, F.E. (ed.) (1975). *Fennoscandian Tundra Ecosystems.* Part 1, *Plants and Microorganisms.* Ecological Studies, vol. 16. Springer-Verlag, New York.

– Bliss, L.C., Svovoda, J., and Doyle, G. (1981). "Primary Production in Tundra." Pp. 187–225 in Bliss, L.C., Heal, O.W., and Moore, J.J. (eds.), *Tundra Ecosystems: A Comparative Analysis*. Cambridge University Press, Cambridge.

Wiken, E. (compiler) (1986). *Terrestrial Ecozones of Canada*. Ecological Land Classification Series, no. 19. Lands Directorate, Environment Canada, Ottawa.

Willms, W.D., and Fraser, J. (1992). "Growth characteristics of rough fescue (*Festuca scabrella* var. *campestris*) after three years of repeated harvesting at scheduled frequencies and heights." *Canadian Journal of Botany*, 70:2125–9.

Wilson, E.O. (1992). *The Diversity of Life*. Belkneys Press of Harvard University, Cambridge, Mass.

Wilson, S.D., and Shay, J.M. (1990). "Competition, fire, and nutrients in a mixed-grass prairie." *Ecology*, 71:1959–67.

Winterhalder, B.P., and Thomas, R.B. (1978). *Geoecology of Southern Highland Peru: A Human Adaptation Perspective*. Occasional Paper, no. 27. Institute of Arctic and Alpine Research, University of Colorado, Boulder.

Wood, D.W., and Morris, W.F. (1990). "Ecological constraints to seedling establishment on the pumice plains, Mount St. Helens, Washington." *American Journal of Botany*, 77: 1411–18.

Woods, K.D. (1984). "Patterns of tree replacement: canopy effects on understory pattern in hemlock–northern hardwood forests." *Vegetatio*, 56:87–107.

Woodward, F.I. (1987). *Climate and Plant Distribution*. Cambridge University Press, Cambridge.

Worbes, M., Klinge, H., Revilla, J.D., and Martius, C. (1992). "On the dynamics, floristic subdivision and geographical distribution of várzea forests in Central Amazonia." *Journal of Vegetation Science*, 3:553–64.

Wright, H.A. (1974). "Effect of fire on southern mixed prairie grasses." *Journal of Range Management*, 27:417–19.

– and Bailey, A.W. (1982). *Fire Ecology: United states and Southern Canada*. John Wiley and Sons, New York.

Wright, H.E., Jr (1971). "Vegetational history of the Central Plains." Pp. 157–72 in Dort, W., Jr, and Jones J.K., Jr (eds.), *Pleistocene and Recent Environments of the Central Great Plains*. University of Kansas, Department of Geology Special Publication, no. 3. University of Kansas Press, Lawrence.

Wylynko, D. (1991). "State of the Industry: How well are we managing our forests?" *Nature Canada*, 20(1):26–34.

Zampella, R.A., Moore, G., and Good, R.E. (1992). "Gradient analysis of pitch pine (*Pinus rigida* Mill.) lowland communities in the New Jersey Pinelands." *Bulletin of the Torrey Botanical Club*, 119:253–61.

Zoladeski, C.A., and Maycock, P.F. (1990). "Dynamics of the boreal forest in northwestern Ontario." *American Midland Naturalist*, 124:289–300.

Zoltai, S.C. (1993). "Cyclic development of permafrost in the peatlands of northwestern Alberta, Canada." *Arctic and Alpine Research*, 25:240–6.

– (1988). "Wetland environments and classification." Pp. 1–26 in National Wetlands

Working Group, *Wetlands of Canada,* Ecological Land Classification Series, no. 24. Sustainable Development Branch, Environment Canada, Ottawa, and Polyscience Publications Inc., Montreal.

– (1975). *The Southern Limit of Coniferous Trees on the Canadian Prairies.* Information Report NOR–X–128. Northern Forest Research Centre, Edmonton.

– (1972). "Palsas and peat plateaus in central Manitoba and Saskatchewan." *Canadian Journal of Forest Research,* 2:291–302.

– and Johnson, J.D. (1985). "Development of a treed bog island in a minerotrophic fen." *Canadian Journal of Botany,* 63:1076–85.

– and Johnson, J.D. (1979). *Vegetation-Soil Relationships in the Keewatin District.* ESCOM Report no. 1–25. Fisheries and Environment Canada, Ottawa.

– and Pollett, F.C. (1983). "Wetlands in Canada: their classification, distribution, and use." Pp. 245–68 in Gore, A.J.P. (ed.), *Mires: Swamps, Bog, Fen and Moor: Regional Studies.* Ecosystems of the World 4B. Elsevier, New York.

– Pollett, F.C., Jeglum, J.K., and Adams, G.D. (1975). "Developing a wetland classification for Canada." Pp. 497–511 in Bernier, B., and Winget, C.H. (eds.), *Proceedings, 4th North American Forest Soils Conference.* Laval University Press, Québec.

– Tarnocai, C., Mills, G.F., and Veldhuis, H. (1988a). "Wetlands of Subarctic Canada." Pp. 58–96 in National Wetlands Working Group, *Wetlands of Canada,* Ecological Land Classification Series. no. 24. Sustainable Development Branch, Environment Canada, Ottawa, and Polyscience Publications Inc., Montreal.

– Taylor, S., Jeglum, J.K., Mills, G.F., and Johnson, J.D. (1988b). "Wetlands of Boreal Canada." Pp. 100–54 in National Wetlands Working Group, *Wetlands of Canada,* Ecological Land Classification Series, no. 24. Sustainable Development Branch, Environment Canada, Ottawa, and Polyscience Publications Inc., Montreal.

– and Vitt, D.H. (1990). "Holocene climatic change and the distribution of peatlands in western interior Canada." *Quaternary Research,* 33:231–40.

Index

Lower-case *t* and *f* following page numbers refer to tables and figures respectively. Index includes selected plant binomials and their respective local names.